NEW REVISED

CAMBRIDGE GED PROGRAM

Mathematics

Jerry Howett

CAMBRIDGE ADULT EDUCATION
A Division of Simon & Schuster
Upper Saddle River, New Jersey

NEW REVISED CAMBRIDGE GED PROGRAM

COVER
Art Director: Josée Ungaretta
Cover Design: Marta Wolchuk, Design Five, NYC
Cover Illustration: Min Jae Hong

PHOTO CREDITS

page	credit
7	Irene Springer
53	Irene Springer
86	Marc Anderson
128	Teri Leigh Stratford
159	Qume: ITT
183	General Precision, Inc.
229	New York Stock Exchange

Copyright © 1993, 1998 by Globe Fearon, Inc. A Simon & Schuster Company. One Lake Street, Upper Saddle River, New Jersey 07458. All rights reserved. No part of this book may be reproduced or transmitted in any form or by any means, electronic, photographic, mechanical, or otherwise, including photocopying, recording, or by any information storage and retrieval system, without permission in writing from the publisher.

Printed in the United States of America

1 2 3 4 5 6 7 8 9 10   01 00 99 98 97

ISBN 0-835-94741-6

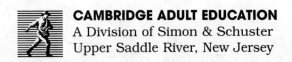

CAMBRIDGE ADULT EDUCATION
A Division of Simon & Schuster
Upper Saddle River, New Jersey

CONTENTS

TO THE STUDENT	xi
PRETEST	1
Pretest Answers	5
CHAPTER 1 WHOLE NUMBERS	7
Level 1 Whole-Number Skills	8
Lesson 1 The Whole-Number System	10
Lesson 2 Basic Operations: Addition and Subtraction	11
Lesson 3 Basic Operations: Multiplication and Division	13
Level 1 Review	*16*
Level 2 Whole Number Applications	17
Formulas	17
Lesson 4 Rounding	18
Lesson 5 Distance and Cost Formulas	20
Lesson 6 Mean and Median	21
Lesson 7 Powers	23
Lesson 8 Square Roots	25
Lesson 9 Perimeter	26
Lesson 10 Area	29
Lesson 11 Volume	30
Lesson 12 Properties of Numbers	32
Level 2 Review	*36*
Level 3 Whole-Number Problem Solving	38
Lesson 13 Word Clues	38
Lesson 14 Estimating	41
Lesson 15 Multistep Problems	43
Lesson 16 Tables	45
Lesson 17 Item Sets	46
Lesson 18 Set-Up Answers	48
Level 3 Review	*50*

CHAPTER 2 DECIMALS — 53

Level 1 Decimal Skills — 54
- Lesson 1 The Decimal Place System — 55
- Lesson 2 Reading and Writing Decimals — 56
- Lesson 3 Addition and Subtraction of Decimals — 58
- Lesson 4 Multiplication and Division of Decimals — 60
- Lesson 5 Comparing and Ordering Decimals — 62
- *Level 1 Review* — *63*

Level 2 Applications of Decimals — 65
- Lesson 6 Rounding Decimals — 65
- Lesson 7 Metric Measurement — 66
- Lesson 8 Reading Metric Scales — 68
- Lesson 9 Powers and Square Roots — 69
- Lesson 10 Circles and Cylinders — 70
- Lesson 11 Formulas and Decimals — 73
- *Level 2 Review* — *74*

Level 3 Problem Solving with Decimals — 77
- Lesson 12 Tables — 77
- Lesson 13 Travel Problems and Other Word Problems — 79
- Lesson 14 Problems Where Not Enough Information Is Given — 81
- *Level 3 Review* — *83*

CHAPTER 3 FRACTIONS — 86

Level 1 Fraction Skills — 87
- Lesson 1 Writing and Reducing Fractions — 88
- Lesson 2 Other Ways to Change Fractions and Mixed Numbers — 90
- Lesson 3 Addition — 93
- Lesson 4 Subtraction — 95
- Lesson 5 Multiplication — 96
- Lesson 6 Division — 98
- Lesson 7 Comparing and Ordering Fractions — 100
- Lesson 8 Changing Fractions and Decimals — 101
- *Level 1 Review* — *103*

Level 2 Fraction Applications — 104
- Lesson 9 Finding What Part One Number Is of Another — 104
- Lesson 10 Standard Units of Measurement — 106
- Lesson 11 Reading Scales — 108
- Lesson 12 Powers and Square Roots — 109
- Lesson 13 Areas of Triangles and Parallelograms — 110
- Lesson 14 Ratio — 112
- Lesson 15 Proportion — 114
- Lesson 16 Probability — 115
- *Level 2 Review* — *117*

Level 3   Problem Solving with Fractions   119
    Lesson 17   Word Problems Using Proportion   119
    Lesson 18   Multistep Proportion Problems   120
    Lesson 19   Unnecessary Information   122
    Lesson 20   Mixed Word Problems   123
    Level 3 Review   *126*

CHAPTER 4   PERCENTS   128

Level 1   Percent Skills   129
    Lesson 1   Percents and Decimals   130
    Lesson 2   Percents and Fractions   132
    Lesson 3   Finding a Percent of a Number (Part)   135
    Lesson 4   Finding What Percent One Number Is of
                 Another (Rate)   137
    Lesson 5   Finding a Number When a Percent of It Is
                 Given (Whole)   138
    Level 1 Review   *139*

Level 2   Percent Applications   141
    Lesson 6   Finding a Percent of a Number (Part)   141
    Lesson 7   Finding What Percent One Number Is of
                 Another (Rate)   142
    Lesson 8   Finding a Number When a Percent of It Is
                 Given (Whole)   143
    Lesson 9   Interest   144
    Level 2 Review   *147*

Level 3   Problem Solving with Percents   149
    Lesson 10   Finding a Percent of a Number in Multistep
                  Problems   149
    Lesson 11   Finding What Percent One Number Is of Another
                  in Multistep Problems   150
    Lesson 12   Finding a Number When a Percent of It Is Given
                  in Multistep Problems   152
    Lesson 13   Solving Percent Problems Using Ratios and
                  Proportions   153
    Lesson 14   Comparing with Percents   156
    Level 3 Review   *157*

CHAPTER 5   GRAPHS   159

    Lesson 1   Pictographs   160
    Lesson 2   Circle Graphs   162
    Lesson 3   Bar Graphs   164
    Lesson 4   Divided-Bar Graphs   166
    Lesson 5   Line Graphs   167
    Review   *170*

REVIEW TEST   175

CHAPTER 6 ALGEBRA 183

Level 1 Algebra Skills 185
- Lesson 1 The Number Line 185
- Lesson 2 Adding Signed Numbers 186
- Lesson 3 Subtracting Signed Numbers 189
- Lesson 4 Multiplying Signed Numbers 190
- Lesson 5 Dividing Signed Numbers 192
- Lesson 6 Substituting Signed Numbers 193
- Lesson 7 Adding Monomials 194
- Lesson 8 Subtracting Monomials 195
- Lesson 9 Multiplying Monomials 196
- Lesson 10 Dividing Monomials 197
- Lesson 11 Evaluating Algebraic Expressions 198
- *Level 1 Review* *200*

Level 2 Algebra Applications 202
- Lesson 12 One-Step Equations 202
- Lesson 13 Multistep Equations 204
- Lesson 14 Equations with Separated Unknowns 205
- Lesson 15 Equations with Parentheses 206
- Lesson 16 Inequalities 207
- Lesson 17 Multiplying Binomials 208
- Lesson 18 Factoring 209
- Lesson 19 Factoring Quadratic Expressions 211
- Lesson 20 Quadratic Equations 213
- Lesson 21 Factoring and Square Roots 215
- *Level 2 Review* *216*

Level 3 Algebra Problem Solving 218
- Lesson 22 Writing Algebraic Expressions 218
- Lesson 23 Writing Multistep Algebraic Expressions 219
- Lesson 24 Translating Words into Algebra 220
- Lesson 25 Writing Equations 222
- Lesson 26 Using Algebra to Solve Formulas 223
- Lesson 27 Algebra Word Problems 225
- *Level 3 Review* *226*

CHAPTER 7 GEOMETRY 229

Level 1 Geometry Skills 230
- Lesson 1 Angles 231
- Lesson 2 Pairs of Angles 233
- Lesson 3 Parallel Lines and Transversals 234
- Lesson 4 Triangles 236
- *Level 1 Review* *238*

Level 2   Geometry Applications **240**
      Lesson 5   Similar Figures 240
      Lesson 6   Congruent Figures 243
      Lesson 7   The Pythagorean Relationship 246
      Lesson 8   Rectangular Coordinates 248
      Lesson 9   Finding the Distance Between Points 251
      Lesson 10 Graphs of Linear Equations 255
      Lesson 11 Graphs and Quadratic Equations 258
      Lesson 12 Slope and Intercepts 262
      Level 2 Review 265

Level 3   Geometry Problem Solving **269**
      Lesson 13 Using Algebra to Solve Formulas 269
      Lesson 14 Algebraic Expressions in Formulas 271
      Lesson 15 Factoring and the Pythagorean Relationship 273
      Level 3 Review 275

POSTTEST **277**

SIMULATED TEST **285**

ANSWERS AND SOLUTIONS **294**

TO THE STUDENT

What Is the Mathematics Test?

The Mathematics Test of the GED Tests examines your ability to solve the kind of math problems you are likely to run into in your daily life.

When you take the test, you will answer questions and solve problems based on brief passages and graphic material. Solving some of the problems will require working through two or more steps. For example, you might be given information about two prospective jobs with different pay and different annual raises. You also might be told the cost of transportation to each job. Then you might be asked to calculate which job would net more money at the end of two years. To answer that question would require performing several different operations.

About two-thirds of the problems are based on written passages. The other one-third will test your ability to use information presented in a graphic format. For example, you might be presented with a circle graph showing in percents how a nursing home's budget is divided. You then might be given the total budget in dollars and be asked to figure the dollar amount of one of the budgeted items.

One-fourth of the items do not require you to solve problems. Instead they ask you to show how you would go about solving the problems. To answer some items correctly, you have to recognize and answer that you are not given enough information.

If a question requires you to use a formula, the formula will be provided for you. You are required only to demonstrate that you can apply mathematical theory to solve the kinds of problems you may face in your work, as a consumer, or in your normal daily family life.

All of the 56 items on the GED Mathematics Test are in multiple-choice format and cover the following content areas: Arithmetic 50% (Measurement 30%, Number Relationships 10%, Data Analysis 10%); Algebra 30%; Geometry 20%.

The Five GED Tests

The GED covers five content areas: writing skills, social studies, science, literature and the arts, and mathematics. The specifics of each test are listed below.

- *Writing Skills, Part I:* You will have 75 minutes to answer 55 questions (Sentence Structure 35%; Usage 35%; Mechanics 30%). Most questions involve detecting and correcting errors.

- *Writing Skills, Part II:* You will have 45 minutes to write a 200-word composition. The topic will be familiar to most people.
- *Social Studies:* You will have 85 minutes to answer 64 questions (History 25%; Geography 15%; Economics 20%; Political Science 20%; Behavioral Science 20%. NOTE: In Canada, Geography is 20% and Behavioral Science is 15%.) Most questions are based on reading passages. About one-third are based on graphic material.
- *Science:* You will have 95 minutes to answer 66 questions (Biology 50%; Physical Sciences 50%). Most questions are based on reading passages. Others are based on graphic material.
- *Literature and the Arts:* You will have 65 minutes to answer 45 questions (Popular Literature 50%; Classical Literature 25%; Commentary 25%). Questions are based on reading passages.
- *Mathematics:* You will have 90 minutes to answer the 56 questions on Arithmetic, Algebra, and Geometry. Most of the questions are word problems.

Actual passing scores differ from area to area, but regardless of your area, you will need to pay attention to two scores. One is the *minimum score* you must get on each test. If your area sets a minimum score of 35, you have to score at least 35 points on each of the five tests. The second score is a *minimum average score* on all five tests. If your area requires a minimum average score of 45, you have to get a total of 225 points to pass ($45 \times 5 = 225$). To pass the GED, you must meet *both* requirements: a) the minimum score on each of the five tests, and b) the minimum average score for all five tests. Failure to meet one or the other score will result in failure to pass the GED.

If you do not pass the test the first time around, you can take one or all five tests again. You will receive a different form of the test each time you take it, but the experience of having taken the test before should improve your score the next time. Of course, you will want to study again to be fully prepared for the test.

Places to contact about taking the GED are the office of the superintendent of schools in your area, vocational education centers, local community colleges, and adult education courses. Or write to:

General Education Development
GED Testing Service of the American Council on Education
One Dupont Circle
Washington, DC 20036

How to Be a Better Test-Taker

Often people say that they are not good test-takers. These people may be intelligent and earn good grades in school, but they do not do well on standardized tests. You may think of yourself as one of these people, but people can take steps to improve their chances of doing well on a test. Here are some helpful hints to make you a better test-taker.

- Study the content areas of the GED.
- Practice taking tests.
- Be well-rested for the actual test.
- Allow yourself enough time to get to the test center.
- Follow directions carefully.
- Pay attention to the time.
- Use your test-taking skills.
- Answer all questions on the GED Test.
- Mark your answers carefully.
- Above all, relax.

How to Use This Book

Take a Glance at the Table of Contents

Before you do anything else, look over the Table of Contents. It's also a good idea to leaf through the book to see what each section looks like.

Take the Pretest

The Pretest is like the actual GED Test in many ways. It will check your skills as you apply them to the kinds of problems you will find on the real test. The questions are similar to those on the actual test.

The Pretest will be most useful if you take it in a manner as close as possible to the way the actual test is given. Try to complete it in one sitting without distractions. Write your answers neatly on a sheet of paper or use an answer sheet provided by your teacher.

As you take the test, don't be discouraged if you find you are having difficulty with some (or even many) of the questions. The purpose of this test is to predict your overall performance on the GED and to locate your strengths and weaknesses. Relax; you will have many opportunities to correct any weaknesses and retest them.

You may want to time yourself to see how long you take to complete the test. When you take the actual Mathematics Test, you will be given 90 minutes. The Pretest is half as long as the actual test, so if you finish within 45 minutes, you are right on target. At this stage, however, you shouldn't worry too much if it takes you longer.

At the end of the test you will find a Pretest Guide (page 5). It will give you a general idea about the areas with which you are most comfortable and those which give you the most trouble. As you work through the instruction in this book, you will complete several short exercises that will help you pinpoint your skill levels even more closely.

Begin Your Instruction

After you have used the Pretest to analyze your strengths and weaknesses, you are ready to begin instruction. The instruction is divided into seven chapters: Whole Numbers, Decimals, Fractions, Percents, Graphs, Algebra, and Geometry.

Most chapters are divided into three levels. The first level concentrates on how to calculate using the kind of numbers covered by the chapter or, in algebra or geometry, on basic principles. The second level of each chapter shows how to apply calculations or basic principles to solving simple problems. The third level shows how to solve more complex problems.

The first step in instruction is to take the Level 1 Preview at the beginning of each chapter. If you score 100%, go ahead to the Level 1 Review. If you score 80% or better on the Level 1 Review, you can be satisfied that you understand the material in that first level. You should always work through the lessons in Level 2 and Level 3. After each level, there is a Level Review. Try to score 80% or better on the Level 2 and Level 3 Reviews.

If you like, you can work through all the Level 1s, then all the Level 2s, and finally all the Level 3s. Or you can work through each chapter in order—Level 1, Level 2, Level 3—before going on to the next chapter. Use the order for studying that is most interesting and effective for you.

A Review Test follows Chapter 5. This test contains problems similar to those you will see on the GED Mathematics Test. All the problems are based on the material in the first five chapters. The Review Test will give you an idea of how well you are doing.

Use the Posttest

This section gives you valuable practice in answering the types of questions you will find on the actual Mathematics Test. In the Posttest, the types of math you need to use vary from item to item, just as on the real test. You can use your results to track your progress and to get an idea of how well prepared you are to take the Mathematics Test.

Take the Simulated Test

Finally, once you have completed the instructions and Posttest, you can take the Simulated Test, which starts on page 285. It contains 56 items, the same number of items on the actual Mathematics Test. This will give you the most-accurate assessment of how ready you are to take the actual GED Mathematics Test.

The Answers and Solutions

At the back of this book you will find a section called Answers and Solutions. This answer key contains the answers to all the questions in the Lesson Exercises, Level Previews and Level Reviews, Review Test, Posttest, and Simulated Test. The answer section is a valuable study tool: It tells you the right answers and shows the process for solving each question. You can benefit a great deal by consulting the answer section as soon as you complete an activity.

PRETEST

Directions: Before you begin to work through the lessons in this book, try the following test. Use any of the formulas on pages 17 and 18 that you need. Choose the one best answer for each item.

When you finish, check your answers. After the test there is a guide to the sections of the book where the skills tested in each problem are presented.

1. Juan makes $7 an hour. In one week his gross income was $245. How many hours did he work that week?
   - (1) 12
   - (2) 20
   - (3) 30
   - (4) 35
   - (5) 40

2. Find, to the nearest kilogram, the weight of six packages each of which weighs 2.95 kg.
   - (1) 19
   - (2) 18
   - (3) 17
   - (4) 16
   - (5) 15

3. Following are the weights of the members of the Ramos family: Mr. Ramos—195 lb, Mrs. Ramos—118 lb, Susana—84 lb, and Alfredo—156 lb. What is the median weight in pounds of the members of the family?
   - (1) 137
   - (2) 140
   - (3) 156
   - (4) 166
   - (5) 174

Items 4 and 5 refer to the following diagram.

4. The diagram shows the plan of a rectangular garden. How many feet of fencing are needed to go around the garden?
   - (1) 36
   - (2) 42
   - (3) 48
   - (4) 84
   - (5) 94

5. How many square feet of sod are needed to cover the garden with grass?
   - (1) 432
   - (2) 216
   - (3) 144
   - (4) 84
   - (5) 18

Items 6 and 7 refer to the following information.

Carol has a large bag that contains 3 glazed doughnuts, 4 cinnamon doughnuts, and 5 plain doughnuts.

6. What is the probability that the first doughnut she pulls from the bag will be cinnamon?
   - (1) $\frac{1}{4}$
   - (2) $\frac{2}{5}$
   - (3) $\frac{1}{2}$
   - (4) $\frac{1}{3}$
   - (5) $\frac{4}{5}$

7. The first doughnut Carol took from the bag was plain, the second was glazed, and the third was cinnamon. What is the probability that the fourth doughnut she takes will be plain?
   (1) $\frac{1}{4}$
   (2) $\frac{1}{3}$
   (3) $\frac{1}{2}$
   (4) $\frac{5}{12}$
   (5) $\frac{4}{9}$

Use the table to answer items 8 and 9.

Yearly Per Capita Expenditures

State	Amount
California	$3,240
Florida	2,555
Massachusetts	3,286
New Jersey	3,297
New York	4,200
Texas	3,240

Source: U.S. Census Bureau.

8. For which state shown on the table were the per capita expenditures by state and local governments the highest?
   (1) New York
   (2) New Jersey
   (3) Massachusetts
   (4) Texas
   (5) California

9. Find the difference in the per capita amount spent by state and local governments in New Jersey and New York.
   (1) $1,197
   (2) $1,097
   (3) $1,103
   (4) $1,003
   (5) $ 903

10. Below are the weights of five boxes.
    Box A—0.15 kg
    Box B—0.9 kg
    Box C—0.85 kg
    Box D—1.05 kg
    Box E—0.955 kg
    Which of the following lists the boxes in order from heaviest to lightest?
    (1) C, B, A, E, D
    (2) E, D, A, C, B
    (3) A, C, B, E, D
    (4) D, B, C, A, E
    (5) D, E, B, C, A

11. Mary McGuire makes $280 a week. Her daughter Margaret, who lives with her, makes $210 a week. Margaret's weekly pay is what fraction of her mother's weekly pay?
    (1) $\frac{1}{8}$
    (2) $\frac{1}{4}$
    (3) $\frac{1}{2}$
    (4) $\frac{2}{3}$
    (5) $\frac{3}{4}$

Items 12 to 14 refer to the following situation.

Max Reade bought 45 acres of farm land from Mr. Brown at a price of $800 an acre. Max made a down payment of 20% of the basic price of the land, and he financed the rest with a bank loan. Besides the basic price of the land, Max had to pay legal fees of $900. Mr. Brown had to pay $1800 to the real estate agent who handled the sale and $700 in legal fees.

12. How much was Max's down payment?
    (1) $3600
    (2) $5400
    (3) $7200
    (4) $8333
    (5) Not enough information is given.

13. The real estate agent's fee was what percent of the basic price Mr. Brown asked for the land?
    (1) 5%
    (2) 7.5%
    (3) 9%
    (4) 12.5%
    (5) Not enough information is given.

14. Find the total amount of interest Max had to pay the first year on his bank loan.
    (1) $ 720
    (2) $1200
    (3) $1440
    (4) $3600
    (5) Not enough information is given.

15. Which of the following expresses the cost of 4.5 meters of cloth at $8 a meter and 2.5 meters of cloth at $10 a meter?
    (1) $4.5 \times 8 \times 2.5 \times 10$
    (2) $4.5 + 8 + 2.5 + 10$
    (3) $(4.5 \times 10) + (2.5 \times 8)$
    (4) $(2.5 \times 4.5) + (10 \times 8)$
    (5) $(4.5 \times 8) + (2.5 \times 10)$

16. A train traveled for $2\frac{1}{2}$ hours at an average speed of 60 mph. Find the distance in miles that the train traveled.
    (1) 120
    (2) 135
    (3) 150
    (4) 165
    (5) 180

17. Find the area of a triangle whose base measures 8.6 meters and whose height measures 6.5 meters. Express the answer to the nearest square meter.
    (1) 56
    (2) 30
    (3) 28
    (4) 24
    (5) 14

18. On this centimeter scale, which of the following expresses the distance between points A and C?

    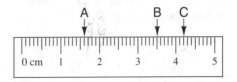

    (1) 0.7 cm
    (2) 1.6 cm
    (3) 1.9 cm
    (4) 2.6 cm
    (5) Not enough information is given.

19. This rectangular container has a length of 20, a width of 3, and a height of 0.75. Find the volume of the container.

    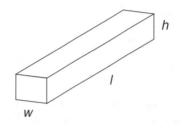

    (1) 23.75
    (2) 45
    (3) 47.5
    (4) 60
    (5) Not enough information is given.

20. Don and his assistant each worked 20 hours on a carpentry job. Don makes $5 an hour more than his assistant. Together they were paid $460 for the job. Find Don's hourly rate.
    (1) $ 7
    (2) $ 8
    (3) $ 9
    (4) $12
    (5) $14

21. Sandy types 80 words per minute. Which of the following tells the number of minutes she needs to type a 250-word letter and a 2000-word report?
    (1) $80 \times 250 + 2000$
    (2) $\frac{80 + 2000}{250}$
    (3) $\frac{250 + 2000}{80}$
    (4) $250 \times 2000 \times 80$
    (5) $\frac{80}{250 + 2000}$

22. David is twice as old as his daughter Catherine. David's wife Mary is three years younger than David. If x represents Catherine's age, which of the following represents Mary's age?
    (1) $2x - 3$
    (2) $2x$
    (3) $2x + 3$
    (4) $3x + 2$
    (5) $3x - 2$

23. A number is one more than three times a smaller number. The larger number is equal to the smaller number increased by 25. Find both numbers.
    (1) 10 and 31
    (2) 10 and 25
    (3) 25 and 37
    (4) 12 and 37
    (5) 15 and 46

24. Which of the following has the same value as 3×10^5?
    (1) $3 \times 10 \times 5$
    (2) $3 \times 3 \times 3 \times 3 \times 3 \times 5$
    (3) 3×50
    (4) $3 \times 10 \times 3 \times 10$
    (5) $3 \times 10 \times 10 \times 10 \times 10 \times 10$

Items 25 and 26 refer to the graph below.

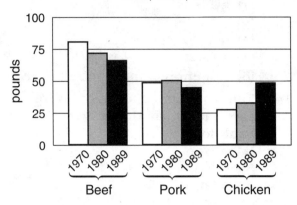

Per Capita Meat Consumption, 1970, 1980, 1989

25. For which product shown in the graph was the per capita consumption more than 75 pounds in one year?
    (1) beef in 1970
    (2) beef in 1989
    (3) pork in 1980
    (4) chicken in 1980
    (5) chicken in 1989

26. For the years and products shown on the graph, which of the following is true?
    (1) While the per capita consumption of beef and pork remained about the same, the consumption of chicken decreased.
    (2) While the per capita consumption of beef decreased, the consumption of pork and chicken increased.
    (3) Of the three products, only chicken increased in per capita consumption.
    (4) The per capita consumption of all three products gradually increased.
    (5) The per capita consumption of all three products gradually decreased.

27. In this triangle $\angle A = 55°$ and $\angle B = 70°$. Find the measurement of $\angle C$.

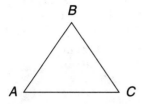

(1) 55°
(2) 65°
(3) 70°
(4) 110°
(5) 125°

28. This diagram shows a rectangle. Which of the following expresses the diagonal distance AC?

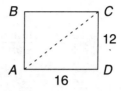

(1) $\sqrt{16^2 - 12^2}$
(2) $\sqrt{16 - 12}$
(3) $\sqrt{16^2 + 12^2}$
(4) $\sqrt{16 + 12}$
(5) $\sqrt{\dfrac{16 + 12}{2}}$

Check your answers below.

Pretest Guide

Below is a list of the problems on the Pretest. After each problem are the chapter and lesson number where you will find similar problems.

1. Whole Numbers/13
2. Decimals/6
3. Whole Numbers/6
4. Whole Numbers/9
5. Whole Numbers/10
6. Fractions/16
7. Fractions/16
8. Whole Numbers/16
9. Whole Numbers/16
10. Decimals/5
11. Fractions/9
12. Percents/6
13. Percents/11
14. Percents/9
15. Decimals/13
16. Fractions/5
17. Decimals/11
18. Decimals/8
19. Decimals/11
20. Algebra/27
21. Whole Numbers/18
22. Algebra/24
23. Algebra/27
24. Whole Numbers/7
25. Graphs/3
26. Graphs/3
27. Geometry/4
28. Geometry/7

PRETEST ANSWERS

1. **(4) 35**
```
        35 hours
$7)$245
     21
     35
     35
```

2. **(2) 18**
```
    2.95
  ×    6
  17.70  to the nearest kilogram = 18
```

3. **(1) 137**
   In order: 84 118 156 195
   Mean of middle two values:
```
   118      137
 + 156    2)274
   274
```

4. **(4) 84**
   $P = 2l + 2w$
   $P = 2(24) + 2(18)$
   $P = 48 + 36$
   $P = 84$ ft

5. **(1) 432**
   $A = lw$
   $A = 24 \times 18$
   $A = 432$ sq ft

6. **(4) $\frac{1}{3}$**

glazed	3	cinnamon	$\frac{4}{12} = \frac{1}{3}$
cinnamon	4	total	
plain	+ 5		
total	12		

7. **(5) $\frac{4}{9}$**

glazed	2	plain	$\frac{4}{9}$
cinnamon	3	total	
plain	+ 4		
new total	9		

8. **(1) New York**

9. **(5) $903**
   New York $4,200
   New Jersey − 3,297
               $ 903

10. **(5) D, E, B, C, A**
    A = 0.15 = 0.150 kg
    B = 0.9 = 0.900 kg
    C = 0.85 = 0.850 kg
    D = 1.05 = 1.050 kg
    E = 0.955 = 0.955 kg

11. **(5) $\frac{3}{4}$**
    $\frac{\$210}{\$280} = \frac{3}{4}$

12. **(3) $7200**
        45   20% = 0.2 $36,000
    $\times$ $800 $\times$ 0.2
    $36,000 $7200.0

13. **(1) 5%**
    $\frac{\text{fee}}{\text{land price}} = \frac{1{,}800}{36{,}000} = \frac{1}{20} \quad \frac{1}{20} \times \frac{100}{1} = 5\%$

14. **(5) Not enough information is given.**
    You do not know the terms of the loan.

15. **(5) (4.5 × 8) + (2.5 × 10)**

16. **(3) 150**
    $d = rt$
    $d = 60 \times 2\frac{1}{2}$
    $d = \frac{60}{1} \times \frac{5}{2} = 150$
    $d = 150$ miles

17. **(3) 28**
    $A = \frac{1}{2}bh$
    $A = \frac{1}{2} \times 8.6 \times 6.5 = 27.95$
    to the nearest m² = 28.

18. **(4) 2.6 cm**
    point C   4.2 cm
    point A   −1.6
                  2.6 cm

19. **(2) 45**
    $V = lwh$
    $V = 20 \times 3 \times 0.75 = 45$

20. **(5) $14**
    assistant = x   20x + 20(x + 5) = 460
    Don = x + 5   20x + 20x + 100 = 460
                            40x + 100 = 460
                                − 100 −100
                            $\frac{40x}{40} = \frac{360}{40}$
                                   x = 9
    x + 5 = 9 + 5 = 14

21. **(3) $\frac{250 + 2000}{80}$**

22. **(1) 2x − 3**
    Catherine's age = x
    David's age = 2x
    Mary's age = 2x − 3

23. **(4) 12 and 37**
    small number = x   3x + 1 = x + 25
    large number = 3x + 1   −x   −x
                               2x + 1 = 25
                                  − 1   −1
                              $\frac{2x}{2} = \frac{24}{2}$
                                  x = 12
    3x + 1 = 3(12) + 1 = 37

24. **(5) 3 × 10 × 10 × 10 × 10 × 10**

25. **(1) beef in 1970**

26. **(3)** Of the three products, only chicken increased in per capita consumption.

27. **(1) 55°**
    ∠A = 55°   180°
    ∠B = +70°   −125°
              125°   55° = ∠C

28. **(3) $\sqrt{16^2 + 12^2}$**
    $c^2 = a^2 + b^2$
    $c^2 = 16^2 + 12^2$
    $c = \sqrt{16^2 + 12^2}$

Chapter 1
WHOLE NUMBERS

Level 1: Whole-Number Skills

Whole numbers are the tools we use to count things. Counting the population of your town or the number of people in your family are examples of how we use whole numbers.

Whole numbers are used to count whole things, not parts of things. Decimals and fractions are used to describe parts of things.

The skills section of this chapter reviews the whole-number system and the basic operations addition, subtraction, multiplication, and division. To find out whether you need to review whole-number skills, try the following preview.

Preview

Choose the correct answer to each question.

1. In the number 18,702 which digit is in the thousands place?
   (1) 1   (2) 8   (3) 7   (4) 0   (5) 2
2. In the number 680,451 which digit is in the hundreds place?
   (1) 6   (2) 8   (3) 0   (4) 4   (5) 5
3. What is the value of 7 in the number 287,356?
   (1) 70   (2) 700   (3) 7000
4. How many digits are in the number 8049?
   (1) 4   (2) 3   (3) 2   (4) 1

Solve each problem.

5. What is the difference between 9000 and 496?
6. Find the product of 473 and 90.
7. Divide: $\frac{7635}{15}$.
8. Find the sum of 49, 207, 5653, and 28.
9. Compute: 16,836 divided by 6.

For the following problems choose the correct setup. Then solve.

10. Subtract 5,207 from 30,050.

    (1) 30,050
        − 5,207

    (2) 30,050
        − 5,207

    (3) 5,207
        − 30,050

11. What is the sum of 12,906, 385, and 4,059?

(1)
```
      12
     906
     385
  + 4059
```
(2)
```
  12,906
     385
  + 4,059
```
(3)
```
   12906
     385
  + 4059
```

12. Divide 9879 by 21.

(1) $9879\overline{)21}$ (2) $\frac{21}{9879}$ (3) $21\overline{)9879}$

Check your answers on page 294. If you have all 12 problems correct, do the Level 1 Review on page 16. If you have fewer than 12 problems correct, study Level 1 beginning with Lesson 1.

Lesson 1

The Whole-Number System

Whole numbers in our number system are written using the **digits** 0, 1, 2, 3, 4, 5, 6, 7, 8, and 9. The number 56 has two digits. The number 28,307 has five digits.

Each position in a whole number gives a **place value**. Below is a chart of the place value of the first seven positions in our whole number system.

Place Names / Places: millions , hundred thousands | ten thousands | thousands , hundreds | tens | units (ones)

As you move left, each position represents ten times the value of the position immediately to its right. The units place has a value of 1. The tens place has a value of $10 \times 1 = 10$. The hundreds place has a value of $10 \times 10 = 100$, and so on.

Think about the number 28,307.

7 is in the units (ones) place. 7 has a value of	$7 \times 1 = 7.$
0 is in the tens place. 0 has a value of	$0 \times 10 = 0.$
3 is in the hundreds place. 3 has a value of	$3 \times 100 = 300.$
8 is in the thousands place. 8 has a value of	$8 \times 1000 = 8000.$
2 is in the ten thousands place. 2 has a value of	$2 \times 10,000 = \underline{20,000}.$
	28,307

Lesson 1 Exercise

Answer each question.

1. Circle the two-digit number(s).   446   89   1,001   16
2. Circle the three-digit number(s).   2,022   980   600
3. Circle the five-digit number(s).   15,000   1,238   90,000

In the number 546,903,

4. What is the value of 9?
5. What is the value of 6?
6. What is the value of 4?
7. What is the value of 5?

In the number 2,750,300,

8. What is the value of 3?
9. What is the value of 5?
10. What is the value of 2?

Check your answers on page 294.

Lesson 2: Basic Operations: Addition and Subtraction

Using the four basic operations, addition, subtraction, multiplication, and division, you can solve many problems on the GED Test. This lesson and the next include hints to help you avoid common mistakes involving these operations.

Addition—The answer to an addition problem is called the **sum** or **total**.

When you set up an addition problem, be sure the numbers are lined up correctly. Put the units under the units, the tens under the tens, the hundreds under the hundreds, and so on. Notice how the examples are set up.

To check an addition problem add the columns in the other direction. For example, if you first added from the top down, check by adding from the bottom up.

Example 1: Find the sum of 95, 5638, and 6.

Step 1. Line up the numbers. Put the units under the units, the tens under the tens, and so on.

Step 2. Add each column from the top down. Carry the left-hand digit from the sum of each column to the next column. The sum is 5739.

```
  1 1
    95
  5638
 +   6
  ----
  5739
```

Step 3. Check by adding each column starting from the bottom.

```
    95
  5638
 +   6
  ----
  5739
```

Subtraction—The answer to a subtraction problem is called the **difference**.

When you subtract one number from another, be sure to put the larger number on top. Line up the numbers with units under units, tens under tens, and so on. Then subtract.

Example 2: Find the difference between 804 and 5946.

Step 1. Put the larger number on the top. Line up the numbers with units under units, tens under tens, and so on.

```
  5946
 - 804
  ----
  5142
```

Step 2: Subtract.

To check a subtraction problem add the answer to the number directly above it. The sum should equal the top number of the problem.

Look carefully at the borrowing in the next two examples. Notice, especially, how the borrowing is handled with zeros.

Example 3: Find the difference between 8,249 and 50,600.

Step 1. Put the larger number, 50,600, on top. Put units under units, tens under tens, and so on.

$$\begin{array}{r} 50,600 \\ -8,249 \end{array}$$

Step 2. Borrow and subtract.

$$\begin{array}{r} 50,600 \\ -8,249 \\ \hline 42,351 \end{array}$$

Step 3. Check by adding the answer to the number directly above it.

$$\begin{array}{r} 8,249 \\ +\,42,351 \\ \hline 50,600 \end{array}$$

On the GED Test you may be asked to choose the correct setup for a problem.

Example 4: Which of the following is the correct setup to find the difference between 12,980 and 7,604?

(1) 7,604   (2) 12,980   (3) 12,980
   − 12,980   − 7,604   − 76,04

Step 1: Remember that to subtract, the larger number must be on top. The larger number is on top in choices **2** and **3**.

Step 2: Also, the units must be under the units and so on. Only choice **2** is correctly lined up, so choice **2** is the correct setup.

Lesson 2 Exercise

Choose the correct setup for each problem. Then solve.

1. Find the sum of 9015, 493, and 76.

    (a)   9015   (b)   9,015   (c)   9015
          493   + 76,493   493
        + 76   + 76

2. What is the difference between 22,500 and 6,087?

    (a)   22,500   (b)   22,500   (c)   6,087
        − 60,87   − 6,087   − 22,500

3. Take 763 from 5030.

    (a)   5030   (b)   763   (c)   5030
        − 763   − 5030   − 763

Chapter 1 Whole Numbers

4. What is the sum of 704, 86, and 10,471?

(a) 704
      86
   + 104,71

(b) 704
      86
   + 10,471

(c) 704
      86
      10
   + 471

Solve each of the following problems.

5. Compute 78 + 4062 + 529.
6. Find the difference between 8,346 and 42,003.
7. Subtract 11,954 from 18,206.
8. Find the difference between 30,005 and 19,472.
9. Find the sum of 428, 61, 593, and 7.
10. What is the difference between 336 and 9048?

Check your answers on page 294.

Lesson 3: Basic Operations: Multiplication and Division

Multiplication—The answer to a multiplication problem is called the **product**.
To find a product it is easier to set up the problem by putting the number with more digits on top. When you multiply by the digit in the units place, be sure to start the **partial product** under the units. When you multiply by the digit in the next column moving left (the tens digit), be sure to start the partial product under the tens.

To check a multiplication problem, repeat each step.

Example 1: Multiply 509 by 28.

Step 1. Put 509 on top and find the partial products.

```
     509
   ×  28
   4 072    ← Partial product for 509 × 8
  10 18     ← Partial product for 509 × 20
  14,252    ← Product
```

Step 2. Add the partial products. The product is 14,252.

Example 2: Find the product of 876 and 204.

Step 1. Find the partial products. Notice how the zero holds a place in the second partial product. The partial product 2 × 876 must begin in the hundreds place since 2 is in the hundreds place.

```
    876
   ×204
   3 504
  175 20    ← Zero is a placeholder
  178,704
```

Step 2. Add the partial products. The product is 178,704.

Division—The answer to a division problem is called the **quotient**.

There are three common ways to write division problems. The problem "20 divided by 4 equals 5" can be written in any of the following ways.

$$20 \div 4 = 5 \qquad \frac{20}{4} = 5 \qquad 4\overline{)20}^{\,5}$$

> To find a quotient repeat these four steps until you finish the problem:
> 1. Divide.
> 2. Multiply.
> 3. Subtract.
> 4. Bring down the next number.

Example 3: Divide 96 by 3.

Step 1. Divide 9 by 3.

Step 2. Multiply 3 × 3.

Step 3. Subtract 9 from 9.

Step 4. Bring down the next number. Repeat the four steps until you cannot bring down any more.

$$\begin{array}{r} 32 \\ 3\overline{)96} \\ \underline{9} \\ 06 \\ \underline{6} \\ 0 \end{array}$$

It is important to line up division problems carefully. When you have located the first digit in a quotient, you must have a digit in the quotient for each remaining digit in the number into which you are dividing.

To check a division problem multiply the answer by the number by which you divided. The product should equal the number into which you divided. If there is a remainder, add the remainder to the product.

In the next two examples be sure you understand how to get each number in the solution.

Example 4: Find 4912 divided by 16. Follow the four steps of division.

Step 1. Divide 16 into 49.

Step 2. Multiply 3 × 16.

Step 3. Subtract 48 from 49.

Step 4. Bring down the next number. The 3 belongs directly above the 9. Notice how the zero in the quotient holds a place. The zero shows that 16 does not divide into 11. The quotient is 307. Check by multiplying 307 by 16. The answer should be 4912.

$$\begin{array}{r} 307 \\ 16\overline{)4912} \\ \underline{48} \\ 11 \\ \underline{0} \\ 112 \\ \underline{112} \\ 0 \end{array} \qquad \begin{array}{r} 307 \\ \times\ 16 \\ \hline 1842 \\ 307 \\ \hline 4912 \end{array}$$

Example 5: Divide: $\frac{965}{32}$.

Step 1. Follow the four steps of division. First divide 32 into 96. Notice how the zero holds a place in the quotient. The zero shows that 32 does not divide into 5. The quotient is 30 with a remainder of 5.

```
      30 r 5         32
32)965             × 30
   96              960
    05            +  5
     0             965
     5
```

Step 2. Check by multiplying 32 by 30. Then add the remainder of 5. The answer should be 965.

Lesson 3 Exercise

Choose the correct setup for each problem. Then solve.

1. Divide 9636 by 12.
   (a) 9636
       × 12
   (b) 9636)12
   (c) 12)9636

2. Find the product of 704 and 18.
   (a) 704
       × 18
   (b) 704)18
   (c) 704
       + 18
   (d) 704
       − 18

3. Find the product of 536 and 800.
   (a) 536)800
   (b) $\frac{800}{536}$
   (c) 536
       × 800
   (d) 536
       + 800

4. Find 7256 divided by 8.
   (a) 7256)8
   (b) 8)7256
   (c) 7256
       × 8
   (d) 8
       × 7256

Solve each of the following problems.

5. Divide: $\frac{3588}{46}$.
6. What is the product of 38 and 506?
7. Find the product of 409 and 817.
8. What is the quotient of 1365 divided by 17?
9. What is the product of 34 times 230?
10. 4424 ÷ 316 = ?
11. Divide 3042 by 78.
12. Find 2392 divided by 523.

Check your answers on page 295.

Level 1 Review

Choose the correct answer to each question.

1. In the number 175,600 which digit is in the ten-thousands place?
   (1) 1   (2) 7   (3) 5   (4) 6   (5) 0
2. Which digit in 38,962 is in the thousands place?
   (1) 3   (2) 8   (3) 9   (4) 6   (5) 2
3. What is the value of 5 in the number 86,540?
   (1) 5   (2) 50   (3) 500   (4) 5000
4. What is the value of 2 in the number 126,300?
   (1) 20   (2) 200   (3) 2000   (4) 20,000

Solve each problem.

5. What is the sum of 890, 23, 4017, and 605?
6. Compute 11,043 divided by 12.
7. What is the difference between 12,050 and 9,947?
8. Find the product of 76 and 308.
9. Divide: $\frac{24,768}{8}$.

Choose the correct setup. Then solve.

10. What is the product of 208 and 675?
    (1) 208 + 675   (2) 675 × 208   (3) $\frac{208}{675}$ (4) 675 − 208
11. Find the difference between 306,471 and 28,295.
    (1) 306,471 − 28,295   (2) 306471 − 28295   (3) 28295 − 306471   (4) 306,471 + 28,295
12. Divide 25,728 by 32.
    (1) $\frac{32}{25,728}$ (2) $32\overline{)25,728}$ (3) $25,728\overline{)32}$

Check your answers on page 295. If you have all 12 problems correct, go on to Level 2. If you answered a question incorrectly, find the item number on the chart below and review that lesson.

Review:	If you missed item number:
Lesson 1	1, 2, 3, 4
Lesson 2	5, 7, 11
Lesson 3	6, 8, 9, 10, 12

Chapter 1: Whole Numbers

Level 2: Whole Number Applications

In this section you will learn to apply basic whole-number skills in a variety of situations. You will learn how to round numbers so they are easier to use. You will use multiplication to find the distance a vehicle travels when you know its rate and time of travel. You will use addition and division to find an average (mean) of a group of numbers. You will also learn about exponents and square roots.

A Word About Formulas

A formula is a mathematical rule written with letters and numbers. The following list contains the formulas you will use as you work through the rest of this text. Refer back to this list as the need arises.

Formulas

Description	Formula
AREA (A) of a	
square	$A = s^2$; where s = side
rectangle	$A = lw$; where l = length and w = width
parallelogram	$A = bh$; where b = base and h = height
triangle	$A = \frac{1}{2} bh$; where b = base and h = height
circle	$A = \pi r^2$; where π = 3.14 and r = radius
PERIMETER (P) of a	
square	$P = 4s$; where s = side
rectangle	$P = 2l + 2w$; where l = length and w = width
triangle	$P = a + b + c$; where a, b, and c are the sides
circumference (C) of a circle	$C = \pi d$; where π = 3.14 and d = diameter
VOLUME (V) of a	
cube	$V = s^3$; where s = side
rectangular container	$V = lwh$; where l = length, w = width, and h = height
cylinder	$V = \pi r^2 h$; where π = 3.14, r = radius, and h = height

Description	Formula
Pythagorean relationship	$c^2 = a^2 + b^2$; where c = hypotenuse and a and b are legs of a right triangle
distance (d) between two points in a plane	$d = \sqrt{(x_2 - x_1)^2 + (y_2 - y_1)^2}$; where (x_1, y_1) and (x_2, y_2) are two points in a plane
slope of a line (m)	$m = \dfrac{y_2 - y_1}{x_2 - x_1}$; where (x_1, y_1) and (x_2, y_2) are two points in a plane
midpoint (M) of a line	$M = \left(\dfrac{x_1 + x_2}{2}, \dfrac{y_1 + y_2}{2}\right)$
mean	mean $= \dfrac{x_1 + x_2 \cdots + x_n}{n}$; where the x's are the values for which a mean is desired and n = number of values in the series
median	median = the point in an ordered set of numbers at which half of the numbers are above and half of the numbers are below this value
simple interest (i)	$i = prt$, where p = principal, r = rate, and t = time
distance (d) as function of rate and time	$d = rt$; where r = rate and t = time
total cost (c)	$c = nr$; where n = number of units and r = cost per unit

Lesson 4

Rounding

Sometimes whole numbers are more exact than they need to be. For example, Mike weighs 178 pounds. We can say that he weighs about 180 pounds because 178 rounded to the nearest ten is 180. Rounding is a way of making numbers easier to read and easier to use. You will also use rounding to estimate answers.

To round whole numbers you must know the place value of every digit in a whole number. If you are unsure of the names of the places, review the chart in Whole-Number System Lesson 1 on page 10.

> To round a whole number:
> 1. Mark the digit in the place to which you want to round.
> 2. If the digit to the right of the marked digit is more than 4, add 1 to the marked digit.
> 3. If the digit to the right of the marked digit is less than 5, do not change the marked digit.
> 4. Replace the digits to the right of the marked digit with zeros.

Example 1: Round 487 to the nearest ten.

Step 1. Mark the digit in the tens place, 8.   48_7_

Step 2. Since the digit to the right of 8 is more than 4, add 1 to 8. Put 9 in the tens place and write 0 in the units place. 487 rounded to the nearest 10 is 490.   490

Example 2: Round 682,394 to the nearest thousand.

Step 1. Mark the digit in the thousands place, 2.   68_2_,394

Step 2. Since the digit to the right of 2 is less than 5, keep the 2 as it is and put zeroes in each place to the right of 2.   682,000

Example 3: Round 196,275 to the nearest ten-thousand.

Step 1. Mark the digit in the ten-thousands place, 9.   1_9_6,275

Step 2. Since the digit to the right of 9 is more than 4, add 1 to 9. Since 1 + 9 = 10, you must carry 1 over to the hundred-thousands column. Put zeros in each place to the right of the marked digit.   200,000

Lesson 4 Exercise

Solve.

1. Round each number to the nearest ten.
   78   164   3198   2433
2. Round each number to the nearest hundred.
   847   1273   6580   351
3. Round each number to the nearest thousand.
   3196   41,826   28,752   149,628
4. Round each number to the nearest hundred-thousand.
   777,500   316,450   567,300   3,470,992
5. Round each number to the nearest million.
   5,648,000   12,387,000   32,479,000   188,750,000

Check your answers on page 296.

Lesson 5: Distance and Cost Formulas

A **formula** is a "short-hand" mathematical instruction in which letters represent numbers. You will use many formulas in this book. A list of the formulas that you will use in the first five chapters of this book appears on pages 17 and 18.

One useful formula is $d = rt$ where d = distance, r = rate, and t = time. When two letters are written next to each other, the two numbers those letters represent must be multiplied together. In words, "$d = rt$" means "Distance equals rate times time." In many distance problems distance is measured in miles, rate is measured in miles per hour, and time is measured in hours.

To use a formula to solve a problem, replace the letters in the formula with the numbers from the problem. Then compute according to the instructions in the formula.

Example 1: Mike drove on a highway for five hours at an average speed of 55 miles per hour. How far did he drive?

Step 1. Replace r with 55 and t with 5 in the formula $d = rt$.

$d = rt$
$d = 55 \times 5$

Step 2. Multiply 55 by 5. Mike drove 275 miles. Notice that distance is measured in miles.

$55 \times 5 = 275$ miles

Another useful formula is $c = nr$ where c = total cost, n = number of units, and r = rate (cost per unit). In words "$c = nr$" means "Cost is equal to the number of units multiplied by the rate." In most cost problems cost is measured in dollars and the rate is measured in dollars per unit.

Example 2: Ruth bought four cans of tuna at the rate (unit cost) of $1.29 per can. Find the total cost of the tuna.

Step 1. Replace n with 4 and r with $1.29 in the formula $c = nr$. Notice that unit cost is the same as rate.

$c = nr$
$c = 4 \times \$1.29$

Step 2. Multiply $1.29 by 4. The total cost is $5.16.

$4 \times \$1.29 = \$5.16$

Lesson 5 Exercise

Use $d = rt$ or $c = nr$ to solve each of the following problems.

1. Alfredo walked for three hours at a speed of 4 miles per hour. How far did he walk?

2. The Johnsons drove for four hours at an average speed of 65 miles per hour. How far did they drive?

3. A plane flew for five hours at an average speed of 475 miles per hour. How far did the plane fly?
4. Sarah drove in city traffic for three hours at an average speed of 15 mph. How far did she drive?
5. Manny bought three shirts at $18 each. Find the total cost of the shirts.
6. At $3.60 per pound what is the cost of five pounds of meat?
7. Find the cost of a dozen reams of typing paper at $6 a ream.
8. The Greenport Community Center bought 30 sets of desks and chairs at the rate of $65 a set. Find the total cost of the furniture.

Check your answers on page 296.

Lesson 6

Mean and Median

The word "average" occurs frequently in the news. There are references to average family income or the average selling price of a home. The average price is usually more than the price of the least expensive home and less than the price of the most expensive.

There are two ways to find a "middle" value for a group of numbers. One way is called the **mean** or **average**. The other way is called the **median**.

To find the **mean** for a group of numbers:
1. Add the numbers.
2. Divide the sum by how many numbers are in the group.

The formula for finding the mean, or average, is

$$\text{mean} = \frac{x_1 + x_2 + \cdots + x_n}{n}$$

where the x's are the values for which a mean is desired and n = the number of values.

Example 1: Sally works three days a week as a waitress. Her tips were $24 on Thursday, $42 on Friday, and $48 on Saturday. Find the average of her tips for the three days.

Step 1. Find the sum of the tips.

$$\begin{array}{r} \$\ 24 \\ 42 \\ +\ 48 \\ \hline \$114 \end{array}$$

Step 2. Divide the sum by the number of days, 3.
The average of her tips for the three days is $38.

$$\begin{array}{r} \$\ 38 \\ 3\overline{)\$114} \end{array}$$

The **median** is the middle value of a group of numbers arranged in order.

> To find the median in a group of numbers:
> 1. Put the numbers in order from smallest to largest.
> 2. The number in the middle is the median.
> 3. If there are two numbers in the middle, find the average (mean) of those two numbers.

Example 2: Find the median amount of Sally's tips in Example 1.

Step 1. Put the numbers in order from smallest to largest. $24 $42 $48

Step 2. The number in the middle, $42, is the median.

Example 3: Find the median for the following set of numbers: 287, 496, 317, and 409.

Step 1. Put the numbers in order from smallest to largest.   287 317 409 496

Step 2. Since two numbers, 317 and 409, are in the middle, find the mean of 317 and 409. The median is 363.

$$\begin{array}{r} 317 \\ + 409 \\ \hline 726 \end{array} \qquad \begin{array}{r} 363 \\ 2\overline{)726} \\ \underline{6} \\ 12 \\ \underline{12} \\ 06 \\ \underline{6} \end{array}$$

Lesson 6 Exercise

Solve each problem.

1. In June Al's electric bill was $14. In July it was $22. In August it was $18. Find the mean electric bill for those months.
2. What was the median electric bill for the months described in problem 1?
3. George Johnson weighs 185 pounds. His wife, June, weighs 138 pounds. Their daughter Jan weighs 97 pounds. Their son Joe weighs 88 pounds. What is the mean weight for the members of the Johnson family?
4. Deborah took five math tests. Her scores were 65, 86, 79, 92, and 88. Find her mean score for the five tests.
5. Paul is a traveling salesman. Monday he drove 284 miles; Tuesday, 191 miles; Wednesday, 297 miles; Thursday, 162 miles; and Friday, 256 miles. Find the mean distance he drove each day.

6. On Monday Don bought 11 gallons of gas. On Wednesday he bought 9 gallons and on Thursday, 13 gallons. Find the mean number of gallons he bought.

7. On Wednesday 213 people went to the Greenport Community Center Talent Show. 191 people went on Thursday, 289 on Friday, and 303 on Saturday. Find the mean attendance.

8. Find the median attendance for the show described in problem 7.

9. Maxine priced cans of tuna at five different stores. She found the following prices for the same brand and can size: $1.29, $1.49, $1.16, $1.25, and $1.39. What was the median price for a can of tuna?

10. In a year Mr. Munro made $24,800. Mrs. Munro made $22,500. Their daughter, Lee, made $6,000, and their son, Nick, made $4,200. What was the mean income of the Munro family?

Check your answers on page 296.

Lesson 7: Powers

Computing **powers** and **roots** are two operations besides addition, subtraction, multiplication, and division that you will use frequently. 7^2 means "Seven to the second power." The number 7 is the **base**. The number 2 is the **exponent**.

> To find a power:
> 1. Write the base the number of times the exponent tells you.
> 2. Multiply the numbers that you wrote.

Example 1: What is the value of 7^2?

Step 1. The exponent is 2. Write 7 two times. $7^2 = 7 \times 7$

Step 2. Multiply 7 by 7. $7^2 = 49$. $7 \times 7 = 49$

Example 2: What is the value of 5^3?

Step 1. The exponent is 3. Write 5 three times. $5^3 = 5 \times 5 \times 5$

Step 2. To multiply 5 by 5 by 5, first find $5 \times 5 = 25$. Then find $25 \times 5 = 125$. $5^3 = 125$. $5 \times 5 \times 5 = 125$.

There are some special cases with powers.

> 1 to any power = 1.

Level 2, Lesson 7: Powers

Example 3: Find the value of 1^4.

Step 1. The exponent is 4. Write 1 four times. $1^4 = 1 \times 1 \times 1 \times 1$

Step 2. Notice that no matter how many times you multiply 1 by itself, the answer is always 1. $1 \times 1 \times 1 \times 1 = 1$

> A number to the first power is that number.

Example 4: Find the value of 20^1.

Step 1. The exponent is 1. Write 20 one time. $20^1 = 20.$

Step 2. Notice that there is nothing to multiply in this problem. $20^1 = 20$.

> A number (except 0) to the zero power equals 1.

Example 5: Find the value of 6^0. $6^0 = 1$
The exponent is 0. According to the rule above, $6^0 = 1$.

Powers can be used in combination with addition and subtraction. First find the value of each power separately. Then add or subtract the values from left to right.

Example 6: What is the value of $3^4 - 6^2 + 2^3$?

Step 1. Find the value of each power.
$3^4 = 3 \times 3 \times 3 \times 3 = 81$
$6^2 = 6 \times 6 = 36$
$2^3 = 2 \times 2 \times 2 = 8$

Step 2. Add or subtract the values from left to right. Subtract 36 from 81. $81 - 36 = 45$. Then add 8. $45 + 8 = 53$.
$81 - 36 + 8 = 53$

Lesson 7 Exercise

Find the value of each of the following.

1. $2^4 =$ $9^2 =$ $3^3 =$ $8^1 =$
2. $13^2 =$ $50^2 =$ $6^3 =$ $12^0 =$
3. $2^5 =$ $1^5 =$ $16^2 =$ $40^2 =$
4. $10^3 =$ $18^0 =$ $25^2 =$ $4^4 =$
5. $5^2 - 2^3 =$ $8^2 + 3^3 =$ $10^2 - 4^2 + 5^2 =$
6. $4^3 + 6^1 - 2^4 =$ $12^2 - 5^0 - 3^2 =$ $10^3 - 10^2 =$

Check your answers on page 296.

Lesson 8: Square Roots

Finding roots is the opposite of finding powers. On the GED Test you will have to solve problems with **square roots**. Finding a square root is the opposite of finding a number to the second power. For example, 8 to the second power is 64. The square root of 64 is 8.

The sign for square root is $\sqrt{}$. To find a square root, find a number that when multiplied by itself gives the number inside the sign.

Example 1: Find the value of $\sqrt{36}$.

Ask yourself, "What number times itself is 36?" $\sqrt{36} = 6$
$6 \times 6 = 36$. Therefore, 6 is the square root of 36.

Following is a list of common square roots. Memorize these before you go on.

$\sqrt{1} = 1$	$\sqrt{49} = 7$	$\sqrt{169} = 13$	$\sqrt{2500} = 50$
$\sqrt{4} = 2$	$\sqrt{64} = 8$	$\sqrt{196} = 14$	$\sqrt{3600} = 60$
$\sqrt{9} = 3$	$\sqrt{81} = 9$	$\sqrt{225} = 15$	$\sqrt{4900} = 70$
$\sqrt{16} = 4$	$\sqrt{100} = 10$	$\sqrt{400} = 20$	$\sqrt{6400} = 80$
$\sqrt{25} = 5$	$\sqrt{121} = 11$	$\sqrt{900} = 30$	$\sqrt{8100} = 90$
$\sqrt{36} = 6$	$\sqrt{144} = 12$	$\sqrt{1600} = 40$	$\sqrt{10,000} = 100$

You can find the square root of a number with a method that uses averages. Suppose that you did not know that $\sqrt{144} = 12$. When you divide a number by its square root, the answer equals the square root. ($144 \div 12 = 12$.)

Guess an answer to $\sqrt{144}$. 10 is a good guess because $10 \times 10 = 100$, which is fairly close to 144.

Divide 144 by 10. $144 \div 10 = 14$ plus a remainder.

Average the guess, 10, and the answer to the division problem, 14. $10 + 14 = 24$. $24 \div 2 = 12$, which is the correct answer.

Check by multiplying the number by itself. $12 \times 12 = 144$.

To find the square root of a number:
1. Guess an answer.
2. Divide the guess into the number.
3. Average the guess and the answer to the division problem.
4. Check.

Example 2: Find the value of $\sqrt{1024}$.

Step 1. Guess. In the list of square roots $\sqrt{900} = 30$. 30 is too small, but it is easy to divide by.

$\sqrt{900} = 30$

Step 2. Divide 1024 by 30. Drop the remainder.

$$\begin{array}{r} 34 \\ 30\overline{)1024} \\ \underline{90} \\ 124 \\ \underline{120} \\ 4 \end{array}$$

Step 3. Find the average of 30 and 34.

$$\begin{array}{r} 30 \\ +\ 34 \\ \hline 64 \end{array} \qquad 2\overline{)64}^{\,32}$$

Step 4. Check. Multiply 32 by 32. 32 is the square root of 1024.

$$\begin{array}{r} 32 \\ \times\ 32 \\ \hline 64 \\ 96 \\ \hline 1024 \end{array}$$

When you use this method to find square roots, always guess a number that ends in zero. It is easier and faster to divide by these numbers. If the average is not the square root of the number, use the average as a new guess and try again.

Lesson 8 Exercise

Find the value of each square root.

1. $\sqrt{289} =$ $\sqrt{784} =$ $\sqrt{1444} =$ $\sqrt{484} =$
2. $\sqrt{1521} =$ $\sqrt{1849} =$ $\sqrt{529} =$ $\sqrt{2704} =$
3. $\sqrt{4096} =$ $\sqrt{8836} =$ $\sqrt{6084} =$ $\sqrt{7056} =$

Check your answers on page 297.

Lesson 9 — Perimeter

There are three geometric figures that you will often see: the **rectangle**, the **square**, and the **triangle**.

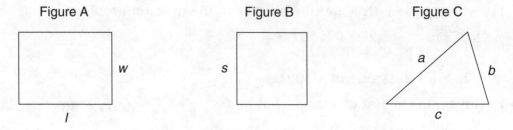

Figure A — rectangle (sides l, w)

Figure B — square (side s)

Figure C — triangle (sides a, b, c)

Chapter 1 Whole Numbers

Figure A is a rectangle. A rectangle has four sides and four **right angles**. A right angle is a square corner. You will learn more about angles later. The sides across from each other are equal. The longer side is usually called the length. The shorter side is usually called the width. In Figure A l is the length and w is the width.

Figure B is a **square**. A square has four equal sides and four right angles. In Figure B s stands for one side.

Figure C is a triangle. A triangle has three sides. All three sides may be the same length. Two of the sides may be the same while the third is different. Or, all three sides may have different lengths. In Figure C the letters a, b, and c stand for the three sides.

Perimeter is a measure of the distance around a flat figure such as a rectangle, a square, or a triangle. To find the perimeter of a flat figure, add the measurements of the sides. You can also use a formula to find the perimeter. Following are the formulas you will use on the GED Test.

Perimeter of a

rectangle: $p = 2l + 2w$ where p = perimeter, l = length, and w = width.

square: $p = 4s$ where p = perimeter and s = one side.

triangle: $p = a + b + c$ where p = perimeter and a, b, and c are the sides of the triangle.

Perimeter is always measured in units such as inches, feet, yards, or meters. In problems where no units are given, the perimeter is simply a number.

Example 1: Find the perimeter of a rectangle with length 20 feet and width 12 feet.

Step 1. In the formula for the perimeter of a rectangle, replace l with 20 and w with 12.	$p = 2l + 2w$ $p = 2 \times 20 + 2 \times 12$
Step 2. Multiply 2 times 20 and 2 times 12 before you add. The perimeter of the rectangle is 64 feet.	$p = 40 + 24$ $p = 64$ feet

Example 2: Find the perimeter of this square.

$s = 15$

Step 1. In the formula for the perimeter of a square, replace s with 15.	$p = 4s$ $p = 4 \times 15$ $p = 60$
Step 2. Multiply. Since there are no units, the perimeter is simply 60.	

Level 2, Lesson 9: Perimeter

Example 3: Find the perimeter of this triangle.

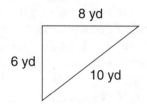

Step 1. In the formula for the perimeter of a triangle, replace *a* with 6, *b* with 8, and *c* with 10.

$p = a + b + c$
$p = 6 + 8 + 10$
$p = 24$ yards.

Step 2. The unit of measurement is yards. The perimeter of the triangle is 24 yards.

Lesson 9 Exercise

Solve each problem.

1. Find the perimeter of a square with a side of 11 feet.
2. What is the perimeter of a rectangular garden with a length of 9 meters and a width of 7 meters?
3. Find the perimeter of this rectangular room.

   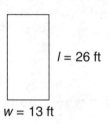

4. What is the perimeter of a square with a side of 20 inches?
5. Find the perimeter of this triangle.

   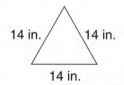

6. Find the perimeter of this square.

   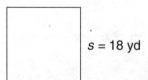

7. What is the perimeter of this rectangle?

   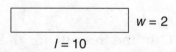

8. Find the perimeter of this triangle.

   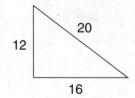

28   Chapter 1   Whole Numbers

9. The sides of a triangle measure 7 centimeters, 8 centimeters, and 9 centimeters respectively. Find the perimeter of the triangle.
10. What is the perimeter of this rectangle?

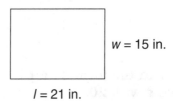

11. What is the perimeter of a square whose side is 21?
12. Find the perimeter of this triangle.

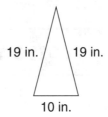

Check your answers on page 298.

Lesson 10

Area

Area is a measure of the amount of space inside a flat figure. Area is always measured in square units such as square inches, square feet, square yards, or square meters. You may see these units abbreviated with an exponent. For example, ft^2 is the same as square feet.

Area is the total number of square units that cover the surface of a flat figure. The rectangle shown at the right has an area of 12 in^2. This means that there are 12 one-inch squares in the rectangle.

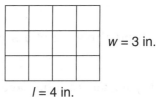

Following are the formulas for the area of a rectangle and a square.

Area of a
rectangle: $A = lw$ where A = area, l = length, and w = width.
square: $A = s^2$ where A = area and s = side.

Example 1: Find the area of a rectangle with length 12 feet and width 9 feet.

Step 1. In the formula for the area of a rectangle, replace l with 12 and w with 9.

$A = lw$
$A = 12 \times 9$
$A = 108 \text{ sq ft}$

Step 2. Multiply. The area is 108 square feet.

Level 2, Lesson 10: Area   29

Example 2: Find the area of this square.

$s = 20$ ft

Step 1. In the formula for the area of a square, replace s with 20.

$A = s^2$
$A = 20^2$
$A = 20 \times 20$

Step 2. Multiply. The area is 400 square feet.

$A = 400$ ft^2

Lesson 10 Exercise

Solve each problem.

1. Find the area of a rectangle with length 16 inches and width 9 inches.
2. What is the area of this square room?

   $s = 8$ ft

3. What is the area of a rectangular countertop with length 20 and width 14?
4. If a garden has length 15 yards and width 3 yards, what is its area?
5. Find the area of a square with side 15 inches.
6. What is the area of a square if one side measures 11 feet?
7. What is the area of this rectangle?

   3
   13

8. Find the area of a square with a side of 16 meters.

Check your answers on page 298.

Lesson 11 Volume

Volume is a measure of the amount of space inside a three-dimensional figure. The unit of measurement for volume is always cubic units such as cubic inches or cubic meters. These units are sometimes abbreviated with exponents. Cubic feet is the same as ft^3.

Volume is used to indicate the capacity of a container. It measures the amount of earth that must be removed to dig a foundation for a house or the amount of water that a pool can hold.

Two of the three-dimensional figures you may see on the GED Test are the cube and the rectangular container.

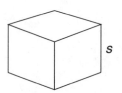
Figure A

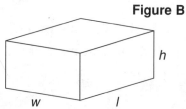
Figure B

Figure A is a cube. A cube is a three-dimensional figure made up of equal edges (sides) and right angles.

Figure B is a rectangular container or rectangular solid. A rectangular solid is also made up of right angles, but the sides are not all the same. The sides are usually called the length, the width, and the height. A shoebox is an example of a rectangular container.

Following are the formulas for the volume of these two figures.

> Volume of a
>
> cube: $V = s^3$ where V = volume and s = side.
>
> rectangular container: $V = lwh$ where V = volume, l = length, w = width, and h = height.

Example 1: Find the volume of a cube with a side of 2 feet.

Step 1. In the formula for the volume of a cube, replace s with 2.

$V = s^3$
$V = 2^3$
$V = 2 \times 2 \times 2$
$V = 8$ cu ft

Step 2. Multiply. The volume of the cube is 8 cubic feet.

Example 2: Find the volume of the rectangular container.

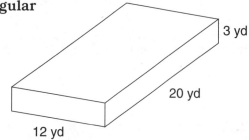

Step 1. In the formula for the volume of a rectangular container, replace l with 20, w with 12, and h with 3.

$V = lwh$
$V = 20 \times 12 \times 3$
$V = 720$ yd^3

Step 2. Multiply. The volume is 720 cubic yards.

Lesson 11 Exercise

Solve each problem.

1. Find the volume of a cube with side 6 feet.
2. Find the volume of this rectangular carton.

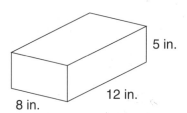

3. Find the volume of a rectangular closet with length 20 feet, width 6 feet, and height 5 feet.
4. What is the volume of this cube?

5. Find the volume of a cube with side 1 yard.
6. What is the volume of a rectangular container that is 100 feet long, 24 feet wide, and 8 feet high?
7. Find the volume of this rectangular solid.

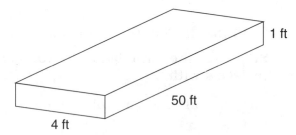

8. What is the volume of a cube with a side that measures 3 feet?

Check your answers on page 298.

Lesson 12: Properties of Numbers

In this section you will learn about three characteristics or **properties** of mathematical operations. On the GED Test you will see these properties applied to answer choices. You do not have to memorize the names of these properties. You simply need to be familiar with how they work.

The Commutative Property for an operation means that you can work forward or backward and get the same answer. For example, when you add two numbers, you can add them in either order. The answers will be the same. 3 + 5 equals 8 and also 5 + 3 equals 8.

> The commutative property for addition of two numbers **a** and **b** is
> $$a + b = b + a$$

Example 1: Which expression below is equal to 12 + 14?
  (1) 12 × 14
  (2) 14 + 12
  (3) 12 × 12
  (4) 12 − 14

Choice **(2)** 14 + 12 is correct. Each expression equals 26.

The commutative property is also correct for multiplication. When you multiply two numbers, you can multiply them in either order and the answers will be the same. For example, 7 × 2 equals 14 and also 2 × 7 equals 14.

> The commutative property for multiplication of two numbers **a** and **b** is
> $$ab = ba.$$

Remember that writing two letters next to each other in a formula indicates multiplication.

Example 2: Which expression equals 15 × 3?
  (1) 3 × 15
  (2) 15 ÷ 3
  (3) 15 + 3
  (4) 15 × 15

Choice **(1)** 3 × 15 is correct. Each expression equals 45.

The Associative Property for an operation means that you can combine the numbers in any order. For example, when you add three numbers, you can add them in any order. Think about adding the numbers 4 + 6 + 9. You can first add 4 + 6 to get 10 and then add 10 + 9 to get 19. You could also first add 6 + 9 to get 15 and then add 4 + 15 to get 19.

> The associative property for addition of numbers **a, b,** and **c** is
> $$(a + b) + c = a + (b + c).$$

The parentheses indicate that the numbers are grouped differently. The expression $(a + b) + c$ means that you should add the first two numbers and then add the third to their sum. The expressions $a + (b + c)$ means that you should add the second and third numbers and then add the first to their sum.

The associative property is also correct for multiplication.

> The associative property for multiplication of numbers **a, b,** and **c,** is
> $$(ab)c = a(bc).$$

Notice that for both addition and multiplication the parentheses are used to group numbers together. To evaluate an expression with parentheses, first do the operation in parentheses. Then do the other operations.

Example 3: Choose the expression that equals $7 \times (3 \times 10)$.
 (1) $7 + (3 \times 10)$
 (2) $(7 \times 3) + 10$
 (3) $7 + 3 + 10$
 (4) $(7 \times 3) \times 10$

Choice (4) $(7 \times 3) \times 10$ is correct.
The original expression is $7 \times (3 \times 10) = 7 \times (30) = 210$.
Choice (4) is $(7 \times 3) \times 10 = (21) \times 10 = 210$.
Each expression equals 210.

> **The Distributive Property** for multiplication over addition is
> $$a(b + c) = ab + ac.$$

This means you can add b and c first, then multiply the sum by a. Or, you can first multiply a and b, then multiply a and c, then find the sum of the products.

> The distributive property also works for multiplication over subtraction:
> $$a(b - c) = ab - ac.$$

Example 4: Which expression is equal to $5(6 + 7)$?
 (1) $5 \times 6 + 7$
 (2) $5 \times 6 + 5 \times 7$
 (3) $5 \times 6 \times 7$
 (4) $5 + 6 \times 7$

Choice (2), $5 \times 6 + 5 \times 7$, is correct. If you add the numbers inside the parentheses you get $5(13) = 65$. If you work out the two separate parts of $5 \times 6 + 5 \times 7$, you get $30 + 35 = 65$. Each expression equals 65.

Lesson 12 Exercise

Choose the correct solution to each of the following.

1. Which expression is equal to 9×6?
   - **(1)** $9 + 6$
   - **(2)** $9 - 6$
   - **(3)** 6×9
   - **(4)** 6×6

2. Which of the following is equal to $14 + 20$?
   - **(1)** $14 - 20$
   - **(2)** $20 + 14$
   - **(3)** 20×14
   - **(4)** 14×20

3. Which of the following equals $3 \times (12 \times 8)$?
   - **(1)** $(3 \times 12) + 8$
   - **(2)** $(3 \times 12) \times 8$
   - **(3)** $3 + (12 \times 8)$
   - **(4)** $3 \times 12 + 3 \times 8$

4. Which expression is equal to $(15 + 4) + 10$?
   - **(1)** $15 \times 4 \times 10$
   - **(2)** $15(4 + 10)$
   - **(3)** $15 + (4 + 10)$
   - **(4)** $(15 + 4) \times 10$

5. Choose the expression that equals $7(8 - 1)$.
   - **(1)** $7 \times 8 - 7 \times 1$
   - **(2)** $7 \times 8 \times 1$
   - **(3)** $7 \times 8 - 1$
   - **(4)** $7 \times 8 + 7 \times 1$

6. Which of the following is equal to $20 + 30$?
   - **(1)** 30×20
   - **(2)** $20 - 30$
   - **(3)** $20 + 20$
   - **(4)** $30 + 20$

7. Which expression is equal to $2 \times 19 + 2 \times 7$?
   - **(1)** $2 \times 19 \times 7$
   - **(2)** $7(2 \times 19)$
   - **(3)** $2(19 + 7)$
   - **(4)** $10(2 + 7)$

8. Which expression is equal to $50(3 + 4)$?
   - **(1)** $50 \times 3 \times 4$
   - **(2)** $50 + (3 \times 4)$
   - **(3)** $(50 + 3) + (50 + 4)$
   - **(4)** $50 \times 3 + 50 \times 4$

9. Which expression equals $4 \times 9 - 4 \times 2$?
   - **(1)** $4 \times 9 - 2$
   - **(2)** $4(9 - 2)$
   - **(3)** $4(9 - 4 + 2)$
   - **(4)** $4(9 + 2)$

10. Choose the expression that has the same value as $3(10 - 1)$.
    - **(1)** $3 \times 10 - 3 \times 1$
    - **(2)** $3 \times 10 - 1$
    - **(3)** $3 \times 10 - 10$
    - **(4)** $3 \times 10 - 1 \times 10$

Check your answers on page 298.

Level 2 Review

Solve each problem. Refer to pages 17 and 18 for any formulas that you need.

1. Round 2,386,475 to the nearest hundred thousand.
2. Round 496,273 to the nearest ten-thousand.
3. Find the cost of 12 gallons of gasoline at $1.20 per gallon.
4. A pilot flew for seven hours at an average speed of 435 mph. How far did he fly?
5. Ernie shipped packages with the following weights: 12 pounds, 36 pounds, 19 pounds, and 25 pounds. Find the average (mean) weight of the packages.
6. What is the median weight of the packages in problem 5?
7. Fran works freelance during the summer months. In June she made $2480; in July, $1630; and in August, $1950. What was her average monthly income for the summer?
8. In March Jack sold five used cars. The prices were $8470, $1950, $6075, $3080, and $2155. What was the median price of the cars he sold during the month?
9. Simplify the expression $15^2 - 10^2 + 25^1$.
10. Find the value of $20^2 - 3^3 + 10^0$.
11. Find the value of $\sqrt{3481}$.
12. Simplify $\sqrt{7744}$.
13. The sides of a triangle measure 8 meters, 11 meters, and 14 meters respectively. Find the perimeter of the triangle.
14. Find the perimeter of a rectangle with a length of 35 meters and a width of 21 meters.
15. What is the area of this rectangle?

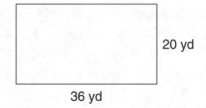

36 yd, 20 yd

16. What is the area of this figure?

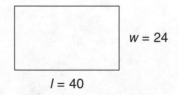

$l = 40$, $w = 24$

17. Find the volume of a rectangular container with length 25 feet, width 12 feet, and height 8 feet.
18. Find the volume of a cube whose side measures 12 inches.

19. Which expression below is the same as 9 × 15 + 9 × 20?
    (1) 9 × 15 + 20
    (2) 9(15 + 20)
    (3) 9 + 15 × 20
    (4) (9 + 15) + 20
20. Which of the following is the same as 7(20 + 1)?
    (1) 7 × 20 + 1
    (2) 20(7 + 1)
    (3) (7 × 20) + (7 × 1)
    (4) (20 × 7) + (20 × 1)

Check your answers on page 298. If you have all 20 problems correct, go on to Level 3. If you answered a question incorrectly, find the item number on the chart below and review that lesson before you go on.

Review:	If you missed item number:
Lesson 4	1, 2
Lesson 5	3, 4
Lesson 6	5, 6, 7, 8
Lesson 7	9, 10
Lesson 8	11, 12
Lesson 9	13, 14
Lesson 10	15, 16
Lesson 11	17, 18
Lesson 12	19, 20

Level 3
Whole-Number Problem Solving

Many problems on the GED Test are word problems. These may involve simple situations in which you need only to substitute numbers from the problem into a formula. You may need to locate information in a table to solve a problem. Other situations may be more complex. You may have to perform several operations to get an answer. You may have to read a long description of a practical situation and then choose the numbers from the description that you need to answer specific questions.

Lesson 13 — Word Clues

This lesson offers hints to help you identify the operations you will need to solve word problems.

Addition problems are easy to recognize. The words **sum**, **total**, **combined**, and **altogether** are word clues that usually mean to **add**.

Subtraction problems are also easy to recognize. The word **difference** and the phrases **how much more?** and **how much less?** mean to **subtract**. The words **gross** and **net** often appear in subtraction problems. **Gross** refers to a total before any deductions are taken from the total. **Net** is the amount that is left after the deductions are made.

Example 1: Jeff earns a gross salary of $280 a week. If his employer deducts $52 for taxes and pensions, what is Jeff's net weekly salary?

To find the net weekly salary, subtract the deductions from the weekly gross salary.

$280 gross salary
− 52 deductions
$228 net salary

The words **product** and **times** do not often appear in multiplication problems. Instead, you may be told information for one thing, such as the distance someone can drive on a gallon of gas. Then you may have to find information for several things, such as the distance the person can drive on several gallons of gas.

Example 2: Celeste can drive her car 18 miles on 1 gallon of gas. How far can she drive with 12 gallons of gas?

38   Chapter 1   Whole Numbers

Multiply the number of miles she can drive on one gallon by 12 gallons.

```
   18
 × 12
 ────
   36
   18
 ────
  216 miles
```

The word **quotient** almost never appears in division problems, but you may see the words **each, per, average, split,** or **share.** These words often mean to divide.

Example 3: John and his four friends were hired to haul 85 pounds of lumber. He and his four friends want to share the work equally. How many pounds will each person haul?

Divide the total weight of the lumber by the number of people who will haul it. Notice that John and his friends total five people.

$$5\overline{)85}\text{17 pounds}$$

Example 4: Sally paid $8 for four pounds of beef. Find the price of one pound of beef.

Divide the total price by the number of pounds. The price is $2 for one pound of beef.

$$4\overline{)\$8}\$2$$

You have learned to solve perimeter, area, and volume problems. Those words may not always appear in problems where you need to use their formulas.

Remember that perimeter is a measure of the distance around a flat figure. A problem that asks you to find the number of meters of fencing needed to enclose a yard is a perimeter problem.

Area is a measure of the amount of surface in a flat shape. A problem that asks you to find the number of square yards of material needed to cover a flat surface is an area problem.

Volume is a measure of the capacity of a three-dimensional object. A problem that asks you to find the number of cubic feet of dirt that a truck can carry is a volume problem.

Lesson 13 Exercise

The questions that follow each problem will help you choose the correct operation needed to solve it. First answer the questions. Then solve.

1. The town of Greenport spent $593,650 for police and fire protection in a year. The town spent $108,212 for health and welfare the same year. How much more did the town spend for police and fire protection than for health and welfare?

    a. The phrase *how much more* suggests which operation?
    b. Solve the problem.

Level 3, Lesson 13: Word Clues

2. John, Jose, Tony, and Mario started a record store. At the end of a year, they had a profit of $27,936. If they shared the profit equally, how much did each person get?
   a. The word *each* suggests which operation?
   b. How many people are sharing the profit?
   c. Solve the problem.

3. Lucia made a gross salary of $18,500 last year. Her empoyer deducted $3,518 from her salary for taxes and insurance. Find Lucia's net income for the year.
   a. To find a *net* amount usually suggests the use of which operation?
   b. Solve the problem.

4. Bill Sutton made $16,456 last year. His wife Connie made $11,294. Their son Henry made $3,367 at his part-time job. Find the combined yearly income for the Sutton family.
   a. What one word in the problem suggests the operation?
   b. Solve the problem.

5. Sandy paid $63 for three pairs of jeans. What was the price of one pair of jeans?
   a. The problem tells you the total price for how many items?
   b. To find the price of one item, what operation should you use?
   c. Solve the problem.

6. It costs $1850 a year to educate one student at the Franklin School. There are 230 students in the school. Find the total cost for educating all the students at the school for a year.
   a. The problem **tells** you the cost of educating how many students?
   b. The problem **asks** you to find the cost for educating how many students?
   c. What operation should you use?
   d. Solve the problem.

7. In 1950 there were 14,273 people living in Elmford. In 1990 9,467 more people lived in Elmford than in 1950. Find the population of Elmford in 1990.
   a. Was the population of Elmford in 1990 more or less than in 1950?
   b. What operation should you use?
   c. Solve the problem.

8. Al wants to cut a board 102 inches long into six equal pieces. If there is no waste, how long will each piece be?
   a. "How long will *each* piece be?" suggests what operation?
   b. Solve the problem.

9. The swimming pool at the Greenport Community Center is 80 feet long, 20 feet wide, and 6 feet deep. Find the capacity of the pool in cubic feet.
   a. To find the capacity do you need the perimeter, the area, or the volume of the pool?
   b. Solve the problem.

10. Oregon became a state in 1859. For how many years had Oregon been a state in 1990?
    a. What operation should you use to solve the problem?
    b. Solve the problem.

11. A large farm in Texas is the shape of a square with each side six miles long. Find the distance around the farm.
    a. To find the distance around a square do you need the perimeter, the area, or the volume?
    b. Solve the problem.

12. In one month the Riveras spent $462 for utilities and mortgage payments, $436 for taxes and social security, $194 for car payments and gasoline, $323 for food, and $245 for everything else. How much did they spend altogether that month?
    a. What word in the problem suggests the operation?
    b. Solve the problem.

13. Jorge makes $295 for a five-day work week. How much does he make each day?
    a. How many days did Jorge work?
    b. You want to know how much he makes in how many days?
    c. What operation should you use?
    d. Solve the problem.

14. The large open office where Vera works is 30 yards long and 15 yards wide. How many square yards of carpet are needed to cover the floor of the office?
    a. To find the number of square yards on the floor, do you need the perimeter, the area, or the volume?
    b. Solve the problem.

Check your answers on page 299.

Lesson 14: Estimating

Estimating or **approximating** means making a good guess. Whenever you solve a word problem, think about the answer. Be sure it makes sense.

You can often decide whether an answer makes sense if you first round the numbers in the problem. (Look back at Lesson 4 if you need help rounding.) Then solve the problem with the rounded numbers. If your exact answer is close to your rounded answer, your exact answer probably makes sense.

Example: Joseph earned a gross salary of $24,763 last year. His employer deducted $5,119 for taxes and social security. What was Joseph's net income last year?

    (1) about $15,000
    (2) about $20,000
    (3) about $23,000

Step 1. Look at the answer choices. Since each answer is expressed in thousands, round each number to the nearest thousand.

$24,763 to the nearest thousand is $25,000.
$5,119 to the nearest thousand is $5,000.

Step 2. To find the estimated net income, subtract $5,000 from $25,000. Joseph's net income was about $20,000, choice (2).

$25,000
− 5,000
$20,000

Step 3. To find the exact net income subtract $5,119 from $24,763. Joseph's exact net income was $19,644. Since the exact net income is close to the estimated income of $20,000, the exact answer makes sense.

$24,763
− 5,119
$19,644

The following word problems are similar to those in Lesson 13. The choices that follow each question will help you estimate the answer.

Lesson 14 Exercise

Choose an approximate answer to each problem. Then find the exact answer.

1. Three partners equally shared a $58,974 profit for their business. How much did each partner get?
    (1) a little less than $10,000
    (2) a little less than $20,000
    (3) a little less than $30,000
    Find the exact answer.

2. Jim Green made $29,260 last year. His wife Karen made $18,420. Their daughter Susan made $9455. Find their combined income.
    (1) a little less than $40,000
    (2) a little less than $50,000
    (3) a little less than $60,000
    Find the exact answer.

3. Tanya paid $112 for four new work uniforms. How much did each uniform cost?
    (1) a little more than $15
    (2) a little more than $25
    (3) a little more than $35
    Find the exact answer.

4. Andrea cut an 87-inch-long copper pipe into three equal pieces. Find the length of each piece.
    (1) about 30 inches
    (2) about 20 inches
    (3) about 15 inches
    Find the exact answer.

5. It cost $2160 a year to educate a student at the Oak Street School. Find the cost of educating 30 students at the school for a year.

   (1) around $600,000
   (2) around $60,000
   (3) around $6000

   Find the exact answer.

6. Ron is a truck driver. Monday he drove 328 miles; Tuesday, 217 miles; Wednesday, 421 miles; Thursday, 186 miles; and Friday, 313 miles. Find the total number of miles he drove that week.

   (1) about 900
   (2) about 1400
   (3) about 1800

   Find the exact answer.

7. Kevin makes $327.50 for a five-day work week. How much does he make in a day?

   (1) between $60 and $70
   (2) between $70 and $80
   (3) between $80 and $100

   Find the exact answer.

8. The floor of Ricardo's basement is 62 feet long and 21 feet wide. What is the area of the floor in square feet?

   (1) between 800 and 1000
   (2) between 1000 and 1200
   (3) between 1200 and 1400

   Find the exact answer.

Check your answers on page 299.

Lesson 15: Multistep Problems

Most of the word problems you have seen so far in this book required only one step to find a solution. Some problems are more complicated. There is no method that guarantees a successful solution to every word problem. Practice is probably the most helpful tool. One goal of this book is to give you a lot of practice with word problems.

Every problem in the next exercise requires at least two steps. Read each problem a couple of times. Think about the operations you will use before you actually perform them. Then estimate an answer. After you have solved each problem, compare your answer to the estimate. Be sure your answer makes sense.

Example: Darryl had a summer job for seven weeks. He worked 35 hours a week for $6 an hour. How much did he make altogether for the seven weeks?

Step 1. Find the amount Darryl made each week.
Multiply the number of hours he worked each week, 35, by the amount he made each hour, $6.
He made $210 each week.

$$\begin{array}{r} 35 \\ \times\ \$6 \\ \hline \$210 \end{array}$$

Step 2. To find the amount he made for the whole summer, multiply the amount he made each week, $210, by the number of weeks he worked, 7.
He made $1470 for the summer.

$$\begin{array}{r} \$210 \\ \times\ 7 \\ \hline \$1470 \end{array}$$

Lesson 15 Exercise

Solve each problem.

1. Louise bought furniture priced at $650. She paid for the furniture in 15 equal payments of $52 each. How much more than $650 did she end up paying for the furniture?

2. Mr. Castro ordered three cases of tomato soup, four cases of chicken soup, and two cases of bean soup for his store. Each case contained 12 cans. Altogether how many cans of soup did he order?

3. Every week Mark takes home $235 and his wife Heather takes home $240. Find their combined income for four weeks.

4. Joaquin's net monthly salary is $1300. He pays $290 a month for rent and $400 a month for child support. How much does he have left each month after these expenses?

5. For the Greenport Community Center Phil bought 20 new baseball gloves at $25 each, eight baseball bats at $12 each, and six softballs at $8 each. Find the total cost of these items.

6. One week Winston worked for 35 hours at his regular wage of $6.50 an hour and for 8 hours overtime at $9.75 an hour. How much did he make altogether that week?

7. Adrienne lived for 18 months in an apartment. For the first year she paid $260 a month rent. Then for the last six months she paid $290 a month. What was her average rent for the months she lived in the apartment?

8. Alex drove for four hours at an average speed of 55 mph and then for two hours at an average speed of 30 mph. Altogether how far did he travel in those six hours?

9. For summer youth projects the town of Greenport received $60,000 from the federal government and $45,000 from the state government. Three projects shared the funds equally. How much did each project receive?

10. Pat's gross monthly income is $1500. Her employer withholds $350 monthly for taxes and insurance. Find Pat's net salary for a year.

11. The picture represents the living room and the dining room of the Garcias' house. At the price of $8 a square foot, how much would it cost the Garcias to carpet both rooms?

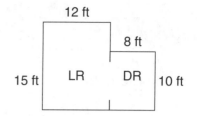

12. Find the area of the shaded part of the figure shown at the right.

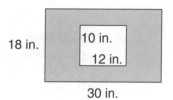

Check your answers on page 300.

Lesson 16

Tables

A **table** is a set of numbers in rows or columns. A table is an orderly way of presenting detailed numerical information.

The table below tells how many calories are used during selected activities.

Calories Used by a 155-Pound Person

Activity	Calories per hour
Walking at 2 mph	140
Mopping floors	270
Swimming	450
Jogging at 5 mph	540

Example 1: Of the activities shown in the table, which uses the most calories per hour?

Step 1. Look at the column of numbers under "Calories per hour."

Step 2. Find the largest number in the column (540) and look at the activity listed to the left. Jogging at 5 mph uses the most calories of the activities shown in the table.

Example 2: How many more calories does a 155-pound person use in one hour of swimming than in one hour of walking at 2 mph?

Step 1. Find the calories per hour that are used swimming (450) and the calories per hour that are used walking at 2 mph (140).

Step 2. Subtract to find how many more calories are used swimming than walking.

$$450 - 140 = 310$$

Level 2, Lesson 16: Tables   45

Lesson 16 Exercise

Use the following table to answer the next questions.

	Per Capita Expenditures	Per Capita Taxes
New York	$1504	$1164
Massachusetts	$1440	1137
Florida	856	694
Illinois	1068	800
Texas	898	705
California	1449	1098
U.S. average	1221	902

Source: U.S. Bureau of Census.

1. Which state in the table had the highest per capita expenditures?
2. What was the difference between the highest amount and the lowest amount in per capita expenditures for the states shown in the table?
3. Find the median per capita expenditures for the states shown in the table.
4. Texas was how far below the U.S. average in per capita expenditures?
5. Central City is in a state where the per capita expenditures were at the national average. The population of Central City is 300,000. Find, to the nearest ten million, the amount spent by the state on Central City.
6. Find the difference between the per capita expenditures and the per capita taxes in Florida.
7. Greenport, with a population of 15,000, is in a state where the taxes are at the U.S. average. To the nearest $100,000, how much did the people of Greenport pay altogether in state taxes?
8. For the states shown on the table, what is the difference between the highest and the lowest per capita state taxes?

Check your answers on page 300.

Lesson 17 — Item Sets

On the GED Mathematics Test, you may see long reading passages. You will have to solve several problems based on information given in these passages.

Read the paragraph below. Then study the examples that follow.

Jeff is trying to decide which of two used cars to buy. The cars he likes are nearly identical. To pay for the car from Dealer A, he must put down $500 cash and then make 24 monthly payments of $120. To pay for the car from Dealer B, he does not have to make a down payment, but he must make 30 monthly payments of $115.

Example 1: What are the total monthly payments for the car from Dealer A?

Multiply the monthly payment, $120, by the number of months, 24.

```
  $120
×   24
   480
   240
 $2880
```

Example 2: What total amount would Jeff have to pay for the car from Dealer A?

Add the total monthly payments to the down payment.

```
 $2880
+  500
 $3380
```

Example 3: Find the total price of the car at Dealer B.

Multiply the monthly payment of $115 by the number of months, 30.

```
  $115
×   30
 $3450
```

Lesson 17 Exercise

Use the following passage to solve problems 1 through 4.

Steve is a trainee on a new job. For six months he will make $950 a month. After the training period he will make $550 more a month for one year. Then, if he does well on his performance review, he will make an additional $250 a month for his second year as a regular employee.

1. After the training period how much will Steve make in his first year as a regular employee?
2. How much will Steve make for his first twelve months at the job including his training period?
3. What is Steve's average monthly income during his first twelve months on the job?
4. If Steve does well on the performance review at the end of his first year of regular employment, what will he make during his second year as a regular employee?

Use the following passage to solve problems 5 through 9.

The yearly payroll for the eight people in management at Paulson's Plastics is $216,000. The payroll for the 35 laborers at the factory is $756,000. The ten people on the clerical staff make a total of $144,000 in a year.

5. Altogether, how many people are on the payroll at Paulson's?
6. What is the total yearly payroll at the factory?

7. Find the total monthly payroll at Paulson's.
8. What is the average yearly income of a laborer at Paulson's?
9. Find the average monthly income of a clerical person at the factory.

Check your answers on page 301.

Lesson 18

Set-Up Answers

In some problems on the GED Mathematics Test you may have to choose from a group of possible methods of solution. In these problems you are not looking for the exact answer, but for the method to find the answer.

Parentheses are important in these "set-up" solutions. Parentheses group together the numbers that must be combined first.

The answers to these problems may look like the answers to the Properties of Numbers Exercise in Lesson 12.

Example: On Friday 650 people attended the Greenport Fair. There were 825 people on Saturday and 940 on Sunday. Everyone paid $5 admission to the fair. Which expression below shows the total dollar receipts for those three days?

(1) $5 \times 650 + 825 + 940$

(2) $\dfrac{650 + 825 + 940}{5}$

(3) $5(650 + 825 + 940)$

(4) $650 + 825 + 940 \times 5$

Choice **(3)** $5(650 + 825 + 940)$ is correct. In choice **(1)** only the Friday attendance is multiplied by the price of a ticket. In choice **(2)** the total attendance is divided by the price of a ticket. In choice **(4)** only the Sunday attendance is multiplied by the price of a ticket.

A Note About Dividing

There are several ways to express division in the language of mathematics. Suppose you want to express the average of the numbers 25 and 31. To find the average you must first add the two numbers and then divide the sum by 2. Look carefully at each of the following ways of expressing this procedure.

Method A: $(25 + 31)/2$
Method B: $(25 + 31) \div 2$
Method C: $\dfrac{25 + 31}{2}$

Notice that Method A and Method B are similar. Both use parentheses to show that 25 and 31 must first be added. In Method A a slash (/) is used to indicate division. In Method C the line under 25 and 31 does two things. It groups the numbers 25 and 31 together. It also means that the sum of 25 and 31 must be divided by the number below the line, 2.

Lesson 18 Exercise

Choose the correct answer to each problem.

1. Tim makes $24,720 a year. Which expression shows the amount he makes in one month?
   (1) $12 \times 24{,}720$
   (2) $12 + 24{,}720$
   (3) $\frac{24{,}720}{12}$
   (4) $24{,}720 - 12$

2. Which of the following expresses the cost of 40 track uniforms if the uniforms cost $29 each?
   (1) 40×29
   (2) $\frac{40}{29}$
   (3) $\frac{29}{40}$
   (4) $40 + 29$

3. Frank drove for two hours at 20 mph and for three hours at 60 mph. Which expression shows the total distance he drove?
   (1) $2(20 + 60)$
   (2) $3(20 + 60)$
   (3) $5(20 + 60)$
   (4) $(2 \times 20) + (3 \times 60)$

4. Eva got scores of 80, 95, and 74 on three math quizzes. Which of the following expresses her mean score on the quizzes?
   (1) $80 + 95 + \frac{74}{3}$
   (2) $\frac{80 + 95 + 74}{3}$
   (3) $\frac{80}{3} + 95 + 74$
   (4) $3(80 + 95 + 74)$

5. Alberto bought four quarts of oil for $7.99 each and paid a total of $1.92 in tax. Which expression shows the amount he paid altogether for the oil?
   (1) $(4 \times 7.99) + 1.92$
   (2) $4 \times 7.99 \times 1.92$
   (3) $4(7.99 + 1.92)$
   (4) $4(7.99 - 1.92)$

6. For a play at the Greenport Community Center, members sold 350 tickets at $8 each and 425 tickets at $6 each. Which of the following represents the total receipts for the tickets?
   (1) $(8 \times 6) + (350 + 425)$
   (2) $8 \times 6 \times 350 \times 425$
   (3) $(8 \times 350) + (6 \times 425)$
   (4) $(8 + 350) \times (6 + 425)$

7. Fred's gross monthly income is $1800. His monthly deductions are $360. Which expression represents his net income for one year?
   (1) $12(1800 + 360)$
   (2) $12 \times 1800 + 360$
   (3) $12(1800 - 360)$
   (4) $12 \times 1800 \times 36$

8. The elementary schools in Greenport had a joint festival for three nights. Receipts on Monday were $2500; on Tuesday, $4850; and on Wednesday, $4200. The total receipts were shared equally by five schools. Which of the following shows the amount each school received?
   (1) $\frac{2500 + 4850 + 4200}{5}$
   (2) $5(2500 + 4850 + 4200)$
   (3) $\frac{2500}{5} + 4860 + 4200$
   (4) $\frac{2500 + 4850 + 4200}{3}$

9. Mr. Vega ordered 20 boxes of shirts containing 12 shirts each for his store. He sent back two of the boxes because they were the wrong color. Which expression shows the total number of shirts he kept from that order?

   (1) $12 \times 20 \times 2$
   (2) $12(20 - 2)$
   (3) $12 + (20 - 2)$
   (4) $12 \times 20 - 2$

10. The length of a rectangle is 100 feet, and the width is 30 feet. Which of the following expresses the perimeter of the rectangle?

    (1) $2 \times 100 \times 30$
    (2) 100×30
    (3) $(2 \times 100) - (2 \times 30)$
    (4) $(2 \times 100) + (2 \times 30)$

Check your answers on page 301.

Level 3 Review

Solve each problem.

1. Elton Electronics shipped 48 boxes that weighed 768 pounds altogether. If each box had the same weight, what was the weight of one box?

2. The town of Greenport received a $67,500 grant for a bilingual program in its elementary schools. If the five elementary schools share the money equally, how much will each school get?

3. Mark used 23 gallons of gasoline on a 437-mile trip. How far could Mark drive on one gallon of gasoline?

4. José takes home $1340 a month. His wife Maria works part time and takes home $610 a month. Find their approximate combined monthly income.

   (1) about $1400
   (2) about $1900
   (3) about $2100

5. Each month Gloria pays rent of $315 and a car payment of $135. She takes home $12,360 a year. How much does she have left at the end of the year after the expenses of rent and car payments?

6. The boiler in the building where Louie is the superintendent uses 760 gallons of fuel oil each month for six months of the year, 1250 gallons a month for three months in winter, and 410 gallons a month for three months in summer. Find the total number of gallons used in a year.

7. For four years the Millers made monthly mortgage payments of $350. Then they paid off $5000 and made $250 a month payments for two more years. Altogether how much did they pay on their mortgage?

Use the following table to answer problems 8 and 9.

Taxi and Bus Licenses Sold in Woods County in 1990

	Taxi Licenses	Bus Licenses
Resident	313	277
Nonresident	55	21
Total	368	298

8. In 1990 the number of resident taxi licenses sold in Woods County was how many more than the number of resident bus licenses sold?

9. The cost of a bus license in Woods County is $30. What was the total value of the bus licenses (both resident and nonresident) sold in Woods County in 1990?

Use the table below to answer question 10.

Average Cost Per Room

Daily charge for semiprivate hospital room, 1990

New York	$339
New Jersey	273
Connecticut	456
United States	297

Source: American Hospital Association.

10. The average cost of a semiprivate hospital room in Connecticut in 1990 was how how much more than the average cost in the entire United States?

Use the following passage to answer problems 11 through 14.

Paul wants to rent a car for four days. Dealer A charges $18 a day with no mileage charge. Insurance costs an additional $12.50 a day. Dealer B charges $12 a day and $.20 a mile. Insurance costs an additional $10.25 a day.

11. How much would a car from Dealer A cost for four days?
12. How much would a car from Dealer B cost for four days excluding mileage?
13. If Paul rents from Dealer B and drives 100 miles, what will the cost of renting the car be?
14. If Paul rents from Dealer B and drives 200 miles, what will the cost of renting the car be?

Use the following passage to answer questions 15 through 18.

The figure at the right shows the floor plan of an office. Store A charges $25 a square yard for carpet. This amount includes installation. Store B charges $20 a square yard plus an installation fee of $500.

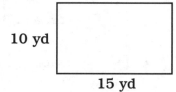

10 yd

15 yd

15. What is the area of the floor?
16. How much will carpet from Store A cost?
17. How much will carpet from Store B cost?
18. What is the difference in the two prices if installation is included?
19. Lois tutors four students in English. Her students are Mai Lee, who is 33 years old, Carlos, who is 25, Ludmilla, who is 47, and Itzhak, who is 19. Which expression represents the average of Lois's students?
    (1) $2(33 + 25) + 2(47 + 19)$ (3) $\frac{33 + 25}{2} + \frac{47 + 19}{2}$
    (2) $33 \times 25 \times 47 \times 19$ (4) $\frac{33 + 25 + 47 + 19}{4}$
20. Eva bought three shirts for her sons for $12 each and four pairs of pants for $15 a pair. Which expression tells the total amount of her purchases?
    (1) $3 + 12 + 4 + 15$ (3) $(3 + 12)(4 + 15)$
    (2) $3 + 4(12 + 15)$ (4) $3 \times 12 + 4 \times 15$

Check your answers on page 301. If you have answered all 20 questions correctly, go on to the next chapter. If you answered a question incorrectly, find the item number on the chart below and review that lesson before you go on.

Review:	If you missed item number:
Lesson 13	1, 2, 3
Lesson 14	4
Lesson 15	5, 6, 7
Lesson 16	8, 9, 10
Lesson 17	11, 12, 13, 14, 15, 16, 17, 18
Lesson 18	19, 20

Chapter 2
DECIMALS

Level 1 Decimal Skills

Decimals are used to describe parts of whole things. Our money system uses dollars to count whole amounts and cents to count parts of dollars. Cents are examples of decimals.

First you will learn to read and write decimals. Then you will learn the basic operations of addition, subtraction, multiplication, and division with decimals and how to compare decimals and arrange them in order of size.

To find out whether you need to review decimals, try the following preview.

Preview

Solve each problem.

1. Which of the following has a value **less** than 1?
   (1) 28.02   **(2)** 1.001   **(3)** 0.708   **(4)** 1.90
2. Which of the following has a value **greater** than 1?
   (1) 5.003   **(2)** 0.987   **(3)** 0.075   **(4)** 0.706
3. Which of the following is the **same** as 0.045?
   (1) four and five hundredths
   (2) forty-five hundredths
   (3) four and five thousandths
   (4) forty-five thousandths
   (5) forty-five ten-thousandths
4. Write eighteen and twenty-three thousandths as a mixed decimal.
5. What is the sum of 0.409, 0.28, and 0.7?
6. What is the difference between 0.82 and 0.197?
7. What is the product of 4.5 and 0.26?
8. Divide 0.468 by 18.
9. Divide 108 by 2.4.
10. Which of the following is greater, 0.067 or 0.06?
11. Which of the following is greater, 0.2 or 0.108?
12. Rewrite the following list of decimals in order from least to greatest: 0.31, 0.013, 0.031, 0.4

Check your answers on page 302. If you have all 12 problems correct, do the Level 1 Review on page 63. If you have fewer than 12 problems correct, study Level 1 beginning with Lesson 1.

Lesson 1: The Decimal Place System

A **decimal** is a kind of fraction. A decimal expresses a part of a whole thing. Decimals get their names from the number of **places** to the right of the **decimal point.** The decimal point itself does not take up a decimal place.

Below is a chart of place names in the decimal system. To the left of the decimal point are the first four whole number places. To the right of the decimal point are the first six decimal places. Notice how the decimal place names all end in **-ths.** Be sure you know these place names before you go on.

thousands	hundreds	tens	units (ones)	.	tenths	hundredths	thousandths	ten thousandths	hundred thousandths	millionths
___	___	___	___	.	___	___	___	___	___	___

One of the most common uses of decimals is our monetary system. For example, $0.19 is a decimal. It represents 19 of the 100 pennies in a dollar—19 of the 100 equal parts of a dollar.

A **mixed decimal** has a whole number to the left of the decimal point and a decimal fraction to the right. For example, $2.35 is a mixed decimal. It represents two whole dollars and 35 of the 100 equal parts of another dollar.

A number has a value **less** than 1 if there are no digits, other than zero, in the whole numbers places. For example, both 0.8 and 0.997 have values less than 1. Notice that a decimal with no whole number is often written with a zero in the units place. Both 0.6 and .6 are acceptable ways to write six tenths.

A number with decimal places has a value **greater** than 1 if there is at least one digit, other than zero, in the whole numbers places. For example, both 1.5 and 80.02 have values greater than 1.

Remember that a decimal gets its name from the number of places to the right of the decimal point. The number 386.4 has one decimal place because there is only one digit to the right of the decimal point. The number 6.08 has two decimal places because there are two digits to the right of the decimal point.

Zeros often cause confusion in decimals. The decimal 0.0807 has two zeros to the right of the decimal point. These zeros keep 8 in the hundredths place and 7 in the ten-thousandths place. The zero to the left of the decimal point has no value. The decimal 0.0310 also has two zeros to the right of the decimal point. The zero to the right of 1 is unnecessary, however, because 0.0310 is the same as 0.031. The zero in the ten-thousandths place does not help keep the other digits in their places.

Lesson 1 Exercise

1. Tell the number of decimal places in each of the following numbers.
   (a) 0.27 (b) 21.486 (c) 6.007
   (d) 0.09546 (e) 1240.06 (f) 250.2

2. Rewrite each of the following numbers and leave out the unnecessary zeros.
   (a) 20.0670 (b) 0.4090 (c) 028.70
   (d) 01.20800 (e) 003.6 (f) 04.500

3. In each number below, circle the digit in the tenths place.
   (a) 0.89 (b) 3.4 (c) 10.307 (d) 5.3681

4. In each number below, circle the digit in the hundredths place.
   (a) 0.125 (b) 4.2679 (c) 5.809 (d) 0.716

5. In each number below, circle the digit in the thousandths place.
   (a) 4.056 (b) 0.0123 (c) 28.3917 (d) 0.9872

6. Which of the following is a mixed decimal?
   (a) 0.8 (b) 0.99 (c) 0.066 (d) 4.02

7. Which of the following has a value **less** than 1?
   (a) 0.9 (b) 5.1 (c) 10.01 (d) 1.99

8. Which of the following has a value **greater** than 1?
   (a) 0.8 (b) 0.99 (c) 0.066 (d) 4.02

9. Which of the following has a value **less** than 1?
   (a) 9.2 (b) 0.879 (c) 2.9 (d) 1.005

10. Which of the following has a value **greater** than 1?
    (a) 0.6 (b) 4.009 (c) 0.01 (d) 0.82

Check your answers on page 302.

Lesson 2 — Reading and Writing Decimals

Remember that a decimal gets its name from the number of places to the **right** of the decimal point.

Reading Decimals

To read a decimal, count the number of places to the right of the decimal point. The last decimal place to the right gives the name of the decimal. In mixed decimals read the decimal point as the word **and**.

Chapter 2 Decimals

Examples:

0.13   is "thirteen hundredths" because the number has two decimal places.

8.007   is "eight and seven thousandths" because the number has three decimal places.

0.0006   is "six ten-thousandths" because the number has four decimal places.

Writing Decimals

When you write a decimal, decide how many decimal places you need. Use zeros in places that are not filled with other digits.

Examples:

Four thousandths = .004 or 0.004
   because for thousandths, three decimal places are needed. Notice that zeros fill the first two decimal places.

Sixteen and two hundredths = 16.02
   because for hundredths, two decimal places are needed. Notice how a zero holds the first decimal place.

Lesson 2 Exercise

For problems 1 to 10, in the blank beside each problem write the letter from the right-hand column that corresponds to each decimal or mixed decimal. Number 1 is done as an example.

1. _c_ 0.006   a. fifteen hundredths
2. ___ 9.03   b. two and eight tenths
3. ___ 0.7   c. six thousandths
4. ___ 0.0015   d. two and eight thousandths
5. ___ 2.008   e. seven tenths
6. ___ 0.07   f. nine and three hundredths
7. ___ 2.8   g. nine and three thousandths
8. ___ 90.3   h. seven hundredths
9. ___ 0.15   i. ninety and three tenths
10. ___ 9.003   j. fifteen ten-thousandths

For problems 11 to 20, in the blank beside each problem write the number in decimal form.

11. _____ four tenths
12. _____ eight and nine hundredths
13. _____ thirty-six thousandths

14. _____ fourteen and three thousandths
15. _____ five hundred nineteen ten-thousandths
16. _____ seventy-two and six ten-thousandths
17. _____ one and five millionths
18. _____ thirty-two hundred-thousandths
19. _____ four and eighteen thousandths
20. _____ four hundred eighteen thousandths.

Check your answers on page 302.

Lesson 3: Addition and Subtraction of Decimals

Adding or subtracting decimals follows the same steps as adding or subtracting whole numbers. The important difference is that you must line up the decimal points. If you do that, everything else falls into place. The thousandths will be under the thousandths, the hundredths under the hundredths and so on. Put in zeros **after** decimal numbers to give each number the same number of decimal places or put the number with the most decimal places on top.

Example 1: Find the sum of 0.17, 0.9, and 0.256.

Step 1. Line up the numbers with decimal points under each other. Do not confuse the period at the end of the sentence with a decimal point. Put in zeros after decimal numbers or write the number with the most decimal places on top.

```
  1 1                1 1
0.170              0.256
0.900      or      0.17
+0.256             +0.9
-----              -----
1.326              1.326
```

Step 2. Add each column. Carry as you would with whole numbers. Notice how the sum of the tenths column carries over to the units. Put the decimal point in the answer in line with the decimal points in the problem.

When you add whole numbers with decimals or mixed decimals, put a point to the right of each whole number. This will help you put each number in the correct column.

Example 2: Find the sum of 18, 2.35, and 0.482.

Step 1. Put a decimal point to the right of the whole number 18 and line up the numbers with the decimal points under each other. Use zeros to give each number the same number of decimal places.

```
 18.000
  2.350
+ 0.482
-------
 20.832
```

Step 2. Add each column. Place the decimal point in the answer under the decimal points in the numbers in the problem.

To subtract decimals put a decimal point to the right of any whole number. Then line up the numbers with the decimal points under each other. Use zeros to give each number the same number of decimal places.

Example 3: Subtract 0.036 from 0.09.

Step 1. Line up the numbers with the decimal points under each other. Then put a zero to the right of 0.09 to give each number the same number of places.

```
   8 10
 0.09̸0̸   ← Use a zero to
-0.036       give each
 0.054       number the
             same number
             of places.
```

Step 2. Borrow and subtract. Place the decimal point in the answer under the decimal points in the numbers in the problem.

Example 4: Take 0.48 from 2.

Step 1. Put a decimal point to the right of 2, and line up the numbers with the decimal points under each other.

```
  2.
- 0.48
```

Step 2. Put two zeros to the right of 2 to give each number the same number of places. Then borrow and subtract.

```
      9
  1 1̸0̸10
  2.0̸0̸
- 0.4̸8
  1.52
```

Lesson 3 Exercise

Solve each problem.

1. $0.36 + 0.5 + 0.607 =$
2. $0.38 + 0.619 + 0.2 =$
3. $0.3 + 0.9 + 0.7 =$
4. $0.006 + 0.05 + 0.8 =$
5. $2.5 + 18 + 1.07 =$
6. $0.506 + 3.1 + 9 =$
7. $38 + 4.078 + 0.0195 =$
8. $9.1 + 0.87 + 0.143 =$
9. $6 - 2.5 =$
10. $8 - 0.19 =$
11. $0.3 - 0.258 =$
12. $5.9 - 2.114 =$
13. $0.015 - 0.009 =$
14. $1 - 0.0865 =$
15. $9 - 0.32 =$
16. $0.6 - 0.24 =$

Check your answers on page 303.

Lesson 4: Multiplication and Division of Decimals

When you multiply or divide decimals, you do not have to line up the numbers with the decimal points under each other. You must be sure, however, that you have the correct number of decimal places in your answer.

Multiplying Decimals

To multiply numbers with decimals, first set up the numbers for easy multiplication. Put the number with more digits on top. After you have multiplied, count the number of decimal places in both numbers. Then count that number of places toward the left from the last digit in the answer to locate the position of the decimal point.

Example 1: Find the product of 3.17 and 8.

Step 1. Since 3.17 has more digits, put it on top and multiply.

Step 2. Count the decimal places in each number. 2.36 has two decimal places, and 8 has none. Put the total number of places, 2 + 0 = 2, in the answer. Start with the last digit and count two decimal places toward the left.

```
  3.17   has 2 decimal places
×    8   has 0 decimal place
 25.36   has 2 decimal places
```

In some problems you may have to add zeros on the left side to get the correct number of decimal places in the answer.

Example 2: Find the product of 0.06 and 0.4.

Step 1. Set up the problem and multiply.

Step 2. Count the decimal places in each number. 0.06 has two decimal places, and 0.4 has one so you need a total of 3 (2 + 1 = 3) decimal places in the answer. Notice that you need to write a zero to the left of 24 to get three decimal places.

```
  0.06   has 2 decimal places
× 0.4    has 1 decimal place
 0.024   has 3 decimal places
```

Dividing Decimals by Whole Numbers

When you divide a decimal (or a mixed decimal) by a whole number, the decimal point in the answer belongs above its position in the problem. Divide as you would with whole numbers.

Example 3: Divide 2.88 by 6.

Set up the problem. Write the decimal point in the answer above its position in the problem. Then divide.

```
   0.48
6)2.88
  2 4
    48
    48
```

In some problems you will have to put zeros in your answer.

Example 4: Divide 0.348 by 4.

Set up the problem. Then write the decimal point in the answer above its position in the problem and divide. Notice the zero above the 3. The zero shows that 4 does not divide into 3. The zero indicates that 8 belongs in the hundredths place.

```
    0.087
4)0.348
```

Dividing by Decimals

Dividing by decimals is more complicated. The idea is to change the problem into one in which the divisor (the number by which you are dividing) is a whole number.

To divide by a decimal first change the divisor into a whole number by moving the decimal point to the right end of the divisor. Next, move the decimal point in the dividend to the right the same number of places.

If you move the point in the divisor one place to the right, move the point in the dividend one place to the right. If you move the point in the divisor three places to the right, move the point in the dividend three places to the right.

These steps are easier to understand with whole numbers. Look at the problem $6 \div 2 = 3$. The answer is the same if the decimal point is moved one place to the right in both 6 and 2. The problem becomes $60 \div 20 = 3$. The problems are different but the answers are the same.

Example 5: Divide 2.52 by 0.4.

Step 1. Set up the problem and make the divisor, 0.4, a whole number by moving the decimal point one place to the right. Then move the decimal point in the dividend the same number of places to the right, one.

```
0.4.)2.5.2
```

Step 2. Write the decimal point up in the answer above its new position in the dividend and divide.

```
         6.3
0.4.)2.5.2
     2 4
       1 2
       1 2
```

It is a good idea to check division problems. Multiply the answer by the divisor. The product should equal the **original** dividend.

```
   6.3
× 0.4
  2.52
```

When you divide a whole number by a decimal, you will have to add zeros to the dividend in order to be able to move the decimal point.

Example 6: Divide 45 by 0.05.

Step 1. Set up the problem and make the divisor, 0.05, a whole number by moving the decimal point two places to the right. Then move the decimal point in the dividend two places to the right. Remember that a whole number is understood to have a decimal point after the units place (at the right end). Notice the two zeros added to the dividend.

$$0.05.\overline{)45.00.}$$

Step 2. Write the decimal point up in the answer above its new position in the dividend and divide.

$$0.05.\overline{)45.00.}^{9\ 00.}$$
$$\underline{45}$$
$$0\ 00$$

Step 3. To check the example, multiply the answer, 900, by the divisor, 0.05.

$$\begin{array}{r}900\\ \times\ 0.05\\ \hline 45.00 = 45\end{array}$$

Lesson 4 Exercise

Solve each problem.

1. $3.5 \times 7 =$
2. $29 \times 0.04 =$
3. $0.06 \times 0.5 =$
4. $0.47 \times 16 =$
5. $0.185 \times 0.4 =$
6. $0.59 \times 0.004 =$
7. $2.09 \times 30 =$
8. $0.0065 \times 0.6 =$
9. $215 \times 0.04 =$
10. $16.2 \div 6 =$
11. $4.32 \div 9 =$
12. $24.195 \div 3 =$
13. $47.5 \div 25 =$
14. $1.152 \div 64 =$
15. $204.6 \div 31 =$
16. $128 \div 0.32 =$
17. $56 \div 0.08 =$
18. $108 \div 1.2 =$
19. $2.4 \div 0.008 =$
20. $261 \div 0.6 =$
21. $0.312 \div 0.026 =$
22. $0.63 \div 0.9 =$
23. $0.144 \div 0.03 =$
24. $0.312 \div 0.052 =$

Check your answers on page 303.

Lesson 5 — Comparing and Ordering Decimals

In Lesson 1 you learned to tell whether decimals had values of more or less than 1. Often it is useful to compare decimals to each other.

To compare decimals use zeros to give each decimal the same number of places.

Example 1: Which is greater, 0.09 or 0.089?

Step 1. Put a zero to the right of 0.09 to give it the same number of places as 0.089

Step 2. Since 0.090 is greater than 0.089, 0.09 is the greater decimal.

0.09 = 0.090
0.089 = 0.089
0.09 is greater

Example 2: Arrange the following in order from least to greatest: 0.4, 0.404, 0.04.

Step 1. Use zeros so that each decimal has three places.

0.4 = 0.400
0.404 = 0.404
0.04 = 0.040

Step 2. Since 0.040 is the least, 0.04 comes first. 0.400 or 0.4 is second, and 0.404 is the greatest.

In order from least to greatest:
0.04, 0.4, 0.404

Lesson 5 Exercise

Solve each problem.

1. Which decimal in each pair is greater?
   a. 0.056 or 0.05
   b. 0.19 or 0.2
   c. 1.08 or 1.082
   d. 0.075 or 0.57

2. Arrange each set of decimals in order from least to greatest.
   a. 0.021, 0.012, 0.21, 0.201.
   b. 0.045, 0.54, 0.5, 0.005
   c. 3.2, 2.33, 3.22, 3.3
   d. 1.008, 0.8, 1.09, 0.9

3. Arrange each set of decimals in order from greatest to least:
   a. 0.38, 0.8, 0.083, 0.308.
   b. 5.0, 0.5, 5.05, 0.055
   c. 0.9, 0.09, 0.999, 9.0
   d. 2.075, 2.75, 2.7, 2.5

4. Which is greatest, 0.705 meter, 0.75 meter, or 0.075 meter?
5. Which is heaviest, 1.2 kg, 1.099 kg, or 1.209 kg?

Check your answers on page 303.

Level 1 Review

Solve each problem.

1. Which of the following has a value **greater** than 1?
   (1) 1.005   (2) 0.9907   (3) 0.098   (4) 0.099
2. Which of the following has a value **less** than 1?
   (1) 8.001   (2) 8.008   (3) 1.808   (4) 0.989

3. Which of the following is the same as 7.06?
   (1) seven and six tenths
   (2) seventy-six hundredths
   (3) seven and six hundredths
   (4) seventy-six thousandths
   (5) seven and six thousandths
4. Write sixty and twelve ten-thousandths as a mixed decimal.
5. What is the sum of 0.385, 0.6, and 0.09?
6. What is the difference between 0.058 and 0.0496?
7. What is the product of 29 and 8.06?
8. What is the product of 12.8 and 0.35?
9. Divide 2.432 by 76.
10. Divide 0.658 by 0.07.
11. Which is greater, 0.085 or 0.07?
12. Arrange in order from least to greatest: 0.44, 4.1, 1.4, 0.4

Check your answers on page 304. If you have 12 problems correct, go on to Level 2. If you have missed a question, find the item number on the chart below and review that lesson before you go on.

Review:	If you missed item number:
Lesson 1	1, 2
Lesson 2	3, 4
Lesson 3	5, 6
Lesson 4	7, 8, 9, 10
Lesson 5	11, 12

Level 2 — Applications of Decimals

In this section you will learn to use decimals in finding powers and square roots. You will learn about the metric measurement system and the formulas for circles and cylinders. Finally you will use decimal numbers in the formulas that you learned earlier in this book.

Lesson 6: Rounding Decimals

As you learned in the Whole Numbers chapter, rounding makes numbers easier to use. You can use rounded decimals to make estimates when you are checking problems.

> To round a decimal:
> 1. Mark the digit in the place to which you want to round.
> 2. If the digit to the right of the marked digit is more than 4, add 1 to the marked digit.
> 3. If the digit to the right of the marked digit is less than 5, leave the digit you marked as it is.
> 4. Drop the digits to the right of the digit you marked.

Example 1: Round 3.18 to the nearest tenth.

Step 1. Mark the digit in the tenths place. 3.1̲8

Step 2. Since the digit to the right of 1 is more than 4, add 1 to the marked digit and drop the 8. 3.2

3.18 rounded to the nearest tenth is 3.2.

Example 2: Round 8.496 to the nearest hundredth.

Step 1. Mark the digit in the hundredths place.   8.4<u>9</u>6

Step 2. Since the digit to the right of 9 is more than 4, add 1 to the marked digit and drop the 6. Notice that you must carry a digit over to the tenths place. 8.496 rounded to the nearest hundredth is 8.50. Since the directions are to round to the nearest hundredth, the zero in the hundredths place is included in the answer.   8.50

Lesson 6 Exercise

Solve each problem.

1. Round each number to the nearest tenth.
   0.634   0.372   0.08   5.16   0.3492
2. Round each number to the nearest hundredth.
   0.527   0.483   2.019   8.296   0.9148
3. Round each number to the nearest thousandth.
   0.1386   0.0577   1.7805   0.1052   6.4326
4. Are 4.2 pounds of potatoes closer to 4 pounds or to 5 pounds?
5. Are 8.7 kilometers closer to 8 kilometers or 9 kilometers?

Check your answers on page 304.

Lesson 7 — Metric Measurement

The metric system of measurement is used throughout the world. You have already seen some of these units in this book.

Following are the three basic units of measure in the metric system.

- The **meter** is the basic unit of length. A meter is a little longer than one yard.
- The **gram** is the basic unit of weight (mass). There are about 30 grams in an ounce and about 450 grams in a pound.
- The **liter** is the basic unit of liquid measure. A liter is a little more than one quart.

66   Chapter 2 Decimals

To read metric measurements, learn the following prefixes.

    kilo-(k)   means 1000.
    deci-(d)   means 0.1 (one tenth).
    centi-(c)   means 0.01 (one hundredth).
    milli-(m)   means 0.001 (one thousandth).

Below is a list of the most common metric measurements. Abbreviations are in parentheses. The list tells how many smaller units are contained in each larger unit. Memorize any units you do not already know.

<u>Length</u>
1 kilometer (km) = 1000 meters (m)
1 meter = 10 decimeters (dm)
1 meter = 100 centimeters (cm)
1 meter = 1000 millimeters (mm)

<u>Weight</u>
1 kilogram (kg) = 1000 grams (g)
1 gram = 100 centigrams (cg)
1 gram = 1000 milligrams (mg)

<u>Liquid Measure</u>
1 liter (l) = 10 deciliters (dl)
1 liter = 100 centiliters (cl)
1 liter = 1000 milliliters (ml)

To change from a **larger** unit to a **smaller** unit, **multiply** by the number of small units that make up one large unit.

Example 1: Change 3.5 liters to milliliters.

Multiply 3.5 l by the number of milliliters in one liter, 1000.

$3.5 \times 1000 = 3500.0$
$3.5 \text{ l} = 3500 \text{ ml}$

To change from a **smaller** unit to a **larger** unit, **divide** by the number of small units that make up one large unit.

Example 2: Change 85 centimeters to meters.

Divide 85 cm by the number of centimeters in one meter, 100.

$85 \div 100 = 0.85$
$85 \text{ cm} = 0.85 \text{ m}$

Lesson 7 Exercise

Solve each problem.

1. Change 2.5 meters to centimeters.
2. Change 3 kilograms to grams.
3. Change 4.8 liters to milliliters.
4. Change 6.5 kilometers to meters.
5. Change 1250 meters to kilometers.
6. Change 8 deciliters to liters.

7. Change 385 grams to kilograms.
8. Change 195 centimeters to meters.
9. Round each of the following measurements to the nearest tenth.
   (a) 1.85 liters   (b) 4.68 kilograms   (c) 0.109 meter
10. Round each of the following measurements to the nearest unit.
    (a) 4.9 kilometers   (b) 28.4 grams   (c) 9.7 liters

Check your answers on page 304.

Lesson 8: Reading Metric Scales

In the last lesson you learned that the basic unit of measuring length in the metric system is the meter.

A centimeter ruler is a tool for measuring the length of small objects such as a nail or a piece of paper. The longest lines on the ruler represent centimeters. The next longest lines represent half centimeters (0.5 centimeters). The shortest lines represent tenth centimeters (0.1 centimeters). The shortest lines also represent millimeters.

To measure a length with a centimeter ruler determine how far from the left end of the centimeter scale a point is.

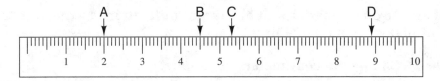

Example 1: How far from the left end of the 10-centimeter ruler pictured above is the point marked A?

A is at the second long line on the ruler.
A is 2 centimeters from the left.

Example 2: How far from the left end of the ruler is point B?

B is at the longest line between 4 and 5 centimeters.
B is 4.5 centimeters from the left.

Example 3: How far from the left end of the ruler is point C?

C is 3 short lines to the right of 5 centimeters.
C is 5.3 centimeters from the left.

68 Chapter 2 Decimals

Example 4: How far from the left end of the ruler is point D?

D is 9 short lines to the right of 8 centimeters.
D is 8.9 centimeters from the left.

Lesson 8 Exercise

Use the ten-centimeter ruler below to answer the following questions.

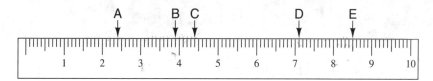

1. How far from the left end of the ruler is point A?
2. How far from the left end of the ruler is point B?
3. How far from the left end of the ruler is point C?
4. How far from the left end of the ruler is point D?
5. How far from the left end of the ruler is point E?
6. What is the distance between points B and D on the ruler?
7. What is the distance between points A and C on the ruler?
8. What is the distance between points D and E on the ruler?

Check your answers on page 304.

Lesson 9: Powers and Square Roots

The operation of raising a decimal to a power or of finding the square root of a decimal is the same as those operations with whole numbers.
To raise a decimal to a power, remember to count the total number of decimal places in the numbers you are multiplying. Notice that when a decimal is raised to the second power the number of decimal places doubles.

Example 1: What is the value of $(0.3)^2$?

Write 0.3 two times with a times sign between them and multiply. Notice that the answer has two decimal places.

$(0.3)^2 = 0.3 \times 0.3 = 0.09$

To find the square root of a decimal notice that the answer will have **half** as many places as the number you are finding the square root of. Remember that when a decimal number is raised to the second power, the number of decimal places doubles.

Example 2: What is the value of $\sqrt{0.0064}$?

The square root of 64 is 8. The square root of a four-place decimal has two places. Check by multiplying $0.08 \times 0.08 = 0.0064$.

$\sqrt{0.0064} = 0.08$

Lesson 9 Exercise

Find the value of each of the following.

1. $(0.5)^2 =$ $(0.02)^3 =$ $(0.4)^2 =$ $(0.12)^2 =$
2. $(0.07)^2 =$ $(0.009)^2 =$ $(0.1)^4 =$ $(1.5)^2 =$
3. $\sqrt{0.16} =$ $\sqrt{0.81} =$ $\sqrt{0.0036} =$ $\sqrt{0.0001} =$
4. $\sqrt{0.0004} =$ $\sqrt{0.0121} =$ $\sqrt{0.000009} =$ $\sqrt{0.0625} =$

Check your answers on page 304.

Lesson 10: Circles and Cylinders

So far in this book you have found the perimeter and the area of figures with straight sides. A **circle** is a closed curved line. Every point on a circle is the same distance from the center of the circle as every other point. Following are definitions of some of the key words related to circles.

The **diameter** (d) is the distance across a circle. The diameter measures the widest part of a circle. The diameter always contains the **center** of the circle.

The **radius** (r) is the distance from the center of the circle to a point on the curved line that is the circle. The radius is exactly one-half of the diameter. There are two formulas for the relationship between the diameter and the radius of a circle:

$d = 2r$ and $r = d/2$ where d stands for the diameter and r stands for the radius.

The **circumference** is the distance around a circle. The circumference is the "perimeter" of a circle.

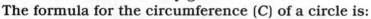

The Greek letter **pi** (π) is the number found by dividing the circumference of a circle by its diameter. The value of π is usually given as 3.14.

The formula for the circumference (C) of a circle is:

$C = \pi d$ where $\pi = 3.14$ and d = diameter.

Circumference is measured in units such as inches, feet, yards, or meters.

Example 1: Find the circumference of the circle.

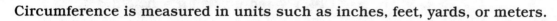

Step 1. Replace π with 3.14 and d with 10 in the formula for the circumference of a circle.

$C = \pi d$
$C = 3.14 \times 10$
$C = 31.4$ ft

Step 2. Multiply.

Example 2: Find the circumference of the circle.

(r = 50 in.)

Step 1. Find the diameter of the circle.

$d = 2r$
$d = 2 \times 50$
$d = 100$ in.

Step 2. Replace π with 3.14 and d with 100 in the formula for the circumference of a circle.

$C = \pi d$
$C = 3.14 \times 100$
$C = 314$ in.

Step 3. Multiply.

The formula for the area (A) of a circle is:

$A = \pi r^2$; where $\pi = 3.14$ and r = radius.

Area is measured in square units such as square inches, square feet, square yards, or square meters.

Example 3: Find the area of the circle.

(r = 3 m)

Step 1. Replace π with 3.14 and r with 3 in the formula for the area of a circle.

$A = \pi r^2$
$A = 3.14 \times 3^2$
$A = 3.14 \times 3 \times 3$
$A = 28.26$ m^2

Step 2. Multiply.

Level 2, Lesson 10: Circles and Cylinders

You learned earlier that volume is a measure of the amount of space or the capacity inside a three-dimensional figure. So far you have found the volume of figures with straight sides.

A **cylinder** is a figure shaped like a tin can. The top and the bottom of a cylinder are circles.

> The formula for the volume of a cylinder is:
> $V = \pi r^2 h$ where $\pi = 3.14$, $r =$ radius, and $h =$ height.

Remember that volume is measured in cubic units such as cubic inches, cubic feet, cubic yards, or cubic meters.

Example 4. Find the volume of the cylinder.

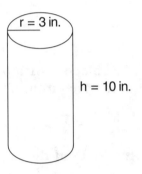

Step 1. Replace π with 3.14, r with 3, and h with 10 in the formula for the volume of a cylinder.

$V = \pi r^2 h$
$V = 3.14 \times 3^2 \times 10$
$V = 3.14 \times 3 \times 3 \times 10$

Step 2. Multiply.

$V = 282.6$ cu in.

Lesson 10 Exercise

Solve each problem.

1. a. What is the radius of this circle?
   b. What is the circumference of the circle?
   c. What is the area of the circle?

   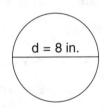

2. The pond in the park has a radius of 30 feet.
   a. What is the diameter of the pond?
   b. What is the circumference of the pond?
   c. What is the area of the pond?

3. a. What is the diameter of this circle?
   b. Find the circumference of the circle to the nearest tenth.
   c. Find the area of the circle to the nearest tenth.

   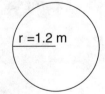

4. A picnic table has a diameter of 40 inches.
   a. Find the radius of the table.
   b. Find the circumference of the table.
   c. Find the area of the table.

5. Find the volume of each of the cylinders pictured below.

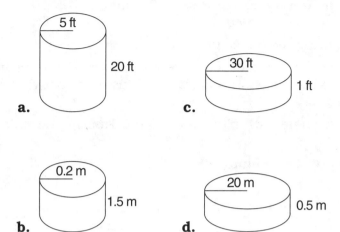

a.   c.

b.   d.

Check your answers on page 304.

Lesson 11 — Formulas and Decimals

The exercise in this lesson gives you a chance to use the formulas on page 17 with decimals.

Example: To the nearest square centimeter what is the area of this rectangle?

$w = 6.2$ cm
$l = 12$ cm

Step 1. Replace l with 12 and w with 6.2 in the formula $A = lw$ and multiply.

$A = lw$
$A = 12 \times 6.2$
$A = 74.4$ sq cm

Step 2. Round 74.4 sq cm to the nearest unit.

74.4 to the nearest sq cm = 74 sq cm

Lesson 11 Exercise

Solve each problem.

1. Sally bought 6.5 yards of lumber at $8.99 per yard. To the nearest cent what was the total cost of the lumber?
2. Kiel drove for 6.25 hours at an average speed of 55 mph. To the nearest mile how far did he drive?

3. Pedro shipped four boxes. One box weighed 8.25 pounds; the second, 4.5 pounds; the third, 7.65 pounds; and the fourth, 3 pounds. What was the average weight of the boxes he shipped?

4. Find the median weight of the boxes in Problem 3.

5. It costs the Quarles $0.036 to run their color television for one hour. In a week they watch television for 45 hours. How much does it cost them to watch television for a week?

6. Find the perimeter of a triangle with sides measuring 1.65 m, 2.4 m, and 1.38 m.

7. a. What is the perimeter of this rectangle?
   b. What is its area?

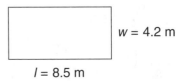

$w = 4.2$ m
$l = 8.5$ m

8. a. What is the perimeter of this square?
   b. What is its area?

$s = 3.4$ cm

9. Find the volume of this cube.

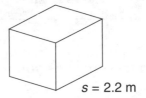

$s = 2.2$ m

10. Find the volume of this figure.

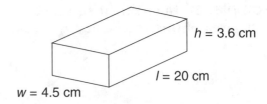

$h = 3.6$ cm
$l = 20$ cm
$w = 4.5$ cm

Check your answers on page 305.

Level 2 Review

Solve each problem. Use any formulas on pages 17 and 18 that you need.

1. What is 6.2975 rounded to the nearest hundredth?
2. What is 0.2836 rounded to the nearest thousandth?
3. Change 2.4 kilograms to grams.
4. Change 655 millimeters to meters.

Use the centimeter scale below to answer items 5 to 7.

5. Point X is how far from the left end of the centimeter scale?

6. Point Y is how far from the left end of the centimeter scale?
7. What is the distance between point X and point Y on the centimeter scale?
8. What is the value of $(0.5)^3$?
9. What is the value of $\sqrt{0.0049}$?
10. Find the circumference of the circle.

11. Find the circumference of a circle with a radius of 0.4 meters. Round the answer to the nearest tenth.
12. Find the area of this circle to the nearest square inch.

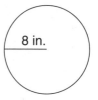

13. Find the volume of this cylinder.

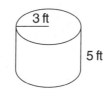

14. Find the cost of using 20 kilowatt hours of electricity at the rate of $0.109 per kilowatt hour.
15. Celia walked for 0.25 hour at the rate of 4 mph. How far did she walk?
16. Find the perimeter of a triangle whose three sides measure 5.4 m, 4.9 m, and 6.2 m.
17. Frank drove an average speed of 60.5 miles per hour for 3.5 hours. To the nearest mile how far did he drive?
18. To the nearest cent what is the cost of 3.6 pounds of beef at $3.48 per pound?
19. Find the volume of a rectangular container with length 10 m, width 3.4 m, and height 2.5 m.
20. To the nearest kilogram what is the average weight of three packages that weigh 2.7 kg, 3.5 kg, and 4.18 kg?

Level 2: Review

Check your answers on page 305. If you have 20 problems correct, go on to Level 3. If you answered a question incorrectly, find the item number on the chart below and review that lesson before you go on.

Review:	If you missed item number:
Lesson 6	1, 2
Lesson 7	3, 4
Lesson 8	5, 6, 7
Lesson 9	8, 9
Lesson 10	10, 11, 12, 13
Lesson 11	14, 15, 16, 17, 18, 19, 20

Level 3: Problem Solving with Decimals

In this section you will learn to read and interpret tables with decimals, to sketch and solve complex distance problems, and to recognize a problem where not enough information is given.

Lesson 12: Tables

In the Whole Numbers chapter (Lesson 16) you used tables to solve word problems. The numbers in tables can also be in decimals. The table below shows the average number of hospital beds for every 1000 people for several years. Notice the use of decimal numbers.

Hospital Beds per 1000 People

1960	1965	1970	1975	1980	1985	1988
9.3	8.9	7.9	6.8	6.0	5.7	5.0

Source: American Hospital Association.

Example: According to the table how many hospital beds did a city with a population of 500,000 have in 1975?

Step 1. The table tells the number of hospital beds per 1000 people. Divide the population by 1000.

$$1000 \overline{)500{,}000} = 500$$

Step 2. Multiply the 1975 rate, 6.8, by 500.

$$\begin{array}{r} 500 \\ \times\ 6.8 \\ \hline 4000 \\ 3000 \\ \hline 3400.0 \end{array}$$

3,400.0 = 3,400 beds

Lesson 12 Exercise

Use the table about hospital beds to answer questions 1 through 5.

1. How many more hospital beds per 1000 people were there in 1960 than in 1985?
2. By how many beds per 1000 did the rate drop from 1965 to 1970?
3. The population of Greenport is 15,000. If Greenport followed the national average, how many hospital beds were there in Greenport in 1988?
4. The population of Central City was 280,000 in 1980. If Central City followed the national average, how many hospital beds were there in Central City in 1980?
5. Which of the following expresses the pattern for the rate per 1000 of hospital beds for the years shown on the table?
   (1) The rate first rose and then fell.
   (2) The rate first fell and then rose.
   (3) The rate gradually increased.
   (4) The rate gradually decreased.

Use the following table to answer questions 6 through 10.

Attendance at Selected Spectator Sports (in millions)

	1975	1980	1985	1989
Major League Baseball	30.4	43.7	47.7	55.7
National Football League	10.8	14.1	14.1	17.4
National Hockey League	9.5	10.5	11.6	12.6
Professional Basketball	7.6	10.7	11.5	16.6

Source: Statistical Abstract of the U.S.

6. For which years shown on the table was the attendance at professional basketball games more than attendance at professional hockey games?
7. The total attendance in 1985 at professional football, hockey, and basketball games was how much less than the attendance at major league baseball games that year?
8. The attendance at major league baseball games was how much greater in 1989 than in 1975?
9. By how much did the attendance at professional hockey games increase from 1980 to 1985?
10. For which sport shown on the table did attendance more than double from 1975 to 1989?

Check your answers on page 306.

Lesson 13: Travel Problems and Other Word Problems

You have already used the formula $d = rt$ to solve distance problems. In problems with more than one vehicle or traveler, it is useful to make a sketch of the trips.

For the examples and the solutions to the next exercise, an arrow to the right indicates travel to the east. An arrow to the left indicates travel to the west. An arrow pointing up indicates north, and an arrow pointing down indicates south.

Example 1: Mark and Heather started hiking at the same time and the same place. Mark walked east at a rate of 5.5 mph, and Heather walked west at a rate of 4.5 mph. How far apart were they at the end of two hours?

Step 1. Substitute each rate and the time of 2 hours in the formula $d = rt$.

$\quad\quad\quad$ ←Heather $\quad\bullet\quad$ Mark→

$d = rt \quad\quad d = rt$
$d = 4.5 \times 2 \quad d = 5.5 \times 2$
$d = 9$ mi $\quad\quad d = 11$ mi

Step 2. Since they walked in opposite directions, the distance between them is the **sum** of the two distances as shown in the diagram.

$9 + 11 = 20$ miles

Example 2: Rick and Janina began driving along a highway at the same time and place and in the same direction. Rick drove at an average speed of 40 mph, and Janina drove at an average speed of 30 mph. How far apart were they after 1.5 hours?

Step 1. Substitute each rate and 1.5 hours into the formula $d = rt$. Notice that the arrows point in the same directions.

$d = rt$
$d = 40 \times 1.5 = 60$ mi
Rick →
Janina →
$d = rt$
$d = 30 \times 1.5 = 45$ mi.

Step 2. Since they drove in the same direction, the distance between them is the **difference** between the two distances.

$60 - 45 = 15$ mi.

The following exercise has travel problems as well as other types of word problems. Some of the problems have set-up answer choices. Other problems are based on an item set or reading passage.

Level 3, Lesson 13: Travel Problems and Other Word Problems

Lesson 13 Exercise

Solve each problem.

1. Horst drove west along a highway at an average speed of 55 mph. Jana began driving at the same time as Horst. She drove east at an average speed of 45 mph. How far apart were they after 0.5 hour?

2. Two friends began driving along a highway at the same time and in the same direction. One drove at an average speed of 40 mph, and the other at an average speed of 30 mph. How far apart were they at the end of 2.5 hours?

3. Miriam rode her bicycle west at an average speed of 12 mph. Her sister Marcia started riding her bicycle at the same time and rode east at an average speed of 10 mph. Which of the following expresses how far apart they were after 1.25 hours?
   - (1) $12 \times 1.25 + 10 \times 1.25$
   - (2) $12 \times 1.25 - 10 \times 1.25$
   - (3) $12 + 10 \times 1.25$
   - (4) $12 \times 10 \times 1.25$

4. Dan used 8.4 gallons of gas on a 200-mile trip. To the nearest whole mile find the average distance he drove on one gallon of gas.

5. An express train started at 10:30 A.M. and traveled at an average speed of 50 mph. At 11:30 A.M. a local train started in the same place and traveled along the same track in the same direction at an average speed of 30 mph. How far apart were the two trains at 1:00 P.M.?

6. Carlos makes $6.50 an hour for the first 40 hours he works each week. He makes $9.75 an hour working overtime. Which expression tells the amount he makes for a week when he works 46 hours?
   - (1) $46 \times \$6.50 + \$9.75$
   - (2) $40 \times \$6.50 + 6 \times \$9.75$
   - (3) $46 \times \$9.75 - \$6.50$
   - (4) $40 \times 6 + \$6.50 \times \$9.75$

7. Before the Pagans went on vacation, the odometer in their car read 3789.6 miles. At the end of their vacation the odometer read 4534.4 miles. How many miles did they travel on their vacation?

8. Nina bought 3.2 pounds of beef for $7.52. Which of the following expresses the price of one pound of the beef?
   - (1) $3.2 \times \$7.52$
   - (2) $3.2 + \$7.52$
   - (3) $\$7.52 \div 3.2$
   - (4) $\$7.52 - 3.2$

Use the following information to answer questions 9 and 10.

A baseball batting average is the number of hits a player gets divided by the number of times the player was at bat. Batting averages are rounded to the nearest thousandth. A batting average multiplied by the number of times a player was at bat tells the number of hits the player makes. Last season Jeff got 17 hits out of 60 times at bat. Hank's batting average for the season was 0.268. Hank was at bat 45 times.

9. What was Jeff's batting average for the season?

10. How many hits did Hank make that season?

11. At 10:00 A.M. Alfredo's temperature was 98.6°. At 12:00 noon it was up 5.8°. By 2:00 P.M. it was down 3.9° from his noon temperature. What was Alfredo's temperature at 2:00 P.M.?

12. Chester wants to arrange the following packages in order from heaviest to lightest.

   Package A weighs 0.65 kg.
   Package B weighs 1.05 kg.
   Package C weighs 0.065 kg.
   Package D weighs 1.65 kg.
   Package E weighs 1.5 kg.

   Which of the following lists the packages in order from heaviest to lightest?

   (1) D, A, E, C, B   **(4)** B, C, E, A, D
   (2) D, E, B, A, C   **(5)** D, B, A, E, C
   (3) B, D, E, C, A

Check your answers on page 306.

Lesson 14 — Problems Where Not Enough Information Is Given

On the GED Test and in the rest of this book you will see multiple-choice questions where the last choice says, "Not enough information is given." This means that at least one number is missing and that there is no way to find a solution.

For practice in this kind of problem try the next exercise.

Lesson 14 Exercise

For problems 1–5 tell what piece of information is missing. The first has been done as an example.

1. Guadalupe mailed four packages. Two of them weighed 4.75 pounds each, and a third weighed 5.6 pounds. Find the average weight of the four packages.
   Missing information: *the weight of the fourth package*

2. Find the perimeter of the triangle.
   Missing information: _____

   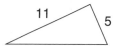

3. Mai Lee bought six pounds of chicken at $0.89 a pound and four pounds of ground beef. What was the total cost of her purchases?
   Missing information: _____

4. According to the table about attendance at spectator sports on page 78, how many of the spectators at National Football League games in 1980 were female?
   Missing information: _____

5. According to the table about hospital beds per 1000 people on page 77, what was the average number of beds per hospital in 1975?
   Missing information: _____

For problems 6 to 10 choose the correct answer.

6. The Millers spend $269.70 a month for rent and $89.90 a month for car payments. The amount the Millers spend each month for food is about how many times the amount they spend for their car payments?

   (1) two
   (2) three
   (3) four
   (4) five
   (5) Not enough information is given.

7. Manny worked 35 hours at his normal rate of $3.90 an hour and five hours at his overtime rate. How much did he make altogether that week?

   (1) $136.50
   (2) $156
   (3) $165.75
   (4) $175.50
   (5) Not enough information is given.

8. From a barrel that contained 20 pounds of nails Hans used 4.5 pounds of nails to build the framing for a remodeled kitchen. Find the weight in pounds of the nails that were left in the barrel when Hans finished.

   (1) 14.5
   (2) 15.5
   (3) 16.5
   (4) 24.5
   (5) Not enough information is given.

9. Mike uses about 30 gallons of gasoline a month to drive to and from work. He pays $0.90 a gallon for the gasoline. Find Mike's total transportation costs including parking for a five-day work week.

   (1) $2.70
   (2) $13.50
   (3) $27
   (4) $135
   (5) Not enough information is given.

10. The Jacksons' living room is 16 feet long, 12 feet wide, and 7.5 feet high. Find the volume of the living room in cubic feet.

    (1) 35.5
    (2) 56
    (3) 112
    (4) 1440
    (5) Not enough information is given.

Check your answers on page 307.

Level 3 Review

Solve each problem. Use formulas as needed from pages 17 and 18.

Use the following table to answer problems 1 through 4.

Smokers and Non-Smokers in the U.S. (in millions)

	Persons who			Total
	never smoked	used to smoke	smoke now	population
Male	28.8	22.2	27.6	78.7
Female	48.0	14.0	25.9	88.0

Source: U.S. National Center for Health Statistics.

1. According to the table how many females in the U.S. used to smoke?
2. According to the table how many more men in the U.S. smoke now than women?
3. The number of males in the U.S. who used to smoke is how many more than the number of females who used to smoke?
4. The number of females who never smoked is how much greater than the number of males who never smoked?

Use the table below to answer questions 5–8.

Per Capita Consumption of Selected Food Items (in pounds)

Item	1970	1975	1980	1985	1989
beef	79.6	83.0	72.1	74.3	65.0
fish	11.8	12.2	12.8	14.4	15.7
cheese	11.4	14.3	17.5	22.5	23.7

Source: U.S. Department of Agriculture.

5. For which item shown in the table was the per capita consumption in 1989 less than in 1970?
6. For which food item shown in the table was the per capita consumption in 1989 more than twice that in 1970?
7. By how many pounds did the per capita consumption of beef drop from 1985 to 1989?
8. Which was the first year in the table when the consumption of cheese exceeded the consumption of fish?

9. Phil hiked north at a rate of 3.5 mph for 1.75 hours. Sue started at the same time and place and hiked south at a rate of 4.5 mph for 1.75 hours. How far apart were they at the end of that time?

10. An express train traveled for 2.25 hours at an average speed of 60 mph. A local train started at the same time as the express and traveled in the same direction on a parallel track at an average speed of 40 mph. How far behind the express was the local after 2.25 hours?

11. Sandy drove east for 3.5 hours at an average speed of 38 mph. Dick started at the same time as Sandy. Dick drove east for 2.5 hours at an average speed of 42 mph. Then he turned around and drove west for one hour at an average speed of 54 mph. How far apart were Sandy and Dick at the end of 3.5 hours?

12. Jack and Manny started driving at the same time and place and in the same direction. Jack drove for 2.5 hours at an average speed of 42 mph and Manny drove for the same length of time at an average speed of 15 mph. How far apart were they at the end of 2.5 hours?

13. Two planes took off at the same time and place and headed in opposite directions. One plane flew east at an average speed of 350 mph and the other flew west at an average speed of 420 mph. How far apart were they at the end of 1.5 hours?

14. Jose drove for 5.5 hours at an average speed of 40 mph. Maria started one hour later and drove in the same direction as Jose at an average speed of 30 mph. How far apart were they at the end of 5.5 hours?

15. Which of the following expresses the perimeter of a rectangle with a length of 6.25 and a width of 4?
    (1) $2 \times 6.25 + 2 \times 4$
    (2) 6.25×4
    (3) $(6.25 + 4)^2$
    (4) $6.25 \times 4 \times 2$
    (5) Not enough information is given.

16. George pays $4.75 a day to park at his job. To drive to work he uses 6.5 gallons of gasoline a week. He pays $1.20 a gallon for the gasoline. For a five-day work week what does George pay altogether for gasoline and parking?
    (1) $ 7.80
    (2) $12.55
    (3) $23.75
    (4) $31.55
    (5) Not enough information is given.

17. Cheryl bought three pounds of pork at $1.29 a pound and four pounds of beef. Which of the following is the total cost of her purchases?
    (1) $9.03
    (2) $7.87
    (3) $5.16
    (4) $3.87
    (5) Not enough information is given.

18. The Graves family spent $124.50 for fuel oil in October, $190 in November, and $310 in December. Which of the following expresses the mean (average) amount they spent during those months?

    (1) $\frac{\$124.50 + \$190 + \$310}{3}$

    (2) 3($124.50 + $190 + $310)

    (3) $124.50 + $190 + $310

    (4) $\frac{\$124.50}{3}$ + $190 + $310

    (5) Not enough information is given.

19. Find the perimeter of this triangle.

    (1) 8.82
    (2) 11.44
    (3) 13.8
    (4) 21.84
    (5) Not enough information is given.

20. Bettina worked one week for 35 hours at her normal hourly wage and six hours at her overtime rate of $6.80 an hour. How much did she make altogether that week?

    (1) $159.80
    (2) $219.30
    (3) $238.00
    (4) $278.80
    (5) Not enough information is given.

Check your answers on page 307. If you have all 20 answers correct, go on to the next chapter. If you missed a question, find the item on the chart below and review that lesson before you go on.

Review:	If you missed item number:
Lesson 12	1, 2, 3, 4, 5, 6, 7, 8
Lesson 13	9, 10, 11, 12, 13, 14
Lesson 14	15, 16, 17, 18, 19, 20

Level 3: Review

Chapter 3
FRACTIONS

Level 1

Fraction Skills

In this section you will review reducing fractions, raising fractions to higher terms, and changing mixed numbers and improper fractions. You will review the basic operations of addition, subtraction, multiplication, and division with fractions. You will also review the steps necessary to compare and order fractions and fractions and decimals.

To find out whether you need a review, try the following preview.

Preview

Solve each problem.

1. Reduce $\frac{24}{64}$ to lowest terms.
2. Change $\frac{15}{4}$ to a mixed number.
3. Find the sum of $4\frac{7}{20}$ and $9\frac{1}{2}$.
4. Add $3\frac{1}{3} + 1\frac{5}{9} + 2\frac{5}{6}$.
5. Find the difference between $9\frac{1}{2}$ and $4\frac{3}{8}$.
6. Take $10\frac{7}{16}$ from $12\frac{1}{4}$.
7. Find the product of $\frac{4}{5}$ and $\frac{5}{6}$.
8. Find $12 \times 4\frac{1}{2} \times 2\frac{2}{3}$.
9. What is $20 \div \frac{4}{5}$?
10. Divide $6\frac{1}{4}$ by $1\frac{7}{8}$.
11. Which is larger, $\frac{5}{16}$ or $\frac{1}{3}$?
12. Change 0.45 to a fraction and reduce.

Check your answers on page 308. If you have all 12 problems correct, do the Level 1 Review on page 103. If you have fewer than 12 problems correct, study Level 1 beginning with Lesson 1.

Level 1: Fraction Skills   87

Lesson 1: Writing and Reducing Fractions

A **fraction**, like a decimal, shows a part of a whole thing. Two days are two of the seven equal parts of a week. Two days are $\frac{2}{7}$ or two sevenths of a whole week.

The top number of a fraction is called the **numerator**. It tells how many parts you have. The bottom number is called the **denominator**. It tells how many equal parts into which the whole is divided.

In the figure at the right, three parts are shaded out of a total of eight equal parts. We say that $\frac{3}{8}$ of the figure is shaded. The numerator, 3, tells how many equal parts are shaded. The denominator, 8, tells how many equal parts there are in the whole figure.

Fractions appear in three forms:

1. In a **proper fraction** the numerator is less than the denominator. For example, $\frac{2}{3}$, $\frac{7}{10}$, and $\frac{1}{50}$ are all proper fractions. The value of a proper fraction is always **less** than one whole.

2. In an **improper fraction** the numerator is equal to or greater than the denominator. For example, $\frac{5}{5}$, $\frac{4}{3}$, and $\frac{24}{6}$ are all improper fractions. When the numerator and the denominator are the same, the value of an improper fraction is one. For example, $\frac{5}{5} = 1$. When the numerator is greater than the denominator, the value of an improper fraction is more than one. For example, $\frac{4}{3}$ and $\frac{24}{6}$ are more than one.

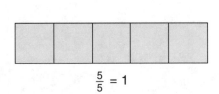

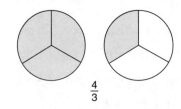

3. In a **mixed number** a whole number is written beside a fraction. For example, $1\frac{1}{4}$, $5\frac{2}{3}$, and $10\frac{5}{6}$ are all mixed numbers. The value of a mixed number is always **more** than one.

 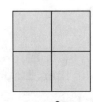

$2\frac{3}{4}$

Reducing Fractions

Reducing means dividing both the numerator and the denominator by a number that divides into each of them evenly. When the numerator and denominator are divided by the same number, **it does not change the value of the fraction.** $\frac{2}{4}$ of a dollar (two quarters) has the same value as $\frac{1}{2}$ a dollar (a fifty-cent piece). $\frac{2}{4}$ and $\frac{1}{2}$ are called **equivalent fractions**.

> To reduce a fraction, divide both numerator and denominator by the same number. Then check to see if any number other than 1 will divide evenly into both. A fraction is in lowest terms when the only number that will divide evenly into both numerator and denominator is 1.

Example 1: Reduce $\frac{14}{16}$.

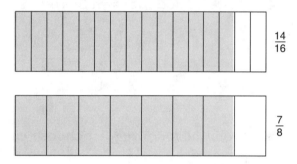

Divide 14 and 16 by a number that divides both evenly. Both 14 and 16 can be divided evenly by 2. Check to see if any other number besides 1 divides evenly into both 7 and 8. Since no other number divides evenly into both, we say $\frac{7}{8}$ is reduced to lowest terms.

$$\frac{14 \div 2}{16 \div 2} = \frac{7}{8}$$

Example 2: Reduce $\frac{36}{48}$.

<u>Step 1.</u> Divide 36 and 48 by a number that goes evenly into both. Try 6. 36 and 48 are divided evenly by 6.

$$\frac{36 \div 6}{48 \div 6} = \frac{6}{8}$$

<u>Step 2.</u> Try another number. 2 divides evenly into both. Since 1 is the only number that goes evenly into both 3 and 4, $\frac{3}{4}$ is reduced to lowest terms.

$$\frac{6 \div 2}{8 \div 2} = \frac{3}{4}$$

Note that the fraction in Example 2, $\frac{36}{48}$, can be reduced to lowest terms in one step by dividing by 12: $\frac{36 \div 12}{48 \div 12} = \frac{3}{4}$.

The solutions in the answer key for the next exercise show each fraction reduced in one step. You may wish to use two or more steps to reduce some of the fractions.

Lesson 1 Exercise

Solve each problem.

1. Choose the proper fractions from the following list.
   $\frac{7}{10}$ $\frac{7}{6}$ $\frac{8}{8}$ $\frac{9}{100}$ $\frac{1}{12}$ $\frac{14}{15}$

2. Choose the improper fractions from the following list.
   $\frac{18}{7}$ $\frac{7}{18}$ $\frac{1}{99}$ $\frac{3}{3}$ $\frac{25}{4}$ $\frac{3}{16}$

3. Choose the mixed numbers from the following list.
   $5\frac{1}{2}$ $\frac{23}{25}$ $\frac{18}{18}$ $6\frac{9}{20}$ $8\frac{1}{2}$

4. Write a fraction that tells the part of each figure that is shaded.

   a.   b.   c.   d.

5. From the following list choose each fraction with a value **less** than 1.
   $\frac{9}{6}$ $\frac{5}{8}$ $\frac{12}{12}$ $\frac{15}{16}$ $\frac{7}{2}$ $\frac{99}{100}$

6. From the following list choose each fraction with a value **more** than 1.
   $\frac{3}{2}$ $\frac{2}{2}$ $\frac{13}{5}$ $\frac{4}{9}$ $\frac{20}{10}$ $\frac{8}{25}$

Reduce each fraction to lowest terms.

7. a. $\frac{5}{20} =$ b. $\frac{32}{56} =$ c. $\frac{25}{30} =$ d. $\frac{90}{200} =$
8. a. $\frac{40}{55} =$ b. $\frac{6}{18} =$ c. $\frac{45}{75} =$ d. $\frac{22}{24} =$
9. a. $\frac{24}{40} =$ b. $\frac{19}{38} =$ c. $\frac{60}{144} =$ d. $\frac{50}{1000} =$
10. a. $\frac{25}{45} =$ b. $\frac{48}{64} =$ c. $\frac{24}{72} =$ d. $\frac{9}{54} =$

Check your answers on page 308.

Lesson 2

Other Ways to Change Fractions and Mixed Numbers

Raising Fractions to Higher Terms

Usually, answers to fraction problems are expressed in lowest terms. When you solve fraction problems, however, you may need to raise fractions to higher terms. You will often raise fractions to higher terms when you add and subtract fractions.

Chapter 3   Fractions

For example, you may want to change $\frac{3}{4}$ to 24ths. In this problem you have a new denominator, 24, and you need to find the new numerator that corresponds to 3.

> To raise a fraction to higher terms:
> 1. Divide the smaller denominator into the larger denominator.
> 2. Then multiply both the numerator and the denominator of the original fraction by the quotient. If you multiply the numerator and denominator by the same number, the result will be an equivalent fraction.

Example 1: Raise $\frac{3}{4}$ to 24ths. $\left(\frac{3}{4} = \frac{}{24}\right)$

Step 1. Divide 24 by 4. The quotient is 6.

$$4\overline{)24} \quad \text{quotient } 6$$

Step 2. Multiply both 3 and 4 by 6. $\frac{3}{4} = \frac{18}{24}$

$$\frac{3 \times 6}{4 \times 6} = \frac{18}{24}$$

To check your work when you have raised a fraction to higher terms, reduce it. The reduced fraction should be the original fraction. For the example, reduce $\frac{18}{24}$ by dividing by 6.

$$\frac{18 \div 6}{24 \div 6} = \frac{3}{4}$$

Changing Improper Fractions to Whole or Mixed Numbers

The answers to many fraction problems are improper fractions. These answers are easier to read if you change them to whole or mixed numbers.

<u>A Note about Mathematical Terms:</u>
In this book the word **change** means to put a number in another form. The terms **rewrite, rename,** and **convert** mean the same thing as the word **change**. When you change an improper fraction to a mixed number, you are changing the way a number is written. You are **not** changing its value.

> To change an improper fraction,
> 1. Divide the denominator into the numerator and write the remainder as a fraction over the denominator.
> 2. If necessary reduce the fraction.

Example 2: Change $\frac{32}{10}$ to a whole or mixed number.

Step 1. Divide 10 into 32. Write the remainder as a fraction, $\frac{2}{10}$.

$$\frac{32}{10} = 10\overline{)32} \;\; 3\frac{2}{10}$$
$$\underline{30}$$
$$2$$

Step 2. Reduce $\frac{2}{10}$. $3\frac{2}{10}$ is equivalent to $3\frac{1}{5}$.

$$3\frac{2 \div 2}{10 \div 2} = 3\frac{1}{5}$$

Changing Mixed Numbers to Improper Fractions

When you multiply and divide fractions, you often have to change mixed numbers to improper fractions. This is the opposite of changing an improper fraction to a mixed number.

> To change a mixed number to an improper fraction:
> 1. Multiply the denominator by the whole number.
> 2. Add the numerator.
> 3. Write the total over the denominator.

Example 3: Change $2\frac{3}{8}$ to an improper fraction.

Step 1. Multiply the denominator, 8, by the whole number, 2. $\quad 8 \times 2 = 16$

Step 2. Add the numerator to the product. $\quad 16 + 3 = 19$

Step 3. Write the total, 19, over the denominator, 8. $\quad 2\frac{3}{8} = \frac{19}{8}$

Lesson 2 Exercise

Raise each fraction to higher terms by finding the missing numerator.

1. a. $\frac{9}{10} = \frac{}{40}$ b. $\frac{5}{12} = \frac{}{36}$ c. $\frac{2}{5} = \frac{}{45}$ d. $\frac{2}{3} = \frac{}{12}$
2. a. $\frac{4}{9} = \frac{}{45}$ b. $\frac{3}{20} = \frac{}{200}$ c. $\frac{3}{5} = \frac{}{35}$ d. $\frac{9}{50} = \frac{}{100}$
3. a. $\frac{7}{25} = \frac{}{75}$ b. $\frac{1}{4} = \frac{}{32}$ c. $\frac{2}{9} = \frac{}{36}$ d. $\frac{5}{8} = \frac{}{16}$
4. a. $\frac{6}{7} = \frac{}{42}$ b. $\frac{7}{12} = \frac{}{48}$ c. $\frac{5}{8} = \frac{}{72}$ d. $\frac{1}{2} = \frac{}{36}$

Change each improper fraction to a whole or mixed number. Reduce each remaining fraction to lowest terms.

5. a. $\frac{11}{2} =$ b. $\frac{17}{3} =$ c. $\frac{25}{8} =$ d. $\frac{36}{9} =$
6. a. $\frac{52}{8} =$ b. $\frac{35}{10} =$ c. $\frac{28}{6} =$ d. $\frac{40}{12} =$

7. a. $\frac{8}{5} =$ b. $\frac{30}{6} =$ c. $\frac{18}{4} =$ d. $\frac{26}{10} =$

Change each mixed number to an improper fraction.

8. a. $3\frac{7}{10} =$ b. $2\frac{2}{3} =$ c. $7\frac{1}{2} =$ d. $1\frac{5}{6} =$
9. a. $10\frac{3}{8} =$ b. $2\frac{3}{4} =$ c. $7\frac{2}{5} =$ d. $12\frac{7}{20} =$
10. a. $1\frac{14}{15} =$ b. $16\frac{1}{2} =$ c. $4\frac{3}{5} =$ d. $4\frac{5}{9} =$

Check your answers on page 309.

Lesson 3: Addition

To add (or subtract) fractions, the denominators of the fractions must be the same. Addition and subtraction problems are usually written vertically. That is, the numbers you add or subtract are written in columns. This is the method used most often in the United States. If you learned another method, you do not need to change. Simply compare your answers to the answers for the exercises.

Adding Fractions with the Same Denominators

To add fractions with the same denominators, first add the numerators. Then put the total over the denominator. Rewrite an improper fraction answer as a mixed number and reduce.

Example 1: Find the sum of $\frac{3}{10}$ and $\frac{9}{10}$.

Step 1. Add the numerators, $3 + 9 = 12$. Then write the total over the denominator, 10.

Step 2. Change $\frac{12}{10}$ to a mixed number and reduce to lowest terms.

$$\begin{array}{r} \frac{3}{10} \\ + \frac{9}{10} \\ \hline \frac{12}{10} = 1\frac{2}{10} = 1\frac{1}{5} \end{array}$$

Adding Fractions with Different Denominators

When the fractions in an additional problem have different denominators, first find a **common denominator**. A common denominator is a number into which each denominator in a problem divides evenly. The lowest number that every denominator divides into evenly is called the **lowest common denominator** or **LCD**. In Example 2, 20 is a common denominator for 5 and 10. But 10 is the *lowest* common denominator.

Example 2: Add $\frac{3}{5}$ and $\frac{7}{10}$.

Step 1. Find the LCD. The lowest number both 5 and 10 divide into evenly is 10. Raise $\frac{3}{5}$ to 10ths.

Step 2. Add the new numerators. Put the total over 10, and change the improper fraction to a mixed number.

$$\frac{3}{5} = \frac{6}{10}$$
$$+\frac{7}{10} = \frac{7}{10}$$
$$\frac{13}{10} = 1\frac{3}{10}$$

With mixed numbers first add the fractions and then add the whole numbers. Then combine the whole number total with the fraction total.

Example 3: Add $2\frac{1}{2} + 3\frac{3}{4} + 4\frac{1}{6}$.

Step 1. Find the LCD. The lowest number into which 2, 4, and 6 divide evenly is 12. Raise each fraction to 12ths.

Step 2. Add the new numerators and put the total over 12.

Step 3. Add the whole numbers. Then change $\frac{17}{12}$ to a mixed number, $1\frac{5}{12}$, and add the mixed number to the whole number ($9 + 1\frac{5}{12} = 10\frac{5}{12}$).

$$2\frac{1}{2} = 2\frac{6}{12}$$
$$3\frac{3}{4} = 3\frac{9}{12}$$
$$+4\frac{1}{6} = 4\frac{2}{12}$$
$$9\frac{17}{12} = 9 + 1\frac{5}{12}$$
$$= 10\frac{5}{12}$$

Lesson 3 Exercise

Solve each problem.

1. $\frac{5}{8} + \frac{7}{8} =$
2. $3\frac{1}{6} + 8\frac{5}{6} =$
3. $2\frac{3}{5} + 1\frac{4}{5} =$
4. $4\frac{11}{12} + 4\frac{7}{12} =$
5. $\frac{1}{4} + \frac{3}{5} =$
6. $\frac{5}{12} + \frac{3}{4} =$
7. $8\frac{2}{5} + 1\frac{7}{15} =$
8. $6\frac{1}{2} + 5\frac{3}{8} =$
9. $\frac{7}{8} + \frac{2}{3} =$
10. $\frac{1}{6} + \frac{5}{9} =$
11. $5\frac{3}{4} + 2\frac{2}{9} =$
12. $4\frac{3}{5} + 4\frac{1}{6} =$
13. $\frac{3}{10} + \frac{1}{2} + \frac{3}{4} =$
14. $\frac{3}{8} + \frac{2}{3} + \frac{1}{12} =$
15. $2\frac{5}{6} + 5\frac{3}{8} + 3\frac{1}{4} =$
16. $1\frac{5}{9} + 4\frac{1}{2} + 3\frac{2}{3} =$

Check your answers on page 309.

Lesson 4: Subtraction

To subtract fractions or mixed numbers, the fractions must have the same denominator. If the denominators are not the same, first change each fraction to a new fraction with the common denominator. Subtract the fractions and whole numbers separately.

Example 1: Subtract $\frac{4}{9}$ from $\frac{7}{9}$.

The denominators are the same, so subtract the numerators and write the answer using the common denominator, 9. Reduce the answer to lowest terms.

$$\frac{7}{9} - \frac{4}{9} = \frac{3}{9} = \frac{1}{3}$$

Example 2: Subtract $4\frac{1}{2}$ from $5\frac{2}{3}$.

Step 1. Find the LCD. The LCD for 3 and 2 is 6. Raise each fraction to 6ths.

Step 2. Subtract the numerators and put the difference over 6.

Step 3. Subtract the whole numbers. Reduce to lowest terms if necessary.

$$5\frac{2}{3} = 5\frac{4}{6}$$
$$-4\frac{1}{2} = 4\frac{3}{6}$$
$$\overline{\phantom{-4\frac{1}{2}}\ 1\frac{1}{6}}$$

Subtracting Fractions with Borrowing

In some problems there may not be a fraction in the number you are subtracting from, so you must **borrow**. To borrow, you take 1 from the whole number and rewrite the 1 you borrow as a fraction. Remember that a fraction with the same numerator and denominator is equal to 1. The numerator and denominator should be the same as the denominator of the other fraction in the problem.

Example 3: Subtract $2\frac{3}{8}$ from 7.

Step 1. Borrow 1 from 7, and rename the 1 as $\frac{8}{8}$ since 8 is the LCD. 7 is rewritten as $6\frac{8}{8}$.

Step 2. Subtract the fractions and the whole numbers. Be sure the answer is reduced.

$$7 = 6\frac{8}{8}$$
$$-2\frac{3}{8} = 2\frac{3}{8}$$
$$\overline{\phantom{-2\frac{3}{8}}\ 4\frac{5}{8}}$$

In some subtraction problems the fraction on top is not great enough to subtract the bottom fraction. To get a greater fraction on top, borrow 1 from the whole number on top. Rewrite the 1 as a fraction. Then add this fraction to the original top fraction.

Example 4: Subtract $3\frac{4}{9}$ from $5\frac{1}{3}$.

Step 1. Find the LCD. The LCD for 3 and 9 is 9. Raise each fraction to the 9ths.

Step 2. Since you cannot subtract $\frac{4}{9}$ from $\frac{3}{9}$, borrow 1 from 5 and rewrite the 1 as $\frac{9}{9}$. Add $\frac{9}{9}$ to $\frac{3}{9}$.

Step 3. Subtract the new fractions and the whole numbers. Be sure the answer is reduced.

$$5\frac{1}{3} = 5\frac{3}{9} = 4\frac{3}{9} + \frac{9}{9} = 4\frac{12}{9}$$
$$-3\frac{4}{9} = 3\frac{4}{9} = \phantom{4\frac{3}{9} + \frac{9}{9} =} 3\frac{4}{9}$$
$$\phantom{-3\frac{4}{9} = 3\frac{4}{9} = 4\frac{3}{9} + \frac{9}{9} =} 1\frac{8}{9}$$

Lesson 4 Exercise

Solve each problem.

1. $\frac{11}{12} - \frac{7}{12} =$
2. $\frac{7}{8} - \frac{3}{8} =$
3. $6\frac{9}{10} - 5\frac{7}{10} =$
4. $9\frac{15}{16} - 3\frac{3}{16} =$
5. $\frac{3}{4} - \frac{3}{8} =$
6. $\frac{4}{5} - \frac{3}{4} =$
7. $7\frac{1}{2} - 2\frac{1}{5} =$
8. $4\frac{7}{9} - 1\frac{1}{6} =$
9. $7 - 3\frac{5}{8} =$
10. $9 - 4\frac{7}{12} =$
11. $8 - 2\frac{7}{10} =$
12. $12 - 5\frac{13}{16} =$
13. $5\frac{1}{6} - 2\frac{5}{6} =$
14. $9\frac{1}{3} - 3\frac{2}{3} =$
15. $8\frac{5}{12} - 7\frac{7}{12} =$
16. $4\frac{1}{5} - 1\frac{4}{5} =$
17. $6\frac{1}{5} - 3\frac{3}{4} =$
18. $7\frac{1}{2} - 1\frac{5}{8} =$
19. $9\frac{1}{3} - 5\frac{3}{4} =$
20. $2\frac{1}{2} - 1\frac{2}{3} =$

Check your answers on page 310.

Lesson 5 — Multiplication

To multiply fractions, the numbers you work with must be written in fraction form. This means that you must change every mixed number to an improper fraction. You must also write every whole number as a fraction with a denominator of 1. However, you do not have to find a common denominator when you multiply fractions.

When multiplying fractions, the problems are usually written horizontally. That is, the numbers you work with are written side by side. For example,

$$\frac{1}{4} \times \frac{3}{4} = \frac{3}{16}.$$

Multiplying Fractions

The answers to fraction multiplication problems may seem strange. When you multiply whole numbers (except for 1 and 0), the answer is greater than either of the two numbers you multiply. When you multiply proper fractions, the answer is smaller than either of the two fractions.

When you multiply fractions, you find **a part of a part**. If you multiply $\frac{1}{2}$ by $\frac{1}{2}$, you find $\frac{1}{2}$ **of** $\frac{1}{2}$. You know that $\frac{1}{2}$ of $\frac{1}{2}$ dollar is $\frac{1}{4}$ dollar. The product, $\frac{1}{4}$, is smaller than either of the fractions you multiplied to get it: $\frac{1}{2} \times \frac{1}{2} = \frac{1}{4}$.

To multiply fractions multiply the numerators together and multiply the denominators together. Check to see if the answer can be reduced.

Example 1: Find the product of $\frac{2}{5}$ and $\frac{1}{3}$.

Multiply the numerators together (2 × 1) and the denominators together (5 × 3). $\frac{2}{15}$ is reduced to lowest terms.

$$\frac{2}{5} \times \frac{1}{3} = \frac{2}{15}$$

Canceling is a way of making fraction multiplication easier. Canceling is similar to reducing. To cancel, divide any numerator and any denominator by a number that goes evenly into both of them.

Example 2: Multiply $\frac{3}{4} \times \frac{5}{6} \times \frac{4}{25}$.

Step 1. Cancel the numerator **3** and denominator **6** by dividing by 3. The **1** shows 3 ÷ 3 = 1, and the **2** shows 6 ÷ 3 = 2.

Step 2. Cancel the numerator **4** and the denominator **4** by dividing by 4.

Step 3. Cancel 5 and 25 by dividing by 5.

Step 4. Multiply the resulting numerators together (1 × 1 × 1) and multiply the resulting denominators together (1 × 2 × 5). The product, $\frac{1}{10}$, is in lowest terms.

$$\frac{\cancel{3}}{4} \times \frac{5}{\cancel{6}} \times \frac{4}{25} =$$

$$\frac{\cancel{3}}{\cancel{4}} \times \frac{\cancel{5}}{\cancel{6}} \times \frac{\cancel{4}}{25} =$$

$$\frac{\cancel{3}}{\cancel{4}} \times \frac{\cancel{5}}{\cancel{6}} \times \frac{\cancel{4}}{\cancel{25}} = \frac{1}{10}$$

Remember to write mixed numbers and whole numbers as improper fractions before you try to multiply.

Example 3: Multiply $2\frac{1}{2}$ by 3.

Step 1. Change $2\frac{1}{2}$ and 3 to improper fractions.

$2\frac{1}{2} \times 3 =$

Step 2. Since you cannot cancel, multiply the numerators together and multiply the denominators together.

$\frac{5}{2} \times \frac{3}{1} = \frac{15}{2} = 7\frac{1}{2}$

Step 3. Change $\frac{15}{2}$ to a mixed number.

Lesson 5 Exercise

Solve each problem.

1. $\frac{2}{3} \times \frac{4}{5} =$
2. $\frac{5}{8} \times \frac{3}{4} =$
3. $\frac{3}{4} \times \frac{1}{2} \times \frac{4}{5} =$
4. $\frac{9}{10} \times \frac{2}{3} =$
5. $\frac{3}{20} \times \frac{5}{12} =$
6. $\frac{2}{3} \times \frac{5}{6} \times \frac{7}{10} =$
7. $6 \times \frac{3}{4} =$
8. $\frac{5}{6} \times 15 =$
9. $8 \times \frac{11}{12} =$
10. $\frac{3}{4} \times 3\frac{1}{5} =$
11. $2\frac{1}{4} \times 1\frac{2}{3} =$
12. $5\frac{1}{3} \times 1\frac{5}{16} =$
13. $2\frac{1}{2} \times 1\frac{2}{5} \times 2\frac{2}{3} =$
14. $\frac{5}{6} \times 3\frac{3}{5} \times 2\frac{2}{3} =$
15. $1\frac{1}{6} \times 1\frac{1}{3} \times 3\frac{3}{4} =$

Check your answers on page 311.

Lesson 6 — Division

Multiplying and dividing fractions are similar. The numbers you work with must be in fraction form. To divide by a fraction means to find out how many times a fraction goes into another number. For example, if you divide $\frac{1}{2}$ by $\frac{1}{4}$, you find out how many times $\frac{1}{4}$ goes into $\frac{1}{2}$. You know that $\frac{1}{4}$ dollar goes into $\frac{1}{2}$ dollar two times. Write this problem as $\frac{1}{2} \div \frac{1}{4}$. The $\div$ sign means "divided by."

To divide by a fraction, **invert** the fraction you are dividing by (the fraction on the right) and follow the rules for multiplying fractions.

To invert a fraction means to rewrite the fraction with the numerator on the bottom and the denominator on the top. For example, when you invert $\frac{3}{4}$, you get $\frac{4}{3}$. $\frac{4}{3}$ is sometimes called the **reciprocal** of $\frac{3}{4}$.

Example 1: Divide $\frac{1}{2}$ by $\frac{1}{4}$.

Step 1. Copy the first fraction. Change the sign to ×. Invert the fraction you are dividing by. $\frac{1}{4}$ becomes $\frac{4}{1}$.

$\frac{1}{2} \div \frac{1}{4} =$

Step 2. Cancel by dividing 4 and 2 by 2. Then multiply the numerators and multiply the denominators.

$$\frac{1}{\underset{1}{\cancel{2}}} \times \frac{\overset{2}{\cancel{4}}}{1} = \frac{2}{1} = 2$$

Step 3. Change $\frac{2}{1}$ to a whole number.

Be sure to change mixed numbers or whole numbers to improper fractions.

Example 2: Divide 3 by $\frac{4}{5}$.

Step 1. Write 3 as an improper fraction with a denominator of 1. Change the sign to × and invert $\frac{4}{5}$.

$$3 \div \frac{4}{5} =$$

Step 2. Multiply the numerators together and the denominators together.

$$\frac{3}{1} \times \frac{5}{4} = \frac{15}{4} = 3\frac{3}{4}$$

Step 3. Change $\frac{15}{4}$ to a mixed number.

Dividing by Mixed Numbers and Whole Numbers

Change every mixed number or whole number in a division problem to an improper fraction <u>before</u> you invert.

Example 3: Divide $\frac{3}{4}$ by 2.

Step 1. Write 2 as an improper fraction with a denominator of 1 and change the ÷ sign to ×.

$$\frac{3}{4} \div 2 =$$

Step 2. Copy $\frac{3}{4}$. Invert $\frac{2}{1}$ to $\frac{1}{2}$ and multiply.

$$\frac{3}{4} \div \frac{2}{1} =$$

$$\frac{3}{4} \times \frac{1}{2} = \frac{3}{8}$$

If inverting and multiplying seems strange, think about this example. $8 \div 2 = 4$. The result, 4, is the same as the result of multiplying 8 by the reciprocal of 2 $\left(\frac{1}{2}\right)$. $8 \times \frac{1}{2} = 4$.

Example 4: Divide $4\frac{1}{2}$ by $2\frac{1}{4}$.

Step 1. Rewrite both $4\frac{1}{2}$ and $2\frac{1}{4}$ as improper fractions.

$$4\frac{1}{2} \div 2\frac{1}{4} =$$

Step 2. Copy $\frac{9}{2}$ and change the ÷ to ×. Invert $\frac{9}{4}$ to $\frac{4}{9}$. Cancel and multiply.

$$\frac{9}{2} \div \frac{9}{4} =$$

Step 3. Change $\frac{2}{1}$ to a whole number.

$$\frac{\overset{1}{\cancel{9}}}{\underset{1}{\cancel{2}}} \times \frac{\overset{2}{\cancel{4}}}{\underset{1}{\cancel{9}}} = \frac{2}{1} = 2$$

Level 1, Lesson 6: Division

Lesson 6 Exercise

Solve each problem.

1. $\frac{2}{3} \div \frac{4}{9} =$
2. $\frac{9}{10} \div \frac{2}{5} =$
3. $\frac{3}{4} \div \frac{1}{8} =$
4. $4 \div \frac{2}{3} =$
5. $8 \div \frac{4}{5} =$
6. $5 \div \frac{3}{4} =$
7. $1\frac{1}{9} \div \frac{5}{6} =$
8. $2\frac{5}{8} \div \frac{1}{4} =$
9. $6\frac{1}{2} \div \frac{3}{8} =$
10. $4\frac{1}{3} \div 5 =$
11. $10\frac{1}{2} \div 8 =$
12. $\frac{3}{8} \div 4 =$
13. $\frac{5}{9} \div 1\frac{1}{3} =$
14. $4 \div 1\frac{3}{4} =$
15. $\frac{3}{8} \div 2\frac{2}{3} =$
16. $1\frac{1}{4} \div 2\frac{1}{2} =$
17. $2\frac{3}{4} \div 1\frac{5}{8} =$
18. $7\frac{1}{2} \div 3\frac{3}{4} =$

Check your answers on page 311.

Lesson 7: Comparing and Ordering Fractions

To compare fractions find a common denominator for the fractions you are comparing. Then raise each fraction to higher terms with the new denominator. Compare the fractions written in higher terms.

Example 1: Which is greater, $\frac{3}{8}$ or $\frac{3}{10}$?

The lowest common denominator for 8 and 10 is 40. Raise each fraction to higher terms with a denominator of 40. Since $\frac{15}{40}$ is more than $\frac{12}{40}$, $\frac{3}{8}$ is the greater fraction.

$$\frac{3}{8} = \frac{15}{40} \qquad \frac{3}{10} = \frac{12}{40}$$

$$\frac{3}{8} \text{ is greater}$$

Example 2: Arrange the following in order from least to greatest: $\frac{13}{20}, \frac{3}{5}, \frac{3}{4}$.

The LCD for 20, 5, and 4 is 20. Raise $\frac{3}{4}$ and $\frac{3}{5}$ to 20ths. Since $\frac{12}{20}$ is the least, $\frac{3}{5}$ comes first. $\frac{13}{20}$ is in the middle, and $\frac{3}{4}$ is the greatest.

$$\frac{13}{20} \qquad \frac{3}{5} = \frac{12}{20} \qquad \frac{3}{4} = \frac{15}{20}$$

In order: $\frac{3}{5}, \frac{13}{20}, \frac{3}{4}$

Lesson 7 Exercise

Solve each problem.

1. Find the greater fraction in each pair.
    a. $\frac{8}{15}$ or $\frac{1}{2}$
    b. $\frac{2}{3}$ or $\frac{5}{9}$
    c. $\frac{5}{8}$ or $\frac{4}{5}$
    d. $\frac{5}{12}$ or $\frac{4}{9}$

2. Arrange in order from least to greatest.
$\frac{3}{5}, \frac{1}{2}, \frac{11}{20}, \frac{7}{10}$

3. Arrange in order from greatest to least.
$\frac{5}{12}, \frac{1}{4}, \frac{5}{16}, \frac{1}{3}$

4. Which is longer, $\frac{5}{8}$ inch or $\frac{9}{16}$ inch?

5. Which is heavier, $\frac{3}{4}$ pound or $\frac{13}{16}$ pound?

Check your answers on page 312.

Lesson 8: Changing Fractions and Decimals

Later, especially when you work with percents, you will often need to change between decimals and fractions. A decimal can always be written as a fraction with a denominator that is a power of 10.

Changing Decimals to Fractions

To change a decimal to a fraction, write the digits in the decimal as the numerator. Write the name of the decimal as a power of 10 in the denominator. Then reduce the fraction.

Example 1: Change 0.08 to a fraction.

Write 08 as the numerator. The denominator is 100 because 0.08 is named "hundredths." Reduce $\frac{8}{100}$.

$$\frac{08}{100} = \frac{2}{25}$$

Example 2: Change 9.6 to a mixed number.

Write 9 as a whole number and 6 as the numerator. The denominator is 10 because .6 is named "tenths." Reduce $9\frac{6}{10}$.

$$9\frac{6}{10} = 9\frac{3}{5}$$

Changing Fractions to Decimals

The line in a fraction means "divided by." For example, $\frac{7}{20}$ means 7 divided by 20. To change a fraction to a decimal divide the numerator by the denominator. Put a decimal point and zeros to the right of the numerator.

Example 3: Change $\frac{7}{20}$ to a decimal.

Divide 7 by 20. Put a decimal point and zeros to the right of 7. Write the decimal point in the answer above the decimal point in the dividend.

$$\begin{array}{r} 0.35 \\ 20\overline{)7.00} \\ \underline{6\,0} \\ 1\,00 \\ \underline{1\,00} \\ 0 \end{array}$$

With some fractions, the division will come out even with just a few zeros to the right of the decimal point. With other fractions the division will never come out even. You can stop after two decimal places and make a fraction with the remainder. Or, you can add a third decimal place and divide. If the answer comes out even, you are finished. If it does not, round to two decimal places.

Example 4: Change $\frac{5}{6}$ to a decimal.

Divide 5 by 6. Put a decimal point and two zeros to the right of 5. Write the decimal point in the answer above the decimal point in the dividend. Reduce $0.83\frac{2}{6}$.

$$0.83\frac{2}{6} = 0.83\frac{1}{3}$$
$$\begin{array}{r} 6\overline{)5.00} \\ \underline{4\,8} \\ 20 \\ \underline{18} \\ 2 \end{array}$$

In some cases the problems you are working may tell you to which place to round the decimal.

Lesson 8 Exercise

Change each decimal to a fraction and reduce.

1. **a.** 0.8 =   **b.** 0.04 =   **c.** 0.35 =   **d.** 0.005 =
2. **a.** 0.065 =   **b.** 0.0075 =   **c.** 0.002 =   **d.** 0.875 =

Change each mixed decimal to a mixed number and reduce.

3. **a.** 8.4 =   **b.** 2.85 =   **c.** 1.004 =   **d.** 6.3 =
4. **a.** 10.125 =   **b.** 3.009 =   **c.** 4.80 =   **d.** 12.16 =

Change each fraction to a decimal.

5. **a.** $\frac{9}{10}$ =   **b.** $\frac{4}{5}$ =   **c.** $\frac{9}{50}$ =   **d.** $\frac{3}{8}$ =
6. **a.** $\frac{1}{2}$ =   **b.** $\frac{2}{3}$ =   **c.** $\frac{8}{25}$ =   **d.** $\frac{1}{6}$ =
7. **a.** $\frac{2}{9}$ =   **b.** $\frac{3}{4}$ =   **c.** $\frac{5}{12}$ =   **d.** $\frac{7}{8}$ =

Check your answers on page 312.

Chapter 3 Fractions

Level 1 Review

Solve each problem.

1. Reduce $\frac{35}{84}$ to lowest terms.
2. Change $8\frac{4}{5}$ to an improper fraction.
3. Find the sum of $8\frac{3}{4}$ and $7\frac{9}{16}$.
4. Find the sum of $2\frac{3}{8}$, $4\frac{1}{2}$, and $6\frac{3}{5}$.
5. Subtract $3\frac{5}{12}$ from $7\frac{2}{3}$.
6. Find the difference between $8\frac{1}{5}$ and $3\frac{3}{4}$.
7. Find the product of $3\frac{3}{5}$ and 10.
8. Find $2\frac{1}{3} \times 4\frac{1}{8} \times \frac{1}{2}$.
9. Divide $6\frac{2}{3}$ by 8.
10. Find $9\frac{1}{3}$ divided by $3\frac{1}{5}$.
11. Arrange in order from least to greatest: $\frac{5}{8}$, $\frac{9}{16}$, $\frac{1}{2}$
12. Change $\frac{11}{20}$ to a decimal.

Check your answers on page 313. If you have all 12 problems correct, go on to Level 2. If you answered a problem incorrectly, find the item number on the chart below and review that lesson.

Review:	If you missed item number:
Lesson 1	1
Lesson 2	2
Lesson 3	3, 4
Lesson 4	5, 6
Lesson 5	7, 8
Lesson 6	9, 10
Lesson 7	11
Lesson 8	12

Level 2: Fraction Applications

In this section you will apply your skills with fractions in a variety of situations. You will learn how to solve one of the basic types of fraction word problems. You will review the United States Customary units of measurement and learn to read an inch scale. You will use powers and square roots with fractions. You will also learn to use two or more area formulas. Finally, you will learn the fundamentals of ratio, proportion, and probability.

Lesson 9: Finding What Part One Number Is of Another

One of the most common applications of fractions is very simple. All you have to do is write a fraction and reduce it. Remember that the numerator tells the **part** and the denominator tells the **whole.**

Example 1: During a week in July three workers at the Greenport post office were on vacation. In all 24 people work at the post office. What fraction of the workers vacationed that week?

Write a fraction in which the numerator tells the number of workers on vacation and the denominator tells the total number of workers. Then reduce the fraction.

$$\frac{\text{part}}{\text{whole}} \quad \frac{3}{24} = \frac{1}{8}$$

In some problems you will have to find the part that is referred to in the question.

Example 2: What fraction of the workers in Example 1 worked during that week in July?

Step 1. To find the number of workers at work subtract the number on vacation from the total number of workers.

```
total          24
on vacation   -3
at work       21
```

Step 2. Write a fraction in which the numerator tells the number of workers at work and the denominator tells the total number of workers. Then reduce.

$$\frac{\text{part}}{\text{whole}} = \frac{21}{24} = \frac{7}{8}$$

Chapter 3 Fractions

In other problems you will have to find the whole.

Example 3: Mary got two problems wrong and eight problems right on a quiz. What fraction of the problems did she get right?

Step 1. To find the total number of problems add the number Mary got right to the number she got wrong.

$$\begin{array}{r} \text{right} \quad 8 \\ \text{wrong} \quad +2 \\ \hline \text{total} \quad 10 \end{array}$$

Step 2. Write a fraction in which the numerator tells the number of problems she got right and the denominator tells the total number of problems. Then reduce.

$$\frac{\text{right}}{\text{total}} = \frac{8}{10} = \frac{4}{5}$$

Lesson 9 Exercise

Solve each problem. Be sure your answers are reduced.

1. Of the 15 students in Laura's GED class, 9 are men.
   a. What fraction of the class are men?
   b. What fraction of the class are women?

2. In the shop where Phil works there are 45 union members and 36 non-members.
   a. What fraction of the workers are union members?
   b. What fraction of the workers do not belong to the union?

3. Of the 33 cars in the lot at Dave's Auto Shop 15 are new.
   a. What fraction of the cars in the lot are new?
   b. What fraction of the cars in the lot are used?

4. For his store Felipe ordered 12 cases of cola, 2 cases of lime drink, and 6 cases of orange drink.
   a. What fraction of the cases were cola?
   b. What fraction of the cases were lime drink?
   c. What fraction of the cases were orange drink?

5. In the office where Kathleen works 45 employees voted to join the union, 35 voted not to join, and 20 were undecided.
   a. What fraction of the employees voted to join the union?
   b. What fraction voted against joining the union?
   c. What fraction were undecided?

Check your answers on page 314.

Lesson 10: Standard Units of Measurement

The list below gives units of measurement for length, weight, time, and liquid measure. These are the units commonly used in the United States and Great Britain. The abbreviation for each unit is in parentheses. The list also tells how many smaller units each larger unit is equal to.

Memorize any units you do not already know.

Length
1 foot (ft) = 12 inches (in.)
1 yard (yd = 3 ft = 36 in.
1 mile (mi) = 5280 ft = 1760 yd

Weight
1 pound (lb) = 16 ounces (oz)
1 ton (t) = 2000 lb

Time
1 minute (min) = 60 seconds (sec)
1 hour (hr) = 60 min
1 day (da) = 24 hr
1 week (wk) = 7 days
1 year (yr) = 365 days = 12 months (mo) = 52 weeks

Liquid Measure
1 pint (pt) = 16 oz
1 quart (qt) = 2 pt = 32 oz
1 gallon (gal) = 4 qt

In some problems you may need to change from a smaller unit to a larger unit. Divide the number of smaller units you have by the number of smaller units that make up one large unit.

When you change from one unit of measurement to another, you will often use fractions.

Example 1: Change 18 months to years.

Divide 18 months by the number of months in one year, 12, and reduce the answer.

$$1\frac{6}{12} = 1\frac{1}{2} \text{ yr}$$
$$12\overline{)18}$$

In some problems you may want to express an answer in more than one unit of measure.

Example 2: Change 18 months to years and months.

Divide 18 months by the number of months in one year, 12, and express the remainder as a number of months.

$$1 \text{ yr } 6 \text{ mo}$$
$$12\overline{)18}$$

When the number of smaller units is less than the number in one large unit, write a fraction and reduce.

106   Chapter 3 Fractions

Example 3: Change 8 inches to feet.

Since the number of inches in a foot, 12, is greater than 8, write a fraction with 12 as the denominator and reduce.

$$\frac{8}{12} = \frac{2}{3} \text{ ft}$$

In other problems you may want to change large units to smaller units. In that case multiply the number of large units you have by the number of smaller units that make up one large unit.

Example 4: Change 10 feet to inches.

Multiply 10 feet by the number of inches in one foot, 12.

$$10 \times 12 = 120 \text{ in.}$$

Lesson 10 Exercise

Solve each problem.

1. a. Change 8 feet to yards.
   b. Change 8 feet to yards and feet.
2. a. Change 135 seconds to minutes.
   b. Change 135 seconds to minutes and seconds.
3. a. Change 10 quarts to gallons.
   b. Change 10 quarts to gallons and quarts.
4. a. Change 20 ounces to pounds.
   b. Change 20 ounces to pounds and ounces.
5. a. Change 20 months to years.
   b. Change 20 months to years and months.
6. a. Change 9000 pounds to tons.
   b. Change 9000 pounds to tons and pounds.
7. a. Change 40 inches to feet.
   b. Change 40 inches to feet and inches.
8. a. Change 100 minutes to hours.
   b. Change 100 minutes to hours and minutes.

For problems 9–12 change each to the new unit indicated.

9. a. 10 mo = _____ yr   b. 45 min = _____ hr   c. 9 in. = _____ ft
10. a. 8 oz = _____ lb   b. 20 in. = _____ yd   c. 2 qt = _____ gal
11. a. 1800 lb = _____ t   b. 16 hr = _____ da   c. 1320 ft = _____ mi
12. a. 4 yd = _____ in.   b. $2\frac{1}{2}$ min = _____ sec   c. $1\frac{1}{2}$ t = _____ lb

Check your answers on page 314.

Lesson 11: Reading Scales

A ruler is a common tool that measures length in fractions of an inch. The longest lines on the ruler represent whole inches. The next longest lines are $\frac{1}{2}$-inch lines. The increasingly shorter lines are $\frac{1}{4}$-inch lines, $\frac{1}{8}$-inch lines, and $\frac{1}{16}$-inch lines, respectively.

To measure a length with a ruler, determine how far from the left end of the scale a point is.

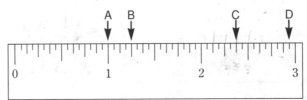

Example 1: How far from the left end of the 3-inch ruler pictured above is the point marked A?

A is at the first long line on the ruler.

A is 1 inch from the left.

Example 2: How far from the left end of the ruler is point B?

B is at the first $\frac{1}{4}$-inch line between 1 and 2 inches.

B is $1\frac{1}{4}$ inches from the left.

Example 3: How far from the left end of the ruler is point C?

C is at the third $\frac{1}{8}$-inch line between 2 and 3 inches.

C is $2\frac{3}{8}$ inches from the left.

Example 4: How far from the left end of the ruler is point D?

D is at the fifteenth $\frac{1}{16}$-inch line between 2 and 3 inches.

D is at $2\frac{15}{16}$ inches from the left.

Lesson 11 Exercise

Use the 5-inch ruler below to answer the following questions.

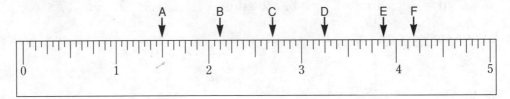

1. How far from the left end of the ruler is point A?
2. How far from the left end of the ruler is point B?
3. How far from the left end of the ruler is point C?
4. How far from the left end of the ruler is point D?
5. How far from the left end of the ruler is point E?
6. How far from the left end of the ruler is point F?
7. What is the distance between point A and point B?
8. What is the distance between point D and point E?

Check your answers on page 315.

Lesson 12: Powers and Square Roots

To raise a fraction to a power, remember to raise both the numerator and the denominator to the power.

Example 1: What is the value of $\left(\frac{3}{4}\right)^2$?

$\left(\frac{3}{4}\right)^2$ means to multiply $\frac{3}{4}$ times itself.

$$\left(\frac{3}{4}\right)^2 = \frac{3}{4} \times \frac{3}{4} = \frac{9}{16}$$

To find the square root of a fraction, remember to find the square root of both the numerator and the denominator.

Example 2: What is the value of $\sqrt{\frac{25}{64}}$?

Find the square root of 25 and the square root of 64.

$$\sqrt{\frac{25}{64}} = \frac{5}{8}$$

Lesson 12 Exercise

Find the value of each of the following.

1. a. $\left(\frac{1}{2}\right)^2 =$ b. $\left(\frac{1}{9}\right)^2 =$ c. $\left(\frac{2}{3}\right)^3 =$ d. $\left(\frac{1}{10}\right)^4 =$
2. a. $\left(\frac{3}{10}\right)^3 =$ b. $\left(\frac{5}{6}\right)^2 =$ c. $\left(\frac{7}{12}\right)^2 =$ d. $\left(\frac{3}{4}\right)^3 =$
3. a. $\sqrt{\frac{25}{36}} =$ b. $\sqrt{\frac{4}{49}} =$ c. $\sqrt{\frac{1}{81}} =$ d. $\sqrt{\frac{9}{100}} =$
4. a. $\sqrt{\frac{64}{81}} =$ b. $\sqrt{\frac{1}{144}} =$ c. $\sqrt{\frac{9}{16}} =$ d. $\sqrt{\frac{36}{49}} =$

Check your answers on page 315.

Lesson 13: Areas of Triangles and Parallelograms

Earlier you learned that a triangle is a figure with three sides. You also learned that to find the perimeter of a triangle you must add the three sides.

> To find the area of a triangle, use the formula
> $$A = \frac{1}{2}bh,$$ where A = area, b = base, and h = height.

The height of a triangle is always **perpendicular** to the base. This means that the height and the base meet to form a square corner. Often the height is not the same as one of the sides.

Look carefully at each of the following triangles to see the positions of each base and height. Notice that the height may be outside the triangle.

Example 1: What is the area of the triangle?

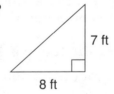

Step 1. In the formula replace b with 8 and h with 7. $A = \frac{1}{2}bh$.

Step 2. Rename the whole numbers as improper fractions and multiply.

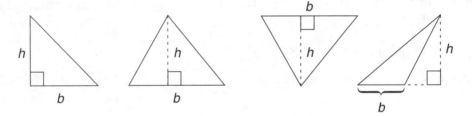

A **parallelogram** is a four-sided figure whose opposite sides are parallel and equal. **Parallel** means that the lines will never cross.

> The formula for the area of a parallelogram is
> $$A = bh,$$ where A = area, b = base, and h = height.

110   Chapter 3   Fractions

Look carefully at each of the following parallelograms to see the positions of each base and height.

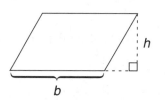 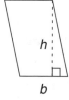 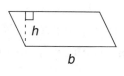

Example 2: Find the area of the figure.

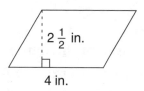

Step 1. In the formula $A = bh$, replace b with 4 and h with $2\frac{1}{2}$.

Step 2. Change the mixed number to an improper fraction and multiply.

$A = bh$

$A = 4 \times 2\frac{1}{2}$

$A = \frac{\overset{2}{\cancel{4}}}{1} \times \frac{5}{\underset{1}{\cancel{2}}} = 10$ sq in.

Remember that area is always measured in **square** units.

Lesson 13 Exercise

Solve each problem.

1. What is the area of the triangle?

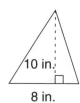

2. What is the area of a triangle with base 30 feet and height 18 feet?

3. Find the area of the triangle.

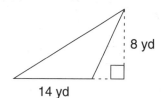

4. What is the area of the triangle?

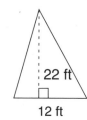

Level 2, Lesson 13: Areas of Triangles and Parallelograms

5. Find the area of the parallelogram.

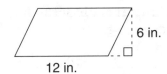

6. What is the area of a parallelogram with base 10 feet and height $1\frac{1}{2}$ feet?

7. What is the perimeter of the figure?

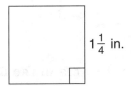

8. What is the area of the figure in problem 7?

9. What is the perimeter of the figure?

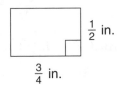

10. What is the area of the figure in problem 9?

Check your answers on page 315.

Lesson 14

Ratio

A **ratio** is a comparison of two numbers. A ratio can compare a part and a whole or a part and a part. A ratio is a fraction. For example, the ratio of an inch to a foot is 1 to 12. The number 1 stands for an inch, and 12 stands for the 12 inches in a foot.

You can write a ratio with the word *to*, with a colon (:), or as a fraction. The ratio of an inch to a foot can be written as 1 to 12, 1 : 12, or $\frac{1}{12}$. Notice that 1 is the numerator and 12 is the denominator when the ratio is written as a fraction.

Ratios, like fractions, should be reduced to lowest terms. It is also important to remember that the numbers in a ratio must be in the same order as the words in the problem.

Example 1: Sally takes home $800 a month and pays $200 for rent. What is the ratio of her rent to her take-home pay?

Step 1. Write a ratio that compares the part to the whole (the rent to take-home pay).

part : whole
rent : take-home

Step 2. Reduce the ratio to lowest terms.

$200 : $800 or $\frac{200}{800}$

$\frac{200}{800} = \frac{1}{4}$ or 1 : 4

In some problems you may have to find one of the numbers in the ratio. Notice that finding a ratio is similar to finding what fraction one number is of another.

Example 2: Mike's softball team won 12 games and lost 8. What is the ratio of the number of games won to the total number played?

Step 1. Add the number of games won to the number lost to find the total number of games played.

won   12
lost + 8
total   20

Step 2. Make a ratio of the number of games won to the total number played.

won : total
12 : 20

Step 3. Reduce the ratio to lowest terms.

$3 : 5$ or $\dfrac{3}{5}$

To compare a part to a part be sure to set up the ratio the same way the numbers are presented. What is the ratio of the number of games lost to the games won?

part : part
lost : won
$8 : 12$ or $\dfrac{8}{12} = \dfrac{2}{3}$

Lesson 14 Exercise

Solve each problem.

1. What is the ratio of 15 inches to a yard?
2. Sandra's living room is 15 feet long and 12 feet wide. What is the ratio of the width to the length?
3. In town Louie gets 16 miles to a gallon of gasoline. In the country he gets 24 miles to a gallon. What is the ratio of his gas mileage in town to his gas mileage in the country?
4. Find the ratio of 4 ounces to a pint. (1 pint = 16 ounces.)
5. The softball team for Ace Electronics has 20 members. 12 of them are men. What is the ratio of the number of men to women on the team?
6. Jack bought a stereo on sale for $450. He saved $150 by buying the stereo on sale. What was the ratio of the sale price to the original price?
7. After paying $220 a month for rent the Santiagos have $880 left for all other expenses. What is the ratio of rent to the total monthly take-home of the Santiagos?
8. Dave weighed 280 pounds in March. By the following September he had lost 70 pounds. Find the ratio of his March weight to his September weight.
9. Denise has paid back $1500 of a $5000 car loan. What is the ratio of the amount she has already paid to the amount she still owes?
10. On a test Geraldine got 40 problems right and 16 problems wrong. What is the ratio of the number she got right to the total number of problems?

Check your answers on page 316.

Lesson 15: Proportion

A **proportion** is a statement that says two ratios are equal. $\frac{6}{8} = \frac{3}{4}$ is a proportion. In a proportion the **cross-products** are equal. This means that the top of the first ratio times the bottom of the other ratio is the same as the bottom of the first ratio times the top of the other.

For the example, one cross-product is $6 \times 4 = 24$. The other cross-product is $8 \times 3 = 24$.

$$\frac{6}{8} = \frac{3}{4}$$

In many proportion problems one number is missing. Usually a letter is written to represent the missing number.

> To find the missing number in a proportion:
> 1. Find the cross-products.
> 2. Write a statement that makes the cross-products equal to each other.
> 3. To find the missing number (letter), divide both products by the number that multiplies the letter.

Example 1: Solve for c in the proportion $\frac{5}{6} = \frac{30}{c}$.

Step 1. Find the cross-products. $5 \times c = 5c$ and $6 \times 30 = 180$. Remember that a number written next to a letter indicates multiplication.

$$\frac{5}{6} = \frac{30}{c}$$

Step 2. Write an equation with the two cross-products.

$$5c = 180$$

Step 3. Divide by 5. $c = 36$ so $\frac{5}{6} = \frac{30}{36}$.

$$\frac{5c}{5} = \frac{180}{5}$$
$$c = 36$$

When the two ratios in a proportion are written with colons, first rewrite the ratios in fraction form.

Example 2: Solve for s in $s:9 = 5:6$.

Step 1. Rewrite the proportion in fraction form.

$$s:9 = 5:6$$
$$\frac{s}{9} = \frac{5}{6}$$

Step 2. Write an equation with the cross-products.

$$6s = 45$$

Step 3. Divide by 6. s is $7\frac{1}{2}$, so $7\frac{1}{2}:9 = 5:6$.

$$\frac{6s}{6} = \frac{45}{6}$$

$$s = 7\frac{3}{6}$$

$$s = 7\frac{1}{2}$$

Lesson 15 Exercise

Find the value of the missing number (letter) in each proportion.

1. **a.** $\frac{c}{6} = \frac{7}{30}$ **b.** $\frac{9}{a} = \frac{3}{2}$ **c.** $\frac{5}{8} = \frac{m}{20}$ **d.** $\frac{4}{w} = \frac{18}{6}$
2. **a.** $\frac{9}{10} = \frac{27}{e}$ **b.** $\frac{x}{14} = \frac{4}{7}$ **c.** $\frac{1}{s} = \frac{8}{5}$ **d.** $\frac{8}{15} = \frac{r}{2}$
3. **a.** $3:n = 6:11$ **b.** $2:5 = t:60$ **c.** $h:4 = 5:6$ **d.** $9:2 = 1:d$
4. **a.** $d:12 = 7:48$ **b.** $4:15 = 24:s$ **c.** $6:v = 9:10$ **d.** $3:11 = y:5$

Check your answers on page 316.

Lesson 16: Probability

Probability is the chance of an event happening. If you toss a coin, there are two possibilities. The coin may land showing its head or it may land showing its tail. The probability of an event is written as a fraction.

The numerator tells the number of possible ways that the event in question can happen. The denominator tells the total number of possibilities.

The probability that a coin will land showing its tail is $\frac{1}{2}$. The denominator, 2, means that there are two possibilities when you toss a coin—heads or tails. The numerator, 1, means that there is just one possibility that the coin will land showing its tail.

Example 1: Tom has two white T-shirts, one gray T-shirt, one blue T-shirt, and one red T-shirt in a laundry bag. What is the probability that the first shirt Tom picks from the bag will be white?

Make a fraction with a numerator that tells the number of white T-shirts in the bag, 2, and a denominator that tells the total number of shirts in the bag, $2 + 1 + 1 + 1 = 5$.

$\frac{2}{5}$

Example 2: What is the probability that the first shirt Tom picks will be blue?

Make a fraction with a numerator that tells the number of blue T-shirts, 1, and a denominator that tells the total number of shirts, 5.

$\frac{1}{5}$

Example 3: The first T-shirt Tom picked from the laundry bag was gray. What is the probability that the next shirt he picks from those remaining in the bag will be white?

Make a fraction with a numerator that tells the number of white T-shirts left in the bag, 2, and a denominator that tells the total number of shirts left in the bag, 5 − 1 = 4. Reduce $\frac{2}{4}$.

$$\frac{2}{4} = \frac{1}{2}$$

Lesson 16 Exercise

Solve each problem.

1. The Greenport Telephone book contains 7500 names. Fifteen Smiths are listed in the phone book. In a random selection from Greenport telephone numbers, what is the probability that the first number selected will belong to a Smith?

2. Liz works as a cashier in a grocery store. At the end of the day she had 12 quarters, 20 dimes, 18 nickels, and 30 pennies. She put all the coins in a bag.
   a. What is the probability that the first coin she takes from the bag will be a dime?
   b. What is the probability that the first coin she takes from the bag will be a penny?
   c. What is the probability that the first coin she takes from the bag will be either a quarter or a nickel?

3. Carlos received a shipment of sweaters to sell in his store. The shipment contained 10 small size sweaters, 15 medium sweaters, and 8 large sweaters. What is the probability that the first sweater he takes from the box will be a medium size?

4. Stored in a locker at the Greenport Community Center are 5 hardballs, 7 softballs, and 3 tennis balls.
   a. What is the probability that the first ball someone grabs from the locker will be a tennis ball?
   b. Frank began to take balls from the locker. The first ball he took was a hardball and both the second and third were softballs. What is the probability that the next ball he takes from those remaining in the locker will be another hardball?

5. A carton contains 8 cans of tomato sauce and 6 cans of green beans.
   a. What is the probability that the first can Max takes from the carton will be a can of tomato sauce?
   b. In fact the first can Max took was tomato sauce and the second was green beans. What is the probability that the next can Max takes from those remaining will be a can of green beans?

6. Jose Acevedo, his wife Beatrice, and their son Felipe each bought a ticket for a chance to win a color TV. Altogether 540 tickets were sold.
   a. What is the probability that Jose will win the television?
   b. What is the probability that one of the Acevedos will win the TV set?

Check your answers on page 317.

Level 2 Review

Solve each problem.

1. Yuki spends $240 a month for rent. She has $960 left over each month for other expenses. Rent is what fraction of Yuki's take-home pay?
2. Gordon received a shipment containing 8 blue shirts, 6 white shirts, and 6 yellow shirts. Blue shirts made up what fraction of the shipment?
3. Change 40 months to years.
4. Change 12 ounces to pounds.

Use the three-inch ruler below to answer questions 5 and 6.

5. How far from the left end of the ruler is point X?
6. What is the distance between point X and point Y?
7. Find the value of $\left(\frac{3}{8}\right)^2$.
8. Find $\sqrt{\frac{49}{100}}$.
9. Find the area of a triangle with a base of 12 inches and a height of 5 inches.
10. Find the area of a rectangle with a length of $6\frac{5}{8}$ inches and a width of 4 inches.
11. Members of the Central County finance committee made telephone interviews to find out whether people wanted a new community center. Of those telephoned 65 people said yes, 25 people said no, and 10 said they had no opinion. What is the ratio of the number of people who want the new community center to the total number of those interviewed?
12. Four women and six men are on the steering committee for the Ninth Street Block Association. Find the ratio of the number of women to the total number of people on the committee.
13. Solve for x in the proportion $8:x = 5:12$.

14. A bag contains 5 green marbles, 7 blue marbles, and 3 black marbles. What is the probability that the first marble Andy picks from the bag will be green?

15. In fact the first two marbles that Andy, in problem 14, took from the bag were blue. The third was green. What is the probability that the next marble he picks from those remaining will be black?

Check your answers on page 317. If you have all 15 problems correct, go on to Level 3. If you have answered a problem incorrectly, find the item number you missed on the chart below and review that lesson before you go on.

Review:	If you missed item number:
Lesson 9	1, 2
Lesson 10	3, 4
Lesson 11	5, 6
Lesson 12	7, 8
Lesson 13	9, 10
Lesson 14	11, 12
Lesson 15	13
Lesson 16	14, 15

Level 3
Problem Solving with Fractions

In this section you will solve word problems that require you to use your skills with fractions, ratios, and proportions. It is especially important to set up proportion problems carefully. Study each example carefully before you try the exercises.

Lesson 17: Word Problems Using Proportion

You can use proportions to solve many word problems. The key to solving proportion word problems is to label each number in the proportion.

Remember that the parts of a proportion must correspond to each other. In the first example below, if the top number on the left represents lime, the top number on the right should also represent lime.

Example 1: The ratio of lime to sand in a mixture of concrete is $1:3$. Jeff is using 12 pounds of sand in the mixture of concrete. How much lime should he use?

Step 1. Write a proportion with the ratio of lime to sand on the left. For the ratio on the right, put 12, the pounds of sand, on the bottom and put x, the pounds of lime, on top.

$$\frac{\text{lime}}{\text{sand}} \quad \frac{1}{3} = \frac{x}{12}$$

Step 2. Write an equation with the cross products.

$$3x = 12$$

Step 3. Divide 12 by 3. Jeff needs 4 pounds of lime.

$$\frac{3x}{3} = \frac{12}{3}$$

$$x = 4 \text{ lb of lime}$$

In some problems you will not see the word **ratio**.

Example 2: If 12 cans of soda cost $4.50, how much do 30 cans cost?

Step 1. Write a proportion with the ratio of the number of cans to the cost. Put 30 on the top on the right. Put x in the place of the cost.

$$\frac{\text{number}}{\text{cost}} = \frac{12}{\$4.50} = \frac{30}{x}$$

Step 2. Write an equation with the cross products. $12x = \$135.00$

Step 3. Divide $135.00 by 12. The cost of 30 cans is $11.25.

$$\frac{12x}{12} = \frac{135.00}{12}$$
$$x = \$11.25$$

Lesson 17 Exercise

1. Six feet of lumber weigh 15 pounds. What is the weight of 16 feet of the same lumber?
2. The ratio of girls to boys at the Oakdale School is 5:4. There are 120 girls in the school. How many boys are there?
3. Carla makes $52.80 in 8 hours. How much does she make in 20 hours?
4. If 12 acres yield 1440 bushels of corn, how many bushels of corn can a farmer expect from a 50-acre field?
5. The ratio of the width to the length of a snapshot is 3:5. If the snapshot is enlarged to be 15 inches wide, how long will it be?
6. Manny drove 72 miles in $1\frac{1}{2}$ hours. If he drives at the same average speed, how far can he go in 4 hours?
7. The scale on a map is $\frac{1}{2}$ inch = 15 miles. How far apart are two towns that are 2 inches apart on the map?
8. The ratio of blue to white paint in a certain mixture is 3:2. If Isabella uses 6 gallons of blue paint for the mixture, how many gallons of white paint should she use?
9. The ratio of wins to losses for the Greenport Grasshoppers baseball team was 3:5. The team won 15 games. How many games did they lose?
10. In a recent year the divorce rate in the U.S. was 5 for every 1000 people. The population of Greenport is 15,000. If Greenport follows the national rate, how many divorces were there in Greenport that year?

Check your answers on page 318.

Lesson 18 Multistep Proportion Problems

In some proportion problems you must first write new ratios before you can solve the problem. Study the next examples carefully. In each problem notice how the parts of the proportion correspond to words in the question.

Example 1: The ratio of men to women working at Central Hospital is 2:9. Thirty men work at the hospital. In all how many people work there?

Step 1. In the ratio, 2 : 9, 2 stands for men and 9 stands for women. Add 2 and 9 to find the total number of people in the ratio. The problem asks for the number of people in all.

$$\begin{array}{r} 2 \\ +\ 9 \\ \hline 11 \end{array}$$

Step 2. Write a proportion with the ratio of the number of men to the total number of workers, $\frac{2}{11}$. Put 30 on the top of the ratio on the right and put x in the place of the total.

$$\frac{\text{men}}{\text{total}} = \frac{2}{11} = \frac{30}{x}$$

Step 3. Write an equation with the cross products.

$$2x = 330$$
$$x = 165$$

Step 4. Divide 330 by 2. The total number of workers is 165.

Example 2: The ratio of problems Sam got right was 4 out of 7 on a test with 35 questions. How many questions did he get wrong?

Step 1. In the ratio 4 : 7, 4 stands for the number right and 7 stands for the total. Subtract 4 from 7 to find the part that corresponds to the number wrong as the problem asks.

$$\begin{array}{r} 7 \\ -\ 4 \\ \hline 3 \end{array}$$

Step 2. Write a proportion with the ratio of the number wrong to the total. Put x in the place of the number wrong, and put 35 in the place of the total.

$$\frac{\text{wrong}}{\text{total}} \quad \frac{3}{7} = \frac{x}{35}$$

Step 3. Write an equation with the cross products.

Step 4. Divide the product, 105, by 7. Sam got 15 problems wrong.

$$7x = 105$$
$$x = 15$$

Lesson 18 Exercise

Solve each problem.

1. The ratio of the number of problems Shirley got right to the number she got wrong was 7 : 2. There were 36 problems on the test. How many problems did she get right?

2. The ratio of the number of men to the total number of workers at the factory where Dan works is 3 : 5. Altogether 60 people work at the factory. How many women work there?

3. The ratio of domestic cars to imported cars in the lot at Al's Auto Shop is 3 : 2. There are 42 domestic cars in the shop. How many cars are there altogether?

4. Out of every 500 parts that are produced at Southern Steel 497 pass inspection. If the factory produces 15,000 parts each day, how many of the parts are defective?

5. The ratio of the amount Lois spends on car payments to her total monthly income is 2 : 7. She spends $96 a month on car payments. How much does she have left over each month for other expenses?

6. In a recent survey the ratio of the number of people who said that they approved of a tax increase to the total number who were interviewed was 5:8. 120 people said they did not approve of the tax increase. How many people approved?

7. The ratio of the number of men to the number of women who registered to run in the Capital City Marathon was 9 : 4. 360 women registered for the marathon. How many people registered altogether?

8. In a recent interview in Greenport, 7 out of 10 people said they prefer coffee in the morning. The rest said that they prefer tea. 27 people said they prefer tea. How many people were interviewed?

Check your answers on page 318.

Lesson 19: Unnecessary Information

Some math problems provide more information than you need. This means that facts are given that you will not use to solve the problem. Read word problems carefully to find unnecessary information.

Example: Mike takes home $1400 a month. He pays $350 a month for rent and $84 a month for car payments. What fraction of his monthly income goes to rent?

Step 1. The question asks you to compare Mike's rent to his monthly income. The $84 car payment is unnecessary information.

$84 is unnecessary.

Step 2. Write a fraction with the rent on top and the total monthly income on the bottom. Then reduce the fraction.

$$\frac{\text{rent}}{\text{total}} = \frac{350}{1400} = \frac{1}{4}$$

Lesson 19 Exercise

For each problem first identify any unnecessary information. Then solve each problem.

1. Harold drove for $6\frac{1}{2}$ hours at an average speed of 48 mph. He used 24 gallons of gasoline. How far did he drive?

2. At the Third Street Day Care Center there are 32 boys and 28 girls. Altogether 48 children have received vaccinations. Girls make up what fraction of the total number of children at the day care center?

3. A board 12 feet long and 10 inches wide weighs $3\frac{1}{2}$ pounds per linear foot. Find the total weight of the board.

4. Carmen bought $4\frac{1}{4}$ pounds of beef at $3.80 a pound and $5\frac{3}{4}$ pounds of chicken at $1.89 a pound. Find the total weight of her purchases.

5. On a test Rick got 6 problems wrong and 54 problems right. Of the total number of problems he guessed the answers to 8 of them. What fraction of the problems did he get right?

6. Geraldine drove 248 miles. She used 16 gallons of gasoline and paid $1.12 per gallon. On an average how far did she drive on one gallon of gasoline?

7. The basement of Colin's house is 60 feet long, 20 feet wide, and $7\frac{1}{2}$ feet high. What is the area of the floor of the basement?

8. The ratio of gray paint to green paint in a certain color mixture is 2:5. Paul has 10 gallons of green paint and 3 gallons of paint thinner. How much gray paint does he need for the mixture?

Check your answers on page 319.

Lesson 20: Mixed Word Problems

This lesson offers you a chance to apply your problem-solving skills with fractions. You will see tables, set-up solutions, and answer choices that include "not enough information is given."

Multiplication and division problems are sometimes more difficult to recognize than addition and subtraction problems. Remember that a fraction followed by the word **of** usually means to multiply.

Example 1: Tom's employer withholds $\frac{1}{5}$ of his salary for taxes and social security. Tom's gross salary is $1200 a month. How much does his employer withhold each month?

Find $\frac{1}{5}$ of $1200. Multiply $1200 by $\frac{1}{5}$.

$$\frac{1}{\cancel{5}} \times \frac{\cancel{1200}^{240}}{1} = \$240$$

Remember that in division problems the thing being divided must come first. The word **per** (as in miles per gallon or miles per hour) sometimes suggests division. The unit that follows the word **per** is usually the divisor.

Example 2: Paula drove 84 miles in $1\frac{3}{4}$ hours. What was her average speed in miles per hour?

In the phrase *miles per hour*, the word that follows *per* is *hour*. The divisor is $1\frac{3}{4}$ hours.

$$84 \div 1\frac{3}{4} =$$
$$\frac{84}{1} \div \frac{7}{4} =$$
$$\frac{\cancel{84}^{12}}{1} \times \frac{4}{\cancel{7}_1} = 48 \text{ mph}$$

You may see set-up solutions to fraction problems on the GED Test. Remember that these solutions show a method instead of an exact solution.

Example 3: Which of the following expresses $3\frac{1}{2}$ gallons changed to quarts?

(1) $3\frac{1}{2} + 4$ (3) $3\frac{1}{2} \div 4$

(2) $3\frac{1}{2} \times 4$ (4) $4 \div 3\frac{1}{2}$

To change gallons to quarts you must multiply the number of gallons by 4, which is the number of quarts in one gallon. Choice **(2)**, $3\frac{1}{2} \times 4$, is correct.

Lesson 20 Exercise

Solve each problem.

1. In the last election in Greenport $\frac{3}{8}$ of the 9600 registered voters failed to vote. How many voters failed to vote?
2. Denise drove 209 miles on $9\frac{1}{2}$ gallons of gasoline. What was the average distance she drove on one gallon of gasoline?
3. Joe bought a jacket originally selling for $72 for $\frac{1}{4}$ off the original price. Find the sale price of the jacket.
4. The Greenport Development Corporation is selling 24 acres of land in $\frac{3}{4}$-acre lots. How many lots can they get from 24 acres?
5. Kate wants to split $10\frac{1}{2}$ bushels of apples into three equal shares. How many bushels will be in each share?

Use the table below to answer questions 6 to 9.

Wholesale Prices per Pound

	imported	domestic
apples	48¢	64¢
lemons	30¢	45¢
lamb	25¢	150¢
beef	45¢	90¢

Source: U.S. Department of Agriculture.

6. The price per pound of imported lemons is what fraction of the price per pound of domestic lemons?
7. The price per pound of imported apples is what fraction of the price per pound of domestic apples?
8. The price per pound of imported lamb is what fraction of the price per pound of domestic lamb?
9. The price per pound of imported beef is what fraction of the price per pound of domestic beef?

Choose the correct solution to each problem.

10. Jill wants to put 10 pounds of berries into jars. Each jar will hold $\frac{1}{4}$ pound. How many jars can she fill?
    - (1) 2
    - (2) $2\frac{1}{2}$
    - (3) 10
    - (4) 40
    - (5) Not enough information is given.

11. Paul spliced together 5 lengths of wire each $2\frac{1}{2}$ feet long. What was the total length of the wire in inches? (Assume there was no waste.)
    - (1) $12(5 \times 2\frac{1}{2})$
    - (2) $\frac{2\frac{1}{2} \times 12}{5}$
    - (3) $(5 \times 2\frac{1}{2} + 12)$
    - (4) $\frac{5 \times 2\frac{1}{2}}{12}$
    - (5) $5 + 2\frac{1}{2} + 12$

12. Gloria can type 90 words per minute. Which of the following expresses the number of minutes she needs to type a report with 500 words?
    - (1) 90×500
    - (2) 60×500
    - (3) $\frac{500}{90}$
    - (4) $\frac{500}{60}$
    - (5) 90×60

13. Janet types 75 words per minute. How many minutes does she need to type a 10-page report if each page has an average of 275 words?
    - (1) $10 \times 275 \times 75$
    - (2) $\frac{75}{10 \times 275}$
    - (3) $\frac{75 \times 275}{10}$
    - (4) $\frac{10 \times 275}{75}$
    - (5) $\frac{10 \times 75}{275}$

14. Which of the following represents the volume in cubic feet of a rectangular container that is 10 feet long, 6 feet wide, and $2\frac{1}{2}$ feet high?
    - (1) $2(10) + 2(6)$
    - (2) $10 \times 6 + 2\frac{1}{2}$
    - (3) $10 \times 6 \times 2\frac{1}{2}$
    - (4) $(10 + 6 + 2\frac{1}{2})^2$
    - (5) $\frac{1}{2}(10 + 6)$

15. Bill wants to arrange the following packages in order from heaviest to lightest.

    Package A weighs $1\frac{1}{2}$ pounds.   Package D weighs $2\frac{3}{16}$ pounds.
    Package B weighs $1\frac{5}{8}$ pounds.   Package E weighs $2\frac{1}{4}$ pounds.
    Package C weighs $2\frac{3}{8}$ pounds.

    Which of the following lists the packages in order from heaviest to lightest?
    - (1) B, A, C, E, D
    - (2) A, B, C, E, D
    - (3) C, D, E, A, B
    - (4) C, E, D, B, A
    - (5) D, E, C, A, B

Check your answers on page 319.

Level 3 Review

Solve each problem.

1. The ratio of wins to losses for Kevin's basketball team was 8:5 last season. The team lost 15 games. How many games did they win?

2. At Irene's Clothing Store, 280 customers made purchases last week. The ratio of customers who paid cash to the total number was 4:7. How many of the customers paid cash?

3. At Paulson's Plastics the ratio of defective plastic bottles to the total number produced is 3:100. The factory produces 20,000 bottles a day. How many of the bottles produced each day are defective?

4. At the factory where Patty works the ratio of management to workers is 3:14. Altogether there are 340 people employed at the factory. How many people work in management?

5. The ratio of the number of girls to boys at the Morningside Day Care Center is 5:4. 84 boys are enrolled at the center. How many children are there altogether?

6. At Ace Products the ratio of freight sent by rail to the total amount of freight is 2:9. All freight that does not go by rail goes by truck. In March Ace Products shipped 140 tons of freight by truck. Find the weight of freight shipped by rail that month.

7. Find the total cost of 10 sheets of $\frac{3}{4}$-inch wallboard at the price of $12.50 per sheet.

8. Mary drove $3\frac{1}{2}$ hours at an average speed of 36 mph. She used 7 gallons of gasoline. How far did she drive altogether?

9. The Cruz family spends $360 a month for food and $240 a month for rent. What fraction of their total monthly income of $960 goes for food?

10. Carlos's employer deducts $\frac{1}{5}$ of his gross salary for taxes and $\frac{1}{10}$ for a savings fund. If his gross salary is $2400 per month, how much is deducted each month for taxes?

11. Phil bought 4 boards each 10 feet long. Which of the following expresses the combined length of the boards in yards?

    (1) $\frac{4 \times 3}{10}$
    (2) $\frac{4 \times 10}{3}$
    (3) $4 \times 10 \times 3$
    (4) $\frac{10 \times 3}{4}$
    (5) $3(4 + 10)$

12. The assembly line in the glass factory where Enzo works produces 25 bottles per minute. If the assembly line works continuously, which of the following expresses the number of bottles produced in an eight-hour day?

    (1) $\frac{8}{25 \times 60}$
    (2) $\frac{8 \times 60}{25}$
    (3) $25 \times 60 \times 8$
    (4) $\frac{25 \times 60}{8}$
    (5) $\frac{8 \times 25}{60}$

13. Bicycling burns off about 450 calories per hour. Which of the following expresses the number of hours required to burn off a pork chop that contains 310 calories and a glass of milk that contains 165 calories?

    (1) $\frac{310 + 450}{165}$

    (2) $310 + 165 + 450$

    (3) $\frac{310 + 165}{450}$

    (4) $\frac{165 \times 450}{310}$

    (5) $\frac{310 \times 165}{450}$

14. Which of the following represents the area in square feet of a triangle with a base of 6 feet and a height of 4 feet?

    (1) $6 + 4$

    (2) 6×4

    (3) $2(6) + 2(4)$

    (4) $6 \times 6 \times 4$

    (5) $\frac{1}{2} \times 6 \times 4$

15. The workers at Green Gardens Inc. can make 20 lawn mowers an hour. Which of the following expresses the total number of lawn mowers they can make working 8 hours a day for 5 days a week?

    (1) $\frac{20 \times 8}{5}$

    (2) $\frac{20 \times 5}{8}$

    (3) $\frac{5 \times 8}{20}$

    (4) $20 \times 8 \times 5$

    (5) $\frac{20 + 8}{5}$

Check your answers on page 320. If you have all 20 answers correct, go on to the next chapter. If you have answered a problem incorrectly, find the item number in the chart below and review that lesson before you go on.

Review:	If you missed item number:
Lesson 17	1, 2, 3
Lesson 18	4, 5, 6
Lesson 19	7, 8, 9, 10
Lesson 20	11, 12, 13, 14, 15

Level 3: Review

Chapter 4
PERCENTS

Level 1 Percent Skills

In this section you will review the methods for interchanging percents and decimals and for interchanging percents and fractions. You will also review the skills involved in the three basic types of percent problems: finding a percent of a number, finding what percent one number is of another, and finding a number when a percent of it is given.

To find out whether you need this review try the following preview.

Preview

Solve each problem.

1. Change 2% to a decimal.
2. Change the decimal 0.035 to a percent.
3. Change 84% to a fraction and reduce.
4. Change the fraction $\frac{7}{12}$ to a percent.
5. What is 3.5% of 700?
6. Find $37\frac{1}{2}$% of 144.
7. 32 is what percent of 80?
8. 120 is what percent of 48?
9. 75% of what number is 72?
10. $12\frac{1}{2}$% of what number is 20?

Check your answers on page 320. If you have all 10 problems correct, do the Level 1 Review on page 139. If you answered a problem incorrectly, study Level 1 beginning with Lesson 1.

Lesson 1: Percents and Decimals

So far in this book you have used decimals and fractions to describe parts of a whole. Percents are a third way to describe parts of a whole. **Percent** means "out of 100." One whole is 100% or 100 out of 100 parts. 1% means 1 out of 100 parts.

Percents are commonly used in business. Interest rates, sales tax, discounts, and markups are usually expressed as percents.

Percents are different from fractions in two ways. The denominator of a percent is always 100, but the denominator is not written. The percent sign, %, stands for the denominator 100. 9% means 9 out of 100 equal parts and is the same as the fraction $\frac{9}{100}$.

Percents can be written as two-place decimals. 9% is the same as the decimal 0.09. Remember that two-place decimals are hundredths.

Look at the drawings below. Each is divided into 100 equal parts. Notice how each can be described using a fraction, decimal, or percent.

This drawing shows 9 out of 100 parts shaded. 9% is shaded. 9% is the same as $\frac{9}{100}$ or 0.09.

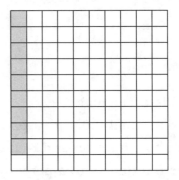

This drawing shows 175% because 175 parts are shaded. It can also be described as $1\frac{3}{4}$ or 1.75.

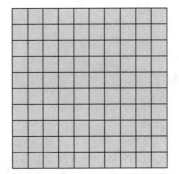
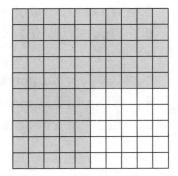

Notice that percents greater than 100% are greater than 1 whole. They are like mixed fractions or mixed decimals.

This drawing shows $\frac{1}{2}$ of a part out of 100 parts shaded or $\frac{1}{2}$%. It can also be described as $\frac{1}{200}$ or 0.005.

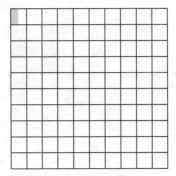

Notice that percents that are less than 1% are less than $\frac{1}{100}$.

When you work with percents, you will usually rewrite the percent in each problem as a decimal or a fraction.

Changing Percents to Decimals

To change a percent to a decimal, remove the percent sign and move the decimal point two places to the **left**.

Example 1: Change 45% to a decimal.

Remove the percent sign. Move the decimal point two places to the left. Notice that 45% is understood to have a decimal point after 45.

$$45\% = 45. = 0.45$$

Example 2: Change 80% to a decimal.

Remove the percent sign and move the decimal point two places to the left. Notice that you can drop the final 0. The final 0 does not affect the value of 0.8.

$$80\% = 80. = 0.8$$

Example 3: Change 3.5% to a decimal.

Remove the percent sign. Move the decimal point two places to the left. Notice that you must add a 0 in order to get two decimal places. This puts the digits 3 and 5 in their correct position.

$$3.5\% = 03.5 = 0.035$$

Example 4: Change $49\frac{1}{3}$% to a decimal.

Remove the percent sign and move the decimal point two places to the left. The decimal point is understood to be after the 9 because the 9 is in the ones place. The fraction $\frac{1}{3}$ does not have a place value.

$$49\frac{1}{3}\% = 49.\frac{1}{3}$$
$$= 0.49\frac{1}{3}$$

Changing Decimals to Percents

To change a decimal to a percent, move the decimal point two places to the **right** and write a percent sign.

Example 5: Change 0.35 to a percent.

Move the decimal point two places to the right and write a percent sign. You do not have to write the decimal point when it comes after a whole number.

$$0.35 = .35 = 35\%$$

Level 1, Lesson 1: Percents and Decimals

Example 6: Change 0.4 to a percent.

Move the decimal point two places to the right and write a percent sign. Here you must add a 0 in order to get two places.

$$0.4 = .40 = 40\%$$

Example 7: Change $0.065\frac{1}{8}$ to a percent.

Move the decimal point two places to the right and write a percent sign. You can drop the 0 because it does not affect the value of the percent.

$$0.065\frac{1}{8} = 0.06\,5\frac{1}{8}$$
$$= 6.5\frac{1}{8}\%$$

Lesson 1 Exercise

Change each percent to a decimal.

1. a. 75% =   b. 4% =   c. 62.5% =   d. 7% =
2. a. 60% =   b. 150% =   c. 1% =   d. 300% =
3. a. $8\frac{3}{4}\%$ =   b. 12.6% =   c. 0.6% =   d. 15% =
4. a. 0.1% =   b. 0.25% =   c. 2.5% =   d. 200% =

Change each decimal to a percent.

5. a. 0.46 =   b. 0.08 =   c. 0.045 =   d. $0.08\frac{1}{3}$ =
6. a. 0.9 =   b. 0.25 =   c. 0.005 =   d. 0.05 =
7. a. 0.0825 = b. 0.4 =   c. 0.675 =   d. 1.2 =
8. a. 4.75 =   b. 0.625 =   c. 8 =   d. $0.66\frac{2}{3}$ =

Check your answers on page 321.

Lesson 2 — Percents and Fractions

When you work percent problems, you will change many of the percents to fractions. Remember that a percent can be thought of as a fraction with a denominator of 100.

Changing Percents to Fractions

To change a percent to a fraction, drop the % sign and write the number as the numerator and write 100 as the denominator. Reduce the fraction to lowest terms.

132   Chapter 4 Percents

Example 1: Change 16% to a fraction.

Drop the % sign. Write the number as the numerator and write 100 as the denominator. Reduce to lowest terms.

$$16\% = \frac{16}{100} = \frac{4}{25}$$

If the percent has a decimal in it, first change the percent to a decimal. Then change the decimal to a fraction.

Example 2: Change 3.5% to a fraction.

Step 1. Move the decimal point two places to the left and drop the % sign.

$$3.5\% = 03.5 = 0.035$$

Step 2. Change 0.035 to a fraction and reduce.

$$\frac{35}{1000} = \frac{7}{200}$$

If the percent has a fraction in it, drop the % sign and write the number as the numerator and 100 as the denominator. Remember that the fraction line means to divide. Divide the numerator by the denominator.

Example 3: Change $16\frac{2}{3}\%$ to a fraction.

Step 1. Drop the % sign. Write the digits in the percent as the numerator and 100 as the denominator.

$$16\frac{2}{3}\% = \frac{16\frac{2}{3}}{100}$$

Step 2. Divide the numerator by the denominator.

$$16\frac{2}{3} \div 100 =$$

$$\frac{50}{3} \div \frac{100}{1} =$$

$$\frac{\cancel{50}^{1}}{3} \times \frac{1}{\cancel{100}_{2}} = \frac{1}{6}$$

Changing Fractions to Percents

To change a fraction to a percent, find a **fraction of** the 100 parts. In other words multiply the fraction by $\frac{100}{1}$ and write a % sign.

Example 4: Change $\frac{9}{20}$ to a percent.

Multiply the fraction by $\frac{100}{1}$ and write a % sign.

$$\frac{9}{\cancel{20}_{1}} \times \frac{\cancel{100}^{5}}{1} = 45\%$$

When you change a fraction to a percent, the answer may include a fraction.

Example 5: Change $\frac{1}{9}$ to a percent.

Multiply the fraction by 100 and write a % sign.

$$\frac{1}{9} \times \frac{100}{1} = \frac{100}{9}$$

$$= 11\frac{1}{9}\%$$

Level 1, Lesson 2: Percents and Fractions

Lesson 2 Exercise

Change each percent to a fraction and reduce.

1. a. 15% = b. 85% = c. 96% = d. 60% =
2. a. 275% = b. 8% = c. 450% = d. 42% =
3. a. 1.5% = b. 4.8% = c. 12.5% = d. 6.25% =
4. a. $6\frac{1}{4}$% = b. $18\frac{3}{4}$% = c. $13\frac{1}{3}$% = d. $56\frac{1}{4}$% =

Change each fraction to a percent.

5. a. $\frac{3}{5}$ = b. $\frac{12}{25}$ = c. $\frac{1}{2}$ = d. $\frac{2}{3}$ =
6. a. $\frac{23}{100}$ = b. $\frac{3}{10}$ = c. $\frac{5}{8}$ = d. $\frac{3}{16}$ =
7. a. $\frac{2}{7}$ = b. $\frac{5}{6}$ = c. $\frac{19}{20}$ = d. $\frac{4}{5}$ =

Check your answers on page 321.

Common Percents, Decimals, and Fractions

When you work a percent problem, you will usually change the percent in the problem to a decimal or to a fraction. The list below includes some of the most common percents along with the decimal form and the fraction form for each.

Take the time now to memorize this chart.

Percent		Decimal		Fraction
25%	=	0.25	=	$\frac{1}{4}$
50%	=	0.5	=	$\frac{1}{2}$
75%	=	0.75	=	$\frac{3}{4}$
$12\frac{1}{2}$%	=	$0.12\frac{1}{2}$	=	$\frac{1}{8}$
$37\frac{1}{2}$%	=	$0.37\frac{1}{2}$	=	$\frac{3}{8}$
$62\frac{1}{2}$%	=	$0.62\frac{1}{2}$	=	$\frac{5}{8}$
$87\frac{1}{2}$%	=	$0.87\frac{1}{2}$	=	$\frac{7}{8}$
$33\frac{1}{3}$%	=	$0.33\frac{1}{3}$	=	$\frac{1}{3}$
$66\frac{2}{3}$%	=	$0.66\frac{2}{3}$	=	$\frac{2}{3}$

Percent		Decimal		Fraction
20%	=	0.2	=	$\frac{1}{5}$
40%	=	0.4	=	$\frac{2}{5}$
60%	=	0.6	=	$\frac{3}{5}$
80%	=	0.8	=	$\frac{4}{5}$
10%	=	0.1	=	$\frac{1}{10}$
30%	=	0.3	=	$\frac{3}{10}$
70%	=	0.7	=	$\frac{7}{10}$
90%	=	0.9	=	$\frac{9}{10}$
$16\frac{2}{3}$%	=	$0.16\frac{2}{3}$	=	$\frac{1}{6}$
$83\frac{1}{3}$%	=	$0.83\frac{1}{3}$	=	$\frac{5}{6}$

Lesson 3: Finding a Percent of a Number (Part)

There are three basic types of percent problems:
- Type 1. Finding a percent of a number (part).
- Type 2. Finding what percent one number is of another (rate).
- Type 3. Finding a number when a percent of it is given (whole).

Each type of problem can be solved using the percent formula: **whole × rate = part**, where the whole is the entire amount, rate is expressed as a percent, and part is a piece of the whole amount. The whole is sometimes called the base. In each type of problem a different element of the formula is missing. In the lessons that follow, you will learn to solve each type of percent problem.

To solve Type 1 problems, you will be finding a percent of a number or the part. In the Fractions chapter you learned that to find a **fraction of** a number you need to multiply. To find a **percent of** a number you also need to multiply. To find the part, use the percent formula:

$$\text{whole} \times \text{rate} = \text{part}.$$

In other words, multiply the whole times the rate (percent) to find the part.

To multiply by a rate (percent) first change the percent to a decimal or to a fraction. Then multiply using the decimal or the fraction.

For some problems, it is easier to change the rate to a decimal.

Example 1: Find 2.5% of 300.

Step 1. To use the formula whole × rate = part, change the rate, 2.5%, to a decimal.

$$2.5\% = .02.5 = 0.025$$

Step 2. Multiply the whole, 300, by the rate, 0.025.

$$\begin{array}{r} 0.025 \\ \times\ 300 \\ \hline 7.500 \end{array} = 7.5$$

With other problems, it is easier to change the rate to a fraction.

Example 2: Find $16\frac{2}{3}\%$ of 48.

Step 1. Use the formula whole × rate = part. Change the rate, $16\frac{2}{3}\%$, to a fraction. $16\frac{2}{3}\%$ is on the list that you should be familiar with.

$$16\frac{2}{3}\% = \frac{1}{6}$$

Step 2. Multiply the whole, 48, by the rate, $\frac{1}{6}$.

$$\frac{1}{\cancel{6}_1} \times \frac{\cancel{48}^8}{1} = 8$$

In some problems you can change the rate to either a decimal or a fraction. Use whichever method is easier for you.

Example 3: Find 60% of 40.

Step 1. Change the rate, 60%, to a decimal.

Using a decimal:
$60\% = 0.60. = 0.6$

Step 2. Multiply the whole, 40, by 0.6.

$$\begin{array}{r} 40 \\ \times\, 0.6 \\ \hline 24.0 = 24 \end{array}$$

Step 1. Change the rate to a fraction.

Using a fraction:
$60\% = \dfrac{3}{5}$

Step 2. Multiply 40 by $\frac{3}{5}$.

$\dfrac{3}{\underset{1}{\cancel{5}}} \times \dfrac{\overset{8}{\cancel{40}}}{1} = 24$

Lesson 3 Exercise

Solve each problem. Use the formula whole × rate = part.
For rows 1 and 2, first change each rate (percent) to a decimal.

1.   **a.** 6% of 150 =   **c.** 125% of 36 =
    b. 40% of 160 =   **d.** 1.9% of 200 =
2.   **a.** 5.4% of 80 =   **c.** 6.25% of 300 =
    b. 0.8% of 50 =   **d.** 10.4% of 500 =

For rows 3 and 4, first change each rate (percent) to a fraction.

3.   **a.** $66\frac{2}{3}$% of 240 =   **c.** $62\frac{1}{2}$% of 48 =
    b. $12\frac{1}{2}$% of 400 =   **d.** $83\frac{1}{3}$% of 120 =
4.   **a.** $33\frac{1}{3}$% of 18 =   **c.** $37\frac{1}{2}$% of 24 =
    b. $16\frac{2}{3}$% of 96 =   **d.** $87\frac{1}{2}$% of 1200 =

For rows 5 and 6, change each rate (percent) to a decimal or to a fraction.

5.   **a.** 75% of 84 =   **c.** 90% of 130 =
    b. 8.5% of 400 =   **d.** 50% of 28 =
6.   **a.** 250% of 36 =   **c.** 25% of 116 =
    b. 60% of 200 =   **d.** 35% of 260 =

Check your answers on page 322.

Lesson 4: Finding What Percent One Number Is of Another (Rate)

In Type 2 percent problems, you must find the rate or percent as an answer. These problems are similar to the fraction problems in which you find what fraction one number is of another.

When you use the percent formula, whole × rate = part, to find what percent one number is of another, you are solving for the rate. To solve for the rate, compare the **part** to the **whole**: $\frac{part}{whole}$ = rate.

First make a fraction with the part as the numerator and the whole as the denominator. Then change the fraction to a percent. In these problems the whole usually follows the word **of**.

Example 1: 18 is what percent of 24?

Step 1. Make a fraction with the part, 18, over the whole, 24, and reduce.

$$\frac{part}{whole} \quad \frac{18}{24} = \frac{3}{4}$$

Step 2. Change $\frac{3}{4}$ to a percent.

$$\frac{3}{\cancel{4}} \times \frac{\cancel{100}^{25}}{1} = 75\%$$

In some problems, the part may be greater than the whole. For these problems, the rate will always be greater than 100%.

Example 2: 30 is what percent of 30?

Step 1. Make a fraction with the part, 30, over the whole, 30, and reduce.

$$\frac{part}{whole} = \frac{30}{30} = 1$$

Step 2. Change 1 to a percent. When the part is the same as the whole, you have 100%

$$1 \times \frac{100}{1} = 100\%$$

Example 3: 42 is what percent of 30?

Step 1. Make a fraction with the part, 42, over the whole 30, and reduce.

$$\frac{part}{whole} = \frac{42}{30} = \frac{7}{5}$$

Step 2. Change $\frac{7}{5}$ to a percent. Think about the answer. In Example 2 you found that 30 is 100% of 30. Since 42 is greater than 30, 42 is greater than 100% of 30.

$$\frac{7}{\cancel{5}} \times \frac{\cancel{100}^{20}}{1} = 140\%$$

Lesson 4 Exercise

Solve each problem.

1. 28 is what percent of 70?
2. 104 is what percent of 160?
3. 16 is what percent of 20?
4. 45 is what percent of 135?
5. 24 is what percent of 32?
6. 30 is what percent of 48?
7. 225 is what percent of 375?
8. 27 is what percent of 72?
9. 36 is what percent of 24?
10. 15 is what percent of 120?
11. 36 is what percent of 40?
12. 150 is what percent of 75?

Check your answers on page 323.

Lesson 5: Finding a Number When a Percent of It Is Given (Whole)

Some percent problems seem backwards. In Type 3 problems, you have a number (part) that is a percent (rate) of a missing number. The missing number is the whole.

Remember, when you find a percent of a number, you must multiply whole × rate = part. Type 3 problems are the opposite. To solve them, you must perform the opposite of multiplication, which is division. You can solve for the whole by dividing: part ÷ rate = whole. To find a number when a percent of it is given, first change the percent to a decimal or to a fraction. Then divide the part by the decimal or the fraction (rate).

Example 1: 35% of what number is 21?

Step 1. Change the rate, 35%, to a decimal.

$35\% = 0.35 = 0.35$

Step 2. Divide the part, 21, by the rate, 0.35. So, 35% of 60 is 21.

$$0.35 \overline{)21.00} \quad \begin{array}{r} 60 \\ \underline{21\ 0} \\ 00 \end{array}$$

Example 2: $16\frac{2}{3}\%$ of what number is 13?

Step 1. Change the rate, $16\frac{2}{3}\%$, to a fraction.

$16\frac{2}{3}\% = \frac{1}{6}$

Step 2. Divide the part, 13, by the rate, $\frac{1}{6}$.

$13 \div \frac{1}{6} =$

$\frac{13}{1} \div \frac{1}{6} =$

$\frac{13}{1} \times \frac{6}{1} = 78$

138 Chapter 4 Percents

It is a good idea to check these problems.

When you find the whole, check your answer using the percent formula: whole × rate = part. The answer should be the original number (part) in the problem.

To check Example 1, find 35% of 60.

$$35\% = .35 = 0.35 \qquad \text{whole} \times \text{rate} = \text{part}$$

$$\begin{array}{r} 0.35 \\ \times\, 60 \\ \hline 21.00 = 21 \end{array}$$

To check Example 2, find $16\frac{2}{3}\%$ of 78.

$$16\frac{2}{3}\% = \frac{1}{6} \qquad \text{whole} \times \text{rate} = \text{part}$$

$$\frac{1}{\cancel{6}_1} \times \frac{\cancel{78}^{13}}{1} = \frac{13}{1} = 13$$

Lesson 5 Exercise

Solve each problem.

1. 60% of what number is 24?
2. 25% of what number is 18?
3. $12\frac{1}{2}\%$ of what number is 15?
4. 4% of what number is 9.6?
5. $66\frac{2}{3}\%$ of what number is 52?
6. $87\frac{1}{2}\%$ of what number is 112?
7. 2.5% of what number is 8?
8. 35% of what number is 140?
9. 90% of what number is 23.4?
10. $62\frac{1}{2}\%$ of what number is 45?
11. 8.5% of what number is 17?
12. $16\frac{2}{3}\%$ of what number is 36?

Check your answers on page 324.

Level 1 Review

In this review the three basic types of percent problems are mixed together. Before you start a problem, decide which type of problem it is. The three types are:

Type 1. Finding a percent of a number (part).

Type 2. Finding what percent one number is of another (rate).

Type 3. Finding a number when a percent of it is given (whole).

Solve each problem.

1. Change 9.6% to a decimal.
2. Change 0.0145 to a percent.
3. Change $8\frac{1}{3}$% to a fraction and reduce.
4. Change $\frac{9}{16}$ to a percent.

For problems 5 to 20, first decide if each problem is Type 1, Type 2, or Type 3 from the list above. Then solve.

5. 24 is what percent of 72?   Type ____
6. Find 4.5% of 2400.   Type ____
7. What is $83\frac{1}{3}$% of 330?   Type ____
8. 65% of what number is 52?   Type ____
9. 105 is what percent of 60?   Type ____
10. $16\frac{2}{3}$% of what number is 25?   Type ____
11. Find 8% of 250.   Type ____
12. 28 is what percent of 42?   Type ____
13. 200 is what percent of 800?   Type ____
14. 20% of what number is 14?   Type ____
15. What is $33\frac{1}{3}$% of 45?   Type ____
16. $12\frac{1}{2}$% of what number is 60?   Type ____
17. Find 1.5% of 700.   Type ____
18. 180 is what percent of 90?   Type ____
19. What is 60% of 110?   Type ____
20. 12 is what percent of 400?   Type ____

Check your answers on page 324. If you have 20 problems correct, go on to Level 2. If you answered a problem incorrectly, find the item number on the chart below and review that lesson before you go on.

Review:	If you missed item number:
Lesson 1	1, 2
Lesson 2	3, 4
Lesson 3	6, 7, 11, 15, 17, 19
Lesson 4	5, 9, 12, 13, 18, 20
Lesson 5	8, 10, 14, 16

Level 2: Percent Applications

In this section you will learn to use your percents skills to solve word problems. You will also learn to compute simple interest.

Lesson 6: Finding a Percent of a Number (Part)

By far the most common application of percents is in problems where you must find a **percent (part) of** a number. Remember that this means to multiply: whole × rate = part.

Percent problems may seem difficult because there is more than one way to solve them. Remember that you can change the percent to a decimal or to a fraction. To build your confidence, try both methods. The solution to each example and exercise shows the method the author finds easier.

Example: Max bought a shirt for $19.80. The sales tax in his state is 6%. How much was the sales tax for the shirt?

Step 1. You need to find 6% of $19.80. Change 6% to a decimal.

6% = .06 = 0.06

Step 2. Multiply the whole (price of the shirt) by the rate, 0.06, and round your answer to the nearest cent.

$19.80
× 0.06
$1.1880 to the nearest cent = $1.19

Lesson 6 Exercise

Solve each problem.

1. Out of 60 questions on a test, Joe got 80% right. How many questions did he get right?

2. Juan's gross salary is $320 a week. His employer deducts 21% of the gross salary for taxes and insurance. How much is the weekly deduction?

3. The Greens made a 6% down payment on a $78,000 house. How much was the down payment?

4. Of 320 employees at the County Hospital, 65% are women. How many women work at the hospital?

5. Jeff sells cars for a 5% commission. One month he sold cars with a total value of $36,400. Find his commissions for the month.

6. Deborah owes $430 on her credit card. She has to pay a fee of 1.5% every month on the amount she owes. Find the monthly fee she must pay.

7. In a state where the sales tax rate is 4.5%, what is the tax on a $119 lawn mower?

8. Of the 420 people who voted in Stone Bridge, $16\frac{2}{3}$% voted in favor of a tax increase to pay for a bigger school budget. How many of the voters voted for the tax increase?

Check your answers on page 325.

Lesson 7: Finding What Percent One Number Is of Another (Rate)

Finding what percent one number is of another is the easiest type of percent problem to recognize. In these problems you are always looking for the rate or percent. Remember the formula for finding the rate: rate = $\frac{part}{whole}$. Make a fraction with the **part** over the **whole**. Then change the fraction to a percent.

Example 1: Of the 20 students in George's math class, 12 are women. What percent of the class are women?

Step 1. Make a fraction with the part (the number of women) over the whole (the total number of students) and reduce.

$$\frac{part}{whole} \text{ or}$$

$$\frac{women}{total} \quad \frac{12}{20} = \frac{3}{5}$$

Step 2. Change $\frac{3}{5}$ to a percent.

$$\frac{3}{\cancel{5}} \times \frac{\cancel{100}\,^{20}}{1} = 60\%$$

In some problems where you must find what percent one number is of another, you will be asked to compare a part that is greater than the whole. The rates for these problems are always greater than 100%

Notice that the amount that follows the phrase **what percent of** usually goes in the denominator.

Example 2: The community of Stone Bridge hoped to raise $5000 to buy new playground equipment. In fact, they raised $7000. The amount they raised was what percent of the amount they had hoped to raise?

Step 1. Make a fraction with the part (the amount the community actually raised) over the whole (the amount they had hoped to raise) and reduce. Notice that the words **the amount they hoped to raise** follow the phrase **what percent of**.

$$\frac{part}{whole} \text{ or}$$

$$\frac{\text{amount raised}}{\text{amount hoped for}} \quad \frac{\$7000}{\$5000} = \frac{7}{5}$$

Step 2. Change $\frac{7}{5}$ to a percent. Since the part is greater than the whole, the rate is greater than 100%

$$\frac{7}{\cancel{5}_1} \times \frac{\overset{20}{\cancel{100}}}{1} = 140\%$$

Lesson 7 Exercise

Solve each problem.

1. On a test with 40 questions, Sandy got 34 right. What percent of the questions did she get right?
2. In the shop where Fred works, 6 of the 48 employees were late one morning because of icy roads. What percent of the employees were late?
3. For a shirt that cost $22.50, Bill paid $1.80 in sales tax. The sales tax was what percent of the price?
4. Of the 320 people who ate at the Riverside Restaurant on Saturday, 280 people paid cash. What percent of the diners paid cash?
5. In 1990 Suzanne's car was worth $6400. By 1992 the car was worth only $2560. The 1992 value of the car was what percent of the 1990 value?
6. Paul takes home $1200 a month and pays $252 a month for rent. His rent is what percent of his take-home pay?
7. The Browns bought a house for $48,000. Fifteen years later they sold it for $84,000. The sale price was what percent of the price they had paid for the house?
8. When John was 20 years old, he weighed 160 pounds. When he was 30, he weighed 200 pounds. John's weight when he was 30 was what percent of his weight when he was 20?

Check your answers on page 325.

Finding a Number When a Percent of It Is Given (Whole)

Finding the whole when a percent of it is given is the hardest type of percent problem to recognize. Remember that these problems are "backwards." You must divide by the percent to solve them. In these problems you have the **part**. You are looking for the **whole** that the part is based on. Use the formula: whole = part ÷ rate.

Again, remember that you can change the percent to a decimal or to a fraction.

Example: Sara got 15 problems right on a test. These problems were 75% of the test. How many problems were on the test?

Step 1. Change 75% to a decimal.

$$75\% = .75 = 0.75$$

Step 2. Divide the part (the number of problems she got right) by the rate, 0.75. There were 20 problems on the test.

$$0.75 \overline{)15.00} = 20 \text{ problems}$$

Lesson 8 Exercise

Solve each problem.

1. On a test Phil got 16 problems right. This was 80% of the total. How many problems were on the test?

2. Alfonso paid $630 as a down payment on a car. This was 9% of the total price. Find the total price.

3. So far the Allens have driven 420 miles on their way to visit their grandchildren. Mr. Allen figures they have driven 75% of the total distance. How far do they have to drive altogether?

4. The Atlas Foundry employs 24 women. Women make up 15% of the total work force. How many people work at the foundry?

5. By April the Fifth Street Block Association had $1440 in a fund to buy trees. This was 60% of their goal. How much do they need altogether?

6. Julio pays 22% of his gross income for taxes. Last year he paid $5280 in taxes. What was his gross income last year?

7. Mark weighs 180 pounds. This is $66\tfrac{2}{3}\%$ of the amount he weighed five years ago. Find his weight five years ago.

8. Of the total number of people who were interviewed 450 said that they had not smoked a cigarette in the last year. They represent $37\tfrac{1}{2}\%$ of the people who were interviewed. How many people were interviewed?

Check your answers on page 326.

Lesson 9 Interest

Interest is money that someone pays for using someone else's money. A bank pays customers interest for using their money deposited in savings accounts. A customer pays a bank interest for using the bank's money as a loan.

The formula for finding simple interest is: $i = prt$ where p = the principal (the money borrowed or loaned), r = rate (expressed as a percent), and t = time (the number of years). Notice that this formula appears on the list on page 18.

Remember that when letters are written next to each other in a formula, you must multiply the values.

Example 1: Find the interest on $700 at 9% annual interest for one year.

Step 1. Change 9% to a fraction.

$$9\% = \frac{9}{100}$$

Step 2. Replace p with $700, r with $\frac{9}{100}$, and t with 1 in the formula $i = prt$ and multiply.

$$i = p \times r \times t$$

$$i = \frac{\cancel{700}^{7}}{1} \times \frac{9}{\cancel{100}_{1}} \times 1 = \$63$$

When the time in an interest problem is less than one year, write it as a fraction of a year.

Example 2: Find the interest on $900 at 11% annual interest for six months.

Step 1. Change 6 months to a fraction of a year.

$$t = 6 \text{ mo} = \frac{6}{12} = \frac{1}{2} \text{ yr}$$

Step 2. Change 11% to a fraction.

$$r = 11\% = \frac{11}{100}$$

Step 3. Replace p with $900, r with $\frac{11}{100}$, and t with $\frac{1}{2}$ in the formula $i = prt$. Then multiply.

$$i = prt$$

$$i = \frac{\cancel{900}^{9}}{1} \times \frac{11}{\cancel{100}_{1}} \times \frac{1}{2} =$$

$$\frac{99}{2} = \$49.50$$

When the time is more than one year, write the time as an improper fraction.

Remember that you can change the percent to either a fraction or a decimal. Use the form that will make the arithmetic easier.

Example 3: Find the interest on $400 at $3\frac{1}{2}$% annual interest for one year and three months.

Step 1. Change 1 year 3 months to a mixed number.

$$t = 1 \text{ yr } 3 \text{ mo}$$

$$= 1\frac{3}{12} = 1\frac{1}{4} = \frac{5}{4} \text{ yr}$$

Step 2. Change $3\frac{1}{2}$% to a fraction.

$$r = 3\frac{1}{2}\% = \frac{3\frac{1}{2}}{100}$$

Level 2, Lesson 9: Interest

Step 3. Replace p with $400, r with $\frac{3\frac{1}{2}}{100}$, and t with $\frac{5}{4}$. Then multiply.

$$i = \frac{\overset{4}{\cancel{400}}}{1} \times \frac{3\frac{1}{2}}{\underset{1}{\cancel{100}}} \times \frac{5}{\underset{1}{\cancel{4}}}$$

$$i = 3\frac{1}{2} \times 5$$

$$i = \frac{7}{2} \times \frac{5}{1} = \frac{35}{2}$$

$$i = \$17.50$$

Lesson 9 Exercise

Find the interest for each of the following.

1. $600 at 4% annual interest for one year.
2. $1500 at 10% annual interest for one year.
3. $400 at 8.5% annual interest for one year.
4. $1600 at 10.5% annual interest for one year.
5. $4000 at 12% annual interest for one year.
6. $1200 at $5\frac{1}{4}$% annual interest for one year.
7. $2000 at $8\frac{3}{4}$% annual interest for one year.
8. $500 at 8% annual interest for six months.
9. $1000 at 9% annual interest for eight months.
10. $400 at $5\frac{1}{2}$% annual interest for one year and three months.
11. $800 at 4.5% annual interest for nine months.
12. $2500 at 15% annual interest for three years.
13. $3000 at 9.5% annual interest for one year and six months.
14. $1600 at 18% annual interest for two years and four months.
15. $1800 at $3\frac{1}{2}$% annual interest for ten months.

Check your answers on page 326.

Chapter 4 Percents

Level 2 Review

In this review the three basic types of percent problems are mixed together. Before you start to solve a problem, decide which type of problem it is. The three types are:

   Type 1. Finding a percent of a number (part).
   Type 2. Finding what percent one numbr is of another (rate).
   Type 3. Finding a number when a percent of it is given (whole).

Decide if each problem is Type 1, Type 2, or Type 3 on the list above. Then solve each problem.

1. Tires usually selling for $55 each were on sale for 15% off. How much can you save by buying a tire on sale?   Type ____

2. In Adrienne's class 20 of the 24 students drive to school. What percent of the students in the class drive to school?   Type ____

3. Sam has saved $1350 for a new motorcycle. This is 75% of what he needs. Find the total that he needs.   Type ____

4. In one week the workers at Stable Steel Products produced 36 defective parts. These represented 1.5% of the total number of parts produced. How many parts were produced that week?   Type ____

5. Isabella made a $700 down payment on a used car selling for $3500. The down payment was what percent of the price?   Type ____

6. Of 500 people who were interviewed only 65 agreed with the president's foreign policy. Of the people who were interviewed what percent agreed with the president's foreign policy?   Type ____

7. Of the 1250 drivers who crossed the Green River toll bridge on Monday 2% put the wrong amount of money in the toll machine. How many drivers put the wrong amount in the machine that day?   Type ____

8. The price of educating one student for a year at the Stone Bridge Elementary School is $6850. 6% of that money is for athletic equipment. Find the yearly cost of athletic equipment for each student.   Type ____

9. When Paul started working at his current job, he made $6.50 an hour. Now, three years later, he makes 18% more. How much more does he make each hour?   Type ____

10. Loisa paid $0.75 sales tax on a book which cost $12.50. The tax was what percent of the price of the book?   Type ____

11. Pat paid $4.96 sales tax on a winter coat. The sales tax in her state is 4%. Find the price of the coat before the tax.   Type ____

12. The Johnsons spend $115 a week for food. Their weekly budget is $345. What percent of their budget goes for food?   Type ____

13. The Salgados have paid off 60% of their mortgage. So far they have paid $46,800. Find the total amount of the mortgage.   Type ____

Level 2: Review

Solve each problem.

14. Find the interest on $2400 at 14.5% annual interest for one year.
15. Find the interest on $360 at 7% annual interest for one year and eight months.
16. What is the interest on $800 at 3% annual interest for six months?

Check your answers on page 327. If you have all 16 problems correct, go on to Level 3. If your have answered a problem incorrectly, find the item number on the chart below and review that lesson before you go on.

Review:	If you missed item number:
Lesson 6	1, 7, 8, 9
Lesson 7	2, 5, 6, 10, 12
Lesson 8	3, 4, 11, 13
Lesson 9	14, 15, 16

Level 3: Problem Solving with Percents

In this section you will learn to solve multistep percent problems. You will also learn to make comparisons using percents.

Lesson 10: Finding a Percent of a Number in Multistep Problems

Problems that involve finding a percent of a number sometimes require more than one step. You may have to calculate the percent of a number and then add or subtract your answer to another number in the problem.

Read the problems carefully. Be sure your result answers the question.

Example 1: An electric drill normally selling for $35 was on sale for 20% off. Find the sale price of the drill.

Step 1. Change 20% to a fraction. $\quad 20\% = \frac{1}{5}$

Step 2. Find $\frac{1}{5}$ of $35. $\quad \frac{1}{\cancel{5}} \times \frac{\cancel{35}^{7}}{1} = \$7$

Step 3. You are looking for the sale price. $7 is the amount that is taken off the regular price. Subtract $7 from the original price to find the sale price.

$$\begin{array}{r} \$35 \\ -7 \\ \hline \$28 \end{array}$$

Lesson 10 Exercise

Solve each problem.

1. The budget for the Stone Bridge Summer Activities Program was $2000 the first year. The budget increased by 40% for the second year. What was the budget the second year?

2. On a sunny Sunday in July 3800 people visited the Capital City Zoo. The next day, which was rainy, the attendance was 35% less. How many people visited the zoo the next day?

3. When José started working as an assistant mechanic he made $4.50 an hour. Now, as a supervisor, he makes 150% more. What is his hourly wage now?

4. Since the Smiths bought their house three years ago, the value of the house has increased 18%. They paid $60,000 for their house. Find the value of the house now.

5. Selma bought a portable cassette player for $49 and cassettes for $12. She had to pay 7% sales tax. Find the total price of the player and the cassettes including tax.

6. Jack pays $28 for each pair of shoes that he sells in his store. He adds a 40% markup on each pair of shoes. To the nearest dollar how much does he charge for a pair of shoes?

7. A computer that cost $5000 in 1980 cost 65% less in 1990. What was the cost of the computer in 1990?

Read the following passage and then answer questions 8–10.

> George and Margie have decided to rent an apartment for $390 a month. They must pay 10% of the first year's rent to the agent who helped them find the apartment, and they must pay the landlord 15% of the first year's rent as a security deposit. If they choose to renew their lease at the end of the year they will have two choices. They can renew for one year at an increase of 6% or they can renew for two years with an increase of 8%.

8. How much did they have to pay the agent who helped them to find the apartment?

9. How much was the security deposit?

10. Find the amount of rent for one month if they choose to renew for two years.

Check your answers on page 327.

Lesson 11: Finding What Percent One Number Is of Another in Multistep Problems

Earlier in this chapter you learned to solve word problems in which you had to find what percent one number is of another. The key to solving these problems is to first make a fraction with the **part** over the **whole**. Sometimes one of the pieces you need is missing.

Example 1: Joyce did 16 problems right and 4 problems wrong on a test. What percent of the problems did she do right?

Chapter 4 Percents

Step 1. Find the **whole**.

$$\begin{array}{r}16 \text{ right} \\ +4 \text{ wrong} \\ \hline 20 \text{ total}\end{array}$$

Step 2. Make a fraction with the part (the number right) over the whole (the total) and reduce the fraction.

$$\frac{16}{20} = \frac{4}{5}$$

Step 3. Change $\frac{4}{5}$ to a percent.

$$\frac{4}{5} = 80\%$$

In problems where an amount changes over a period of time, think of the **part** over the **whole** as the **change** over the **original**. Remember that the original is always earlier in time.

These problems are **percent of increase** or **percent of decrease** problems.

Example 2: The membership in the Southside Tenants' Union rose from 250 to 300. By what percent did the membership increase?

Step 1. Find the amount of change. Subtract the old membership from the new membership.

$$\begin{array}{r}300 \\ -250 \\ \hline 50 \text{ new members}\end{array}$$

Step 2 Make a fraction with the change (the number of new members) over the original and reduce.

$$\frac{\text{change}}{\text{original}} = \frac{50}{250} = \frac{1}{5}$$

Step 3. Change $\frac{1}{5}$ to a percent.

$$\frac{1}{5} = 20\%$$

You will often see this kind of problem in business applications. When an item is on sale there is a percent of decrease or **discount**. When a store owner determines a price for an item, he includes a percent of increase or a **markup** from his original cost on the item.

Lesson 11 Exercise

Solve each problem.

1. The price of gas dropped from $1.32 to $1.21 a gallon. Find the percent of decrease.
2. A television that originally sold for $360 was on sale for $306. Find the discount rate (percent of decrease).
3. The population of Greenport was 12,000 ten years ago. Now it is 15,000. By what percent did the population increase?
4. Frank pays $16 each for the shirts he sells for $24 in his store. Find the markup rate (percent of increase) on each shirt.
5. In the factory where Gladys works there are 84 women and 126 men. What percent of the workers are women?

6. Sally spends $270 a month for food for her family. She has $630 left each month to cover all other expenses. What percent of her monthly budget goes for food?

7. Steve bought a car for $4800 and sold it three years later for $1800. By what percent did the value of the car drop?

8. There were 720 unemployed people in Greenport in October. By November the number had increased to 792. By what percent did the number of unemployed people increase from October to November?

Use the following passage to answer questions 9 and 10.

Carlos made a down payment of $900 on a two-year-old car selling for $6000. The payments on the loan that Carlos got from his credit union will be $115 a month. The insurance policy will cost $1020 a year.

9. The down payment was what percent of the price of the car?

10. Carlos will pay for the car insurance in equal monthly installments. What percent of his monthly take-home pay of $1600 is the combined loan payment and insurance payment?

Check your answers on page 328.

Lesson 12: Finding a Number When a Percent of It Is Given in Multistep Problems

Earlier in this chapter you learned that problems where you must find a number when a percent of it is given are "backwards." You have the **part**, and you are looking for the **whole**. These problems may seem more difficult when the question asks for something other than the whole. Be sure you answer the question asked.

Think carefully about the next example.

Example: Alvaro has driven 240 miles. This is 80% of his goal. How much farther does he have to drive?

Step 1. Change 80% to a decimal. $80\% = 0.8$

Step 2. To find out how much farther he has to go, first find his goal. Divide the part (distance he has driven) by the rate 0.8. His goal is 300 miles.

$$0.8 \overline{)240.0} = 300.0 \text{ mi}$$

Step 3. Subtract the distance he has driven from the total distance he has to drive. He has to drive 60 miles farther.

$$\begin{array}{r} 300 \\ -240 \\ \hline 60 \text{ miles} \end{array}$$

152   Chapter 4   Percents

Lesson 12 Exercise

Solve each problem.

1. Jack has saved $270, which is 60% of the down payment he wants to make on a car. How much more does he need?
2. The Farmers' Party of Greenport now has 900 signatures on a petition to get their candidate on the ballot for mayor. This represents 75% of the number of signatures needed. How many more signatures do they need?
3. There are 112 workers in a factory who have agreed to join a union. This is 40% of the number the organizers need. How many more workers do they need?
4. Celeste paid $102 for a coat on sale. This was 85% of the original price. How much did Celeste save by buying the coat on sale?
5. Last month 420 students in Capital City's night school program passed the GED Test. This represents 70% of the number who took the test. How many students failed?
6. Herb weighs 216 pounds, which is 90% of his weight one year ago. How much did Herb lose in a year?
7. The Greens have driven 338 miles. This is 65% of the total distance they need to drive. How much farther do they have to drive?
8. Bill's employer deducts $3465 a year for taxes and social security. This is 21% of Bill's gross salary. Find his net salary for the year.

Check your answers on page 328.

Lesson 13: Solving Percent Problems Using Ratios and Proportions

This lesson gives you a chance to use proportions to solve percent problems. Remember that a proportion is a statement that two ratios are equal.

Each part of a percent problem can fit into a proportion. The percent can be written as a ratio. For example, 25% means 25 out of 100 or $\frac{25}{100}$. You can also write a ratio comparing the part to the whole. The proportion for solving a percent problem is:

$$\frac{\text{part}}{\text{whole}} = \frac{\%}{100}$$

To find the missing number in the proportion write an equation with the cross-products and solve.

Example 1: A clock radio normally selling for $28 was on sale for 15% off. Find the sale price of the clock radio.

Step 1. Set up a proportion. The whole is $28. 15% is written as $\frac{15}{100}$. You are solving for the part.

$$\frac{p}{\$28} = \frac{15}{100}$$

Step 2. Write an equation with the cross-products. Solve the equation.

$$\$420 = 100\,p$$
$$\frac{\$420}{100} = \frac{100\,p}{100}$$
$$\$4.20 = p$$

Step 3. You are looking for the sale price. The amount taken from the original price is $4.20. Subtract $4.20 from the original price to find the sale price.

$$\begin{array}{r} \$28.00 \\ -\ 4.20 \\ \hline \$23.80 \end{array}$$

Example 2: In the office where Wei works there are 126 women and 84 men. What percent of the workers are women?

Step 1. Find the whole, the total number of workers.

$$\begin{array}{r} 126 \\ +\ 84 \\ \hline 210 \end{array}$$

Step 2. Set up a proportion. The rate is unknown. You need to compare the part, the number of women, to the whole, the total number of workers.

$$\frac{126}{210} = \frac{r}{100}$$

Step 3. Write an equation with the cross-products. Solve the equation. 60% of those working in the office are women.

$$\frac{12{,}600}{210} = \frac{210\,r}{210}$$
$$60\% = r$$

Example 3: Betsy has saved $240. This is 80% of her goal. How much more does she need to save?

Step 1. Set up a proportion. You need to solve for the whole.

$$\frac{\$240}{w} = \frac{80}{100}$$

Step 2. Write an equation with the cross-products. Solve the equation.

$$\frac{24{,}000}{80} = \frac{80w}{80}$$
$$\$300 = w$$

Step 3. Subtract the amount Betsy has saved from her goal. She needs $60 more to reach her goal.

$$\begin{array}{r} \$\ 300 \\ -240 \\ \hline \$\ 60 \end{array}$$

To solve percent of increase or percent of decrease problems, use the following proportion:

$$\frac{\text{change}}{\text{original}} = \frac{\%}{100}$$

Chapter 4 Percents

Example 4: Marcy pays $20 for each skirt that she sells for $31 in her store. Find the markup (percent of increase) on each skirt.

Step 1. Find the amount of markup.

$$\begin{array}{r} \$\ 31 \\ -20 \\ \hline \$11 \end{array}$$

Step 2. Set up a proportion. Write an equation with the cross-products.

$$\frac{\$11}{\$20} = \frac{r}{100}$$

Step 3. Solve the equation. There is a 55% markup on each skirt.

$$\$1100 = \$20\ r$$
$$55\% = r$$

Lesson 13 Exercise

Use a proportion to solve each problem.

1. On Saturday 1200 people saw the performance at Newberry Theater. On Sunday attendance was 15% less. How many people saw the Sunday performance?

2. Matt budgets $64 a month for telephone. He has $1216 left each month to cover all other expenses. What percent of his monthly income is budgeted for telephone?

3. At a doctor's office 380 patients have health insurance. This represents 80% of all the patients. How many patients do not have insurance?

4. Clark paid $104 for a suit that was on sale. This was 65% of the original price. How much did Clark save by buying the suit on sale?

5. Sean bought an electric drill for $49 and drill bits for $9. He had to pay 6.5% sales tax. Find the total price of the drill and drill bits.

6. The number of subscribers to the *Edgewood Daily* in 1990 was 12,000. Now it is 13,500. Find the percent of increase in subscribers.

Use a proportion or the percent formula to solve each problem.

7. In 1980 Gemini Express Company had 150 drivers. Today they have 120% more drivers. How many drivers do they have now?

8. Brian's insurance company paid 70% of the cost of fixing his car. If they paid $437.50 for the repairs, how much did Brian have to pay?

9. The Taylors made a down payment of $1200 on a used car selling for $8000. The down payment is what percent of the price of the car?

10. Tony has car insurance that costs $1224 a year which he pays in equal monthly installments. He also has a monthly car loan payment of $118. What percent of his monthly take-home pay of $1600 is the combined car payment and insurance?

Check your answers on page 329.

Lesson 14

Comparing with Percents

On the GED Test you may see a type of problem that takes a while to solve. In these problems you must compare four or five choices in order to find the best solution. These problems usually include some percent calculations in which you must find the percent of a number.

The next exercise gives you a chance to try this type of problem. Be sure to work through each choice before you make your decision.

Lesson 14 Exercise

Solve each problem.

1. Mr. and Mrs. Cooke want to rent a cottage for a month in the summer. They have looked at four similar cottages. Below are the rent schemes for each cottage. Which one is least expensive?
   A. $35 a day for 30 days.
   B. 15% of the yearly rent of $6000.
   C. $25 a day plus a non-refundable deposit of $250.
   D. $285 a week for four weeks.

2. The Parents' Association of the Fourth Street School has raised $10,000 for after-school programs. They need more money for the next six months. Which of the following plans will give them the most?
   A. 50% more than they currently have.
   B. $800 a month for six months.
   C. Three gifts of $1500 each.
   D. $250 a week for half a year.

3. The town of Eastport has a population of 25,000. Four different projections have been made about the growth of the town over the next three years. Which of the following would result in the largest growth in population?
   A. An increase of 15% over the entire three years.
   B. 1300 more people per year for three years.
   C. A 5% increase for the first year followed by an increase of 1000 people per year for each of the next two years.
   D. 80 more people per month for the next three years.

4. Ovidio wants to ship 500 pounds of freight from Boston to Cali, Colombia. Four different shippers have given him the following rates. Which is the least expensive?
   A. $300 for the first 100 pounds and $12.50 for each additional ten pounds.
   B. $100 down and $65 for every 50 pounds plus a 5% tax on the total including the down payment.

C. $150 for each 100 pounds.
D. $1.25 a pound plus a $150 handling charge.

Check your answers on page 330.

Level 3 Review

Solve each problem.

1. On a test with 56 problems Adrian did 21 problems wrong. What percent of the problems did he do right?
2. The Santanas paid $25,000 for their house in 1960. In 1990 the house was worth $85,000. By what percent did the value of the house increase from 1960 to 1990?
3. Gail paid $192 for a television that was on sale. The sale price was 80% of the original price. How much did Gail save by buying the television on sale?
4. Chris is spending his vacation hiking. So far he has walked 54 miles, which is 75% of his goal. How much farther does he have to walk?
5. A shirt that originally cost $24 was on sale for $20.40. Find the percent of decrease in the price of the shirt.
6. Mike has saved $390 toward a down payment on a car. He still needs $260. What percent of the down payment has he saved so far?
7. Janet weighs 136 pounds, which is 85% of her weight two years ago. How much weight has she lost in two years?
8. At the Civic Auditorium on amateur night there were 675 occupied seats. This represents 75% of the capacity of the auditorium. How many seats were empty on amateur night?

Use the next passage to answer questions 9–12.

Gambol's Department Store had a three-month clearance sale when they went out of business. In July every item in the store was on sale for 15% off. In August every item was on sale for 25% off. For the first three weeks of September, every item was on sale for 50% off. Then, during the last week of September, every item was on sale for 80% off. Bill and Caroline have been thinking of buying a carpet originally selling at Gambol's for $700 and a stereo set originally selling for $400.

9. What was the price of the stereo at Gambol's in July?
10. What was the price of the carpet at Gambol's in August?
11. Find the difference in the price of the carpet between August and early September.
12. What was the combined price of the stereo and the carpet, including 5% sales tax, during the last week of September?

13. Danny has a job selling farm equipment. He gets a combination of a $20,000 base salary and 6% commissions. How much will he make altogether in a year if he sells $100,000 worth of farm equipment?

14. Randolph and Shirley want to buy a set of furniture listed for $600. They have a choice of four payment plans. Which of these plans will be the least expensive?
    A. $33\frac{1}{3}$% down and $40 a month for a year.
    B. $250 every four months for a year.
    C. $\frac{1}{4}$ down and $50 a month for ten months.
    D. $45 a month for one year and six months.

15. Arturo now makes $1200 a month. He has four choices for raises. Which of the following will give him the greatest yearly raise?
    A. $40 a week more.
    B. A raise of $2000 for the year.
    C. A 10% raise.
    D. $550 more each quarter.

Check your answers on page 330. If you have all 15 problems correct, go on to the next chapter. If you answered a question incorrectly, find the item number on the chart below and review that lesson before you go on.

Review:	If you missed item number:
Lesson 10	9, 10, 11, 12, 13
Lesson 11	1, 2, 5, 6
Lesson 12	3, 4, 7, 8
Lesson 13	1–13
Lesson 14	14, 15

Chapter 5
GRAPHS

A **graph** is a diagram that compares the relative sizes of numbers. You often see graphs in newspapers and magazines. In this chapter you will learn to read and interpret pictographs, circle graphs, bar graphs, and line graphs.

The questions based on graphs on the GED Test range from locating information on a graph to drawing conclusions from the information shown on a graph.

This chapter is not divided into levels. Work through the entire chapter to get practice with the types of graphs you may see on the GED Test.

Lesson 1: Pictographs

A **pictograph** uses symbols to represent numbers. On every pictograph a **key** or **legend** explains what each symbol stands for.

To understand a graph, first read the title. Then read the labels for the columns and rows of information.

The title of the pictograph below is "Pigs Raised in One Year." The first column contains state names. The legend at the end of the graph tells that each symbol represents 1,000,000 pigs.

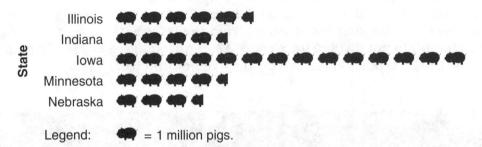

Example 1: How many pigs were produced in Nebraska in one year?

Count the number of symbols to the right of the label *Nebraska*. There are $3\frac{1}{2}$ symbols to the right of Nebraska. Since each symbol stands for one million pigs, multiply $3\frac{1}{2}$ by 1 million.

$3\frac{1}{2} \times 1$ million

$= 3\frac{1}{2}$ million

Some questions give information that you must apply to the numbers on the graphs.

Example 2: At a value of $200 a pig, find the total value of the pigs produced in Nebraska in one year.

Multiply the number of pigs produced in one year in Nebraska by the value of one pig.

```
  3.5 million
× 200
-----------
700.0 = $700 million
```

Lesson 1 Exercise

Use the graph about pig production to answer questions 1–6.

1. Tell the number of pigs produced in one year in each of the following states.
   a. Illinois   b. Indiana   c. Iowa   d. Minnesota
2. Which state shown on the graph produces the most pigs in one year?

3. The states of Minnesota and Nebraska together produce how many pigs in one year?
4. The combined production of pigs in the states of Illinois, Indiana, and Minnesota is which of the following?
   (1) about the same as the production in Nebraska
   (2) slightly more than the production in Iowa
   (3) about one-third of the production in the U.S.
   (4) less than half the production in Iowa
5. The number of pigs produced in a year in Nebraska is what percent of the number produced in Iowa?
6. At a value of $175 per pig, what is the total value of the pigs produced in Minnesota in one year?

Use the following graph to answer questions 7–15.

CORN PRODUCTION AND EXPORTS IN THE UNITED STATES

Year	Production	Exports
1970	4 ears	1 ear
1975	6 ears	2 ears
1980	6½ ears	3 ears
1985	8 ears	1 ear
1990	8 ears	2 ears

Legend: 🌽 = 1 billion bushels.

7. What does the symbol of an ear of corn represent?
8. Tell the total amount of corn produced in the following years.
   a. 1970   b. 1980   c. 1985   d. 1990
9. Tell the total amount of corn exported in the following years.
   a. 1970   b. 1980   c. 1985   d. 1990
10. For which year shown on the graph was production the highest?
11. For which year shown on the graph were exports the highest?

Lesson 1: Pictographs   161

12. In 1970 exports were what fraction of production?

13. The 1980 production of corn was how much greater than the 1970 production?

14. The amount of corn produced in 1970 was what percent of the amount produced in 1990?

15. For what year shown on the graph was the export of corn the highest percentage of the amount produced?

Check your answers on page 332.

Lesson 2

Circle Graphs

A **circle graph** shows the parts of a whole. A pie-shaped piece of a circle graph usually stands for the percent of a whole. It could also represent the number of cents that are part of a whole dollar.

When the pieces of a circle graph are expressed as percents, the pieces total 100%. When the pieces of a circle graph are measured in cents, the pieces total one dollar.

To understand a circle graph, read the title and the names of each category. The circle graph below shows how a dollar of the federal budget is spent in a year.

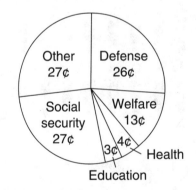

HOW A DOLLAR OF
THE FEDERAL BUDGET IS SPENT

Example: What percent of federal expenditures goes for health?

Notice that all the parts of the circle graph add up to $1.00. 4¢ of each dollar goes to health. 4¢ is the same as $\frac{4}{100}$ or 4%.

4%

162   Chapter 5   Graphs

Lesson 2 Exercise

Use the circle graph on page 162 to answer the questions 1–6.

1. What percent of federal expenditures goes for each of the following categories?
   a. defense   b. welfare   c. social security   d. education
2. Which category shown on the graph represents the smallest federal expenditure?
3. The health category makes up what fraction of the total yearly federal expenditures?
4. Together, health, education, and welfare make up what fraction of the total federal expenditures?
5. Together, defense and social security make up which of the following parts of the total federal expenditures in a year?
   (1) a little over $\frac{1}{4}$
   (2) a little over $\frac{1}{2}$
   (3) a little over 80%
   (4) approximately 95%
6. The total federal expenditures for a year were $960 billion. How much was spent on health?

Use the following two circle graphs to answer questions 7–13.

MARITAL STATUS OF PERSONS OVER 18

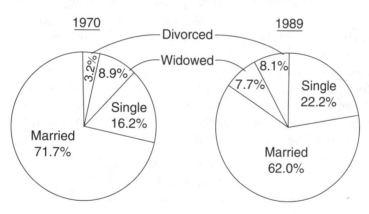

7. The graph on the left tells the marital status of persons over 18 for what year?
8. For persons over 18 tell each of the following:
   a. the percent who were married in 1970.
   b. the percent who were married in 1989.
   c. the percent who were divorced in 1970.
   d. the percent who were divorced in 1989.
9. For 1970 which category represents the fewest people?
10. By how many percentage points did the number of adults who were divorced change from 1970 to 1989?

Lesson 2: Circle Graphs

11. In 1970 the people who were either widowed or single made up about what fraction of the adults over 18?

    (1) $\frac{1}{6}$ (2) $\frac{1}{4}$ (3) $\frac{1}{3}$ (4) $\frac{1}{2}$

12. The population of Capital City in 1989 was 300,000. If Capital City follows the national pattern, how many married adults were in Capital City in 1989?

13. Which of the following describes the change from 1970 to 1989 as shown by the two graphs?

    (1) The percent of adults who were married or widowed decreased, and the percent who were single or divorced increased.
    (2) The percent of adults who were married or divorced increased, and the percent who were single or widowed decreased.
    (3) The percent of each category increased.
    (4) The percent of adults who were married or widowed increased, and the percent who were single or divorced decreased.

Check your answers on page 332.

Lesson 3: Bar Graphs

A **bar graph** uses thick lines or bars to compare the sizes of numbers. The size of each number being compared corresponds to the length of its bar. The bars may be vertical (up and down) or horizontal (from left to right).

To estimate the number each bar stands for, you must use the **scale** along the side or across the bottom of the bar graph.

Read the title and the categories on each bar graph carefully.

The bar graph below shows the average price of a semiprivate hospital room for six different years. The vertical scale at the left of the graph shows the price in dollars. The horizontal scale across the bottom of the graph tells the different years. Notice that the bars are vertical.

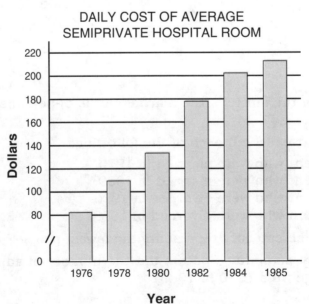

Example: Which of the following was the average cost of a semiprivate hospital room in 1978?

        (1) $203
        (2) $148
        (3) $109
        (4) $ 78

Choice (3) $109 is correct. Find 1978 on the scale at the bottom of the graph. Follow the bar above 1978 to the top and look directly to the left on the vertical scale. The bar above 1978 ends about half-way between $100 and $120 on the vertical scale. $109 is the closest answer.

Lesson 3 Exercise

Use the bar graph on page 164 to answer the following questions.

1. Which of the following tells the average cost of a semiprivate hospital room in 1976?
   - **(1)** $78
   - **(2)** $83
   - **(3)** $88
   - **(4)** $98

2. Which of the following tells the average cost of a semiprivate hospital room in 1982?
   - **(1)** $148
   - **(2)** $158
   - **(3)** $168
   - **(4)** $178

3. During which year was the average cost of a semiprivate hospital room $203?
   - **(1)** 1980
   - **(2)** 1982
   - **(3)** 1984
   - **(4)** 1985

4. Which of the following is closest to the difference in price for a semiprivate hospital room in 1980 and in 1985?
   - **(1)** $50
   - **(2)** $60
   - **(3)** $70
   - **(4)** $80

5. The average price of a semiprivate hospital room in 1978 was about what fraction of the price of a semiprivate room in 1985?
   - **(1)** $\frac{1}{2}$
   - **(2)** $\frac{1}{3}$
   - **(3)** $\frac{1}{4}$
   - **(4)** $\frac{2}{3}$

6. The average price of a room in 1984 was about how many times that of the average price of a room in 1976?
   - **(1)** $1\frac{1}{2}$ times
   - **(2)** $2\frac{1}{2}$ times
   - **(3)** 3 times
   - **(4)** 4 times

7. During which two years shown on the graph did the average price of a semiprivate hospital room rise the most?
   - **(1)** 1976 to 1978
   - **(2)** 1978 to 1980
   - **(3)** 1980 to 1982
   - **(4)** 1982 to 1984

8. Sam was in the hospital for eight days in 1982. If the price of his semiprivate room followed the national average, which of the following is closest to the cost of the room for his total stay in the hospital?

   (1) $1600   (3) $1100
   (2) $1400   (4) $ 900

Check your answers on page 333.

Lesson 4: Divided-Bar Graphs

A **divided-bar graph** separates the amounts represented by each bar into parts. The divided-bar graph below shows the number of new houses sold in various parts of the United States during four different years. Notice that the bars in this graph are horizontal.

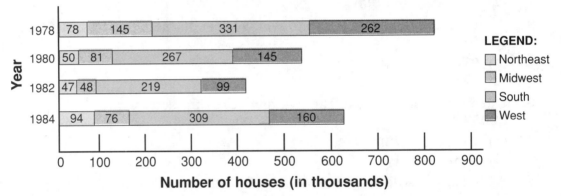

Example: Find the total number of new houses sold in the United States in 1980.

The total for 1980 lies between 500 thousand and 600 thousand. To get the exact total, add the subtotals for each region in 1980.

```
   50
   81
  267
+ 145
  543 thousand
```

166   Chapter 5: Graphs

Lesson 4 Exercise

Use the divided-bar graph on page 166 to answer the following questions.

1. Tell the number of new houses that were sold in each of the following categories.
   a. the Midwest in 1978
   b. the South in 1980
   c. the Northeast in 1982
   d. the West in 1984
2. What was the total number of new houses sold in the United States in 1984?
3. For which year shown on the graph was the number of new houses sold the greatest?
4. For which year shown on the graph was the number of new houses sold in the Northeast the greatest?
5. For which year shown on the graph was the number of new houses sold in the West the smallest?
6. The total number of houses sold in the United States in 1982 was about what fraction of the total number sold in 1978?
   (1) $\frac{1}{5}$ (3) $\frac{1}{3}$
   (2) $\frac{1}{4}$ (4) $\frac{1}{2}$
7. For which year shown on the graph was the number of new houses sold in the Northeast more than the number sold in the Midwest?
8. In 1984 the number of new houses sold in the South was about what percent of the total number of new houses sold?
   (1) 75% (3) 25%
   (2) 50% (4) 10%
9. For the years shown on the graph, in which region was the number of new houses sold consistently the highest?

Check your answers on page 333.

Lesson 5

Line Graphs

A **line graph** is a way to show changing amounts over a period of time. The vertical scale on the left can be in almost any unit of measurement—dollars, numbers of people, percent, pounds, and so on. The horizontal scale across the bottom usually shows units of time.

A line that **rises** from left to right shows an increasing or upward **trend**. A line that **falls** from left to right shows a decreasing or downward trend.

To understand a line graph, read the title and the labels on both scales carefully.

The line graph below shows the unemployment rate in Capital City from 1975 to 1990. Notice that the vertical scale is measured in percent. The horizontal scale shows the year.

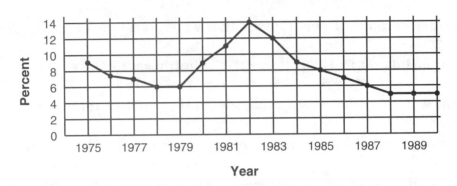

Example: What was the unemployment rate in Capital City in 1981?

Step 1. Find 1981 on the horizontal scale at the bottom of the graph. Then follow the line marked 1981 straight up until you reach the changing line.   11%

Step 2. Look straight across to the left to the vertical scale. You should be half-way between 10% and 12%. The unemployment rate in 1981 was 11%.

Lesson 5 Exercise

Use the line graph above to answer questions 1–9.

1. What was the unemployment rate in Capital City in:
   a. 1975?   b. 1982?   c. 1985?   d. 1990?
2. For what year shown on the graph did the unemployment rate in Capital City first get as low as 6%?
3. For which year on the graph was the unemployment rate highest?
4. What is the difference in percentage points between the lowest and the highest unemployment rates shown on the graph?
5. The 1979 unemployment rate was what fraction of the 1983 rate?
6. By how many percentage points did the unemployment rate in Capital City drop from 1982 to 1985?
7. The labor force in Capital City in 1985 was 200,000 people. How many of those people were unemployed that year?
8. For which of the periods shown below did the unemployment rate show a constant upward trend?
   (1) 1975 to 1979   (3) 1979 to 1982
   (2) 1977 to 1981   (4) 1981 to 1984

9. For what three consecutive years was the unemployment rate the same?
   (1) 1977, 1978, 1979
   (2) 1979, 1980, 1981
   (3) 1985, 1986, 1987
   (4) 1988, 1989, 1990

The next line graph compares farm income to expenses from 1974 to 1984. Notice that there are two lines. One shows the changing income, and the other shows the changing expenses. Use this graph to answer questions 10–17.

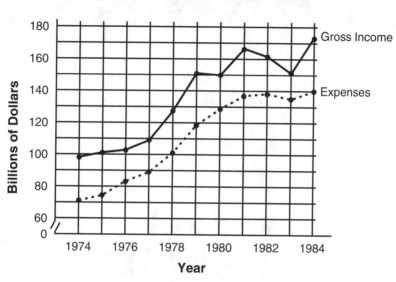

Source: U.S. Department of Agriculture.

10. What is the unit of measurement of the vertical scale on the graph?

11. What was the gross income from farms in 1978?
    (1) $128 billion
    (2) $120 billion
    (3) $112 billion
    (4) $108 billion

12. What were the farm expenses in 1979?
    (1) $ 89 billion
    (2) $ 99 billion
    (3) $109 billion
    (4) $119 billion

13. What were the farm expenses in 1983?
    (1) $145 billion
    (2) $135 billion
    (3) $125 billion
    (4) $115 billion

14. The net income for farms is the difference between the gross income and the expenses. Approximately what was the net income for farms in 1975?
    (1) $40 billion
    (2) $25 billion
    (3) $10 billion
    (4) $ 5 billion

15. What was the net income for farms in 1981?
    (1) $10 billion
    (2) $15 billion
    (3) $20 billion
    (4) $30 billion

16. In what year did the gross income for farms first reach $150 billion?
    (1) 1979
    (2) 1981
    (3) 1983
    (4) 1984

Lesson 5: Line Graphs   169

17. Between what two consecutive years did the expenses on farms drop?
    (1) 1976–1977   (3) 1982–1983
    (2) 1980–1981   (4) 1983–1984

Check your answers on page 333.

Review

Solve each problem.
Use the following pictograph to answer questions 1–3.

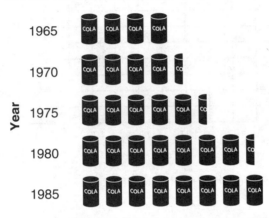

PER CAPITA CONSUMPTION OF SOFT DRINKS

Legend: = 5 gallons of soft drinks.

1. What was the per capita consumption of soft drinks in 1970?
   (1) $22\frac{1}{2}$ gallons
   (2) $27\frac{1}{2}$ gallons
   (3) $32\frac{1}{2}$ gallons
   (4) $37\frac{1}{2}$ gallons

2. The per capita consumption of soft drinks in 1985 was how many times that of the per capita consumption in 1965?
   (1) 8 times
   (2) 4 times
   (3) 2 times
   (4) $1\frac{1}{2}$ times

3. The 1980 per capita consumption of soft drinks was how much greater than the per capita consumption in 1975?
   (1) 20 gallons
   (2) 15 gallons
   (3) 12 gallons
   (4) 10 gallons

Chapter 5: Graphs

Use the following circle graph to answer questions 4–7.

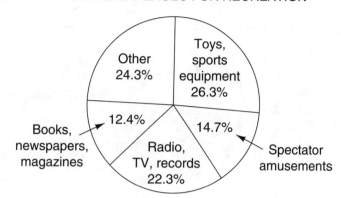

PERSONAL EXPENSES FOR RECREATION

4. What is the difference in percentage points between the amount spent on toys and sports equipment and the amount spent on books, magazines, and newspapers?
   (1) 15.9%
   (2) 13.9%
   (3) 11.9%
   (4) 9.9%

5. The amount spent on spectator amusements and the amount spent on books, magazines, and newspapers together make up about what fraction of the total expenditures for recreation?
   (1) a little over $\frac{1}{2}$
   (2) a little over $\frac{1}{3}$
   (3) a little over $\frac{1}{4}$
   (4) a little over $\frac{1}{5}$

6. The amount spent on toys and sports equipment and the amount spent on the category called "other" together make up about what percent of the total?
   (1) 20%
   (2) 30%
   (3) 40%
   (4) 50%

7. The Jones family spends $3000 a year on recreation. If they spend at the same rate as the national average, how much do they spend each year on books, magazines, and newspapers?
   (1) $372
   (2) $441
   (3) $669
   (4) $744

Review 171

Use the bar graph below to answer questions 8–10.

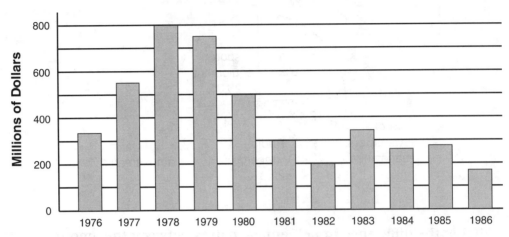

8. The amount spent in 1982 on parks and preserves was about what fraction of the amount spent in 1978?
   (1) $\frac{1}{8}$
   (2) $\frac{1}{4}$
   (3) $\frac{3}{8}$
   (4) $\frac{3}{4}$

9. Between what two consecutive years shown on the graph was there the largest drop in expenditures on parks and preserves?
   (1) 1979–1980
   (2) 1980–1981
   (3) 1981–1982
   (4) 1983–1984

10. For which period shown below did the yearly amount spent on parks and preserves rise each year?
    (1) 1984–1986
    (2) 1982–1984
    (3) 1978–1981
    (4) 1976–1978

Use the following divided-bar graph to answer questions 11 and 12.

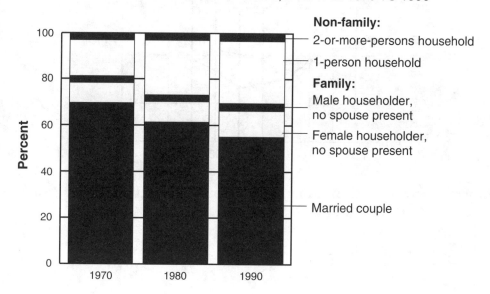

11. Which category showed a steady percent decrease from 1970 to 1990?
    (1) non-family 2-or-more-persons household
    (2) non-family 1-person household
    (3) female householder, no spouse present
    (4) married couple

12. In 1990 1-person non-family households were approximately what fraction of total households?
    (1) $\frac{1}{10}$
    (2) $\frac{1}{4}$
    (3) $\frac{1}{2}$
    (4) $\frac{2}{5}$

Use the following line graph to answer questions 13–15.

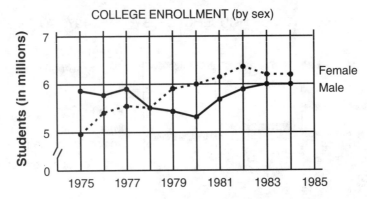

13. In what year was the number of female college students the same as the number of male college students?

   (1) 1978
   (2) 1980
   (3) 1982
   (4) 1984

14. In 1980 the number of female college students was about how much greater than the number of male college students?

   (1) a little over 1 million
   (2) a little over 500,000
   (3) a little over 100,000
   (4) a little over 50,000

15. Which of the following best describes college enrollment from 1975 to 1980?

   (1) Both the number of male students and the number of female students decreased.
   (2) The number of male students and the number of female students increased gradually.
   (3) The number of female students increased steadily and the number of male students gradually decreased.
   (4) The number of male students decreased and the number of female students decreased more slowly.

Check your answers on page 333. If you have all 15 problems correct, go on to the Review Test. If you answered a question incorrectly, find the item number on the chart below and review that lesson before you go on.

Review:	If you missed item number:
Lesson 1	1, 2, 3
Lesson 2	4, 5, 6, 7
Lesson 3	8, 9, 10
Lesson 4	11, 12
Lesson 5	13, 14, 15

REVIEW TEST

This exercise has problems similar to the problems you will see on the GED Mathematics Test. The problems are all based on material you have studied so far in this book.

Choose the correct answer to each problem. Use any formulas on pages 17 and 18 that you need.

1. Herb bought a suit for $128. Find the cost of the suit including 6.5% sales tax.
   (1) $121.50
   (2) $134.50
   (3) $136.32
   (4) $141.00
   (5) Not enough information is given.

2. Donna weighs 120 pounds and her husband Steve weighs 180 pounds. Donna's weight is what percent of Steve's weight?
   (1) 40%
   (2) $66\frac{2}{3}$%
   (3) 75%
   (4) 80%
   (5) $83\frac{1}{3}$%

3. Find the volume in cubic inches of a cube with a side of 9 inches.
   (1) 27
   (2) 270
   (3) 729
   (4) 810
   (5) 900

4. Of the 150 members in the First Street Tenants' Association, 85 voted to go on a rent strike. What is the ratio of the association members who voted to strike to the number of members who voted not to strike?
   (1) 13:30
   (2) 17:13
   (3) 1:85
   (4) 17:30
   (5) 85:1

5. A player's batting average is the number of hits divided by the number of times the player is at bat. Andrew got 23 hits in 75 times at bat. Find his batting average to the nearest thousandth.
   (1) 0.230
   (2) 0.299
   (3) 0.307
   (4) 0.350
   (5) 0.750

6. Which of the following is equal to $(\frac{2}{3})^3$?
   (1) 2
   (2) $\frac{4}{9}$
   (3) $\frac{2}{3}$
   (4) $\frac{8}{9}$
   (5) $\frac{8}{27}$

7. The three highest mountains in Alaska are McKinley (20,320 ft), St. Elias (18,008 ft), and Foraker (17,400 ft). Find the average (mean) height of these three mountains.
   (1) 17,704 ft
   (2) 18,008 ft
   (3) 18,576 ft
   (4) 18,860 ft
   (5) 19,164 ft

Review Test 175

8. The ratio of flour to sugar in a certain recipe is 5:2. If a cook uses 8 cups of flour, how many cups of sugar should be used?
   (1) $3\frac{1}{5}$
   (2) 10
   (3) $12\frac{1}{2}$
   (4) 15
   (5) 20

9. Lois is a real estate broker. In April she sold a house for $62,400. In May she sold another house for $76,600, and in June she sold a house for $59,300. Find the average (mean) price of the houses she sold from April through June.
   (1) $50,400
   (2) $59,300
   (3) $60,800
   (4) $62,400
   (5) $66,100

10. Simplify the expression $14^2 - 8^0 + 16^1$.
    (1) 172
    (2) 179
    (3) 204
    (4) 211
    (5) 213

Use the following information to answer questions 11 and 12.

Three families bought raffle tickets. The Johnsons bought 8 tickets, the Millers bought 12, and the Smiths bought 10. Altogether 500 raffle tickets were sold.

11. What is the probability that the Smiths will win the raffle?
    (1) $\frac{1}{10}$
    (2) $\frac{1}{3}$
    (3) $\frac{1}{2}$
    (4) $\frac{1}{50}$
    (5) $\frac{3}{500}$

12. What is the chance that one of the three families will win the raffle?
    (1) $\frac{1}{30}$
    (2) $\frac{3}{50}$
    (3) $\frac{1}{3}$
    (4) $\frac{3}{500}$
    (5) $\frac{3}{25}$

13. The expression 9×10^4 is the same as which of the following?
    (1) $9 \times 10 \times 4$
    (2) $10 \times 9 \times 9 \times 9 \times 9$
    (3) $9 \times 4 \times 4 \times 4 \times 4$
    (4) $9 \times 10 \times 10 \times 10 \times 10$
    (5) $4 \times 9 \times 9 \times 9 \times 9$

Item 14 refers to the following diagram.

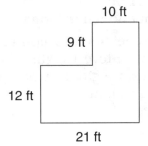

14. Find the total area in square feet of this room.
    (1) 84
    (2) 309
    (3) 318
    (4) 342
    (5) 441

15. Find the circumference of a circle whose diameter measures 1.5 m.
    (1) 0.75 m
    (2) 2.355 m
    (3) 3.0 m
    (4) 4.71 m
    (5) 9.42 m

Items 16 and 17 refer to the following table.

Per Capita Expenditures in a Year

State	Education	Welfare	Health
New York	$500	$392	$153
Illinois	371	278	69
California	553	397	90

Source: U.S. Bureau of the Census.

16. How much more does California spend per capita for education than Illinois?
    (1) $212
    (2) $192
    (3) $182
    (4) $153
    (5) $ 53

17. How much more does New York spend per capita on education, welfare, and health than Illinois?
    (1) $129
    (2) $198
    (3) $229
    (4) $243
    (5) $327

18. What is the distance from point X to point Y on this 5-centimeter scale?

    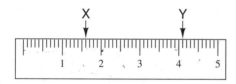

    (1) 2.5 cm
    (2) 2.7 cm
    (3) 2.9 cm
    (4) 3.5 cm
    (5) 3.7 cm

Use the passage below to answer questions 19–21.

On Monday morning Mike's odometer (mileage gauge) read 38,526. He then drove for three days in the city where he lives and used a total of 16 gallons of gasoline. On Wednesday night his odometer read 38,734. On Thursday and Friday he drove in the country and used another 14 gallons of gasoline. His odometer read 39,028 on Friday night.

19. How many miles did Mike get on a gallon of gasoline for the three days he drove in the city?
    (1) 11
    (2) 12
    (3) 13
    (4) 15
    (5) 20

20. How many miles did Mike get on a gallon of gasoline for the two days he drove in the country?
    (1) 18
    (2) 21
    (3) 24
    (4) 25
    (5) 26

21. Mike paid $1.15 a gallon for the gasoline he used. Altogether how much did he spend on gasoline for those five days?
    (1) $18.40
    (2) $24.15
    (3) $33.35
    (4) $34.50
    (5) $35.40

22. Sharon made $8.50 an hour for 35 hours and then $12.75 an hour for 6 hours of overtime. Which expression tells the total amount she made in dollars?
    (1) 41 × 8.50
    (2) 35 × 12.75 + 6 × 8.50
    (3) 35(12.75 + 8.50)
    (4) 35 × 8.50 + 6 × 12.75
    (5) 41(8.50 + 12.75)

23. What is the area in square inches of a square with a side that measures $4\frac{1}{2}$ inches?
    (1) $24\frac{1}{4}$
    (2) $20\frac{1}{4}$
    (3) 18
    (4) $16\frac{1}{4}$
    (5) 9

Review Test   177

24. What is the volume in cubic centimeters of this rectangular container?

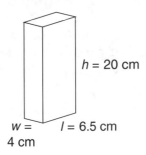

(1) 46
(2) 134
(3) 460
(4) 502
(5) 520

25. Arrange the following lengths of metal tubing in order from <u>shortest</u> to <u>longest</u>.

   Tube A is 0.85 m long.
   Tube B is 0.095 m long.
   Tube C is 1.2 m long.
   Tube D is 0.9 m long.
   Tube E is 1.07 m long.

(1) E, B, C, A, D
(2) B, A, D, E, C
(3) D, A, C, B, E
(4) C, E, B, A, D
(5) C, A, E, B, D

Items 26 and 27 refer to the following table.

Turkeys Raised in a Year (in millions)

Arkansas	12.9
California	20.2
Minnesota	27.0
Missouri	13.0
North Carolina	29.4
Virginia	11.4

Source: U.S. Department of Agriculture.

26. Which state shown on the table raised the <u>third</u> largest number of turkeys?

(1) California
(2) Minnesota
(3) Missouri
(4) North Carolina
(5) Virginia

27. The total number of turkeys raised in the U.S. for the year shown in the table was 170.7 million. The four states that raised the most turkeys raised about what fraction of the national total?

(1) about $\frac{1}{10}$
(2) about $\frac{1}{4}$
(3) less than $\frac{1}{3}$
(4) over $\frac{1}{2}$
(5) Not enough information is given.

Items 28 and 29 are based on the following information.

A box contains 15 green tennis balls, 12 orange tennis balls, and 9 white tennis balls.

28. What is the probability that the first ball someone picks from the box will be green?

(1) $\frac{2}{5}$
(2) $\frac{3}{5}$
(3) $\frac{4}{5}$
(4) $\frac{5}{12}$
(5) $\frac{7}{12}$

29. The first tennis ball Alan took from the box was orange, and the next two he took were green. What is the probability that the fourth tennis ball he takes from those remaining will be white?

(1) $\frac{9}{11}$
(2) $\frac{8}{11}$
(3) $\frac{3}{11}$
(4) $\frac{2}{3}$
(5) $\frac{1}{3}$

30. Carmela pays $20 for each pair of shoes she sells in her store. She charges her customer $29 for the shoes. Find her percent of markup.
    (1) 20%
    (2) 25%
    (3) 30%
    (4) $33\frac{1}{3}$%
    (5) 45%

31. In the Central High School adult education program there are 165 women and 135 men. What percent of the students are women?
    (1) 35%
    (2) 45%
    (3) 55%
    (4) 65%
    (5) 75%

32. For a circus performance the ratio of the number of child's tickets sold to the number of adults' tickets was 5:2. 630 child's tickets were sold. Find the total number of tickets sold.
    (1) 882
    (2) 1260
    (3) 1470
    (4) 1890
    (5) 2205

33. Faye types an average of 50 words per minute. Which of the following tells the number of minutes Faye will need to type a 6-page report if each page has an average of 380 words?
    (1) $6 \times 380 \times 50$
    (2) $\frac{6 \times 380}{50}$
    (3) $\frac{380}{6 \times 50}$
    (4) $\frac{6 \times 50}{380}$
    (5) $\frac{50 \times 380}{6}$

34. Jeff is selling an old car for which he paid $4500. Which of the following offers is the best?
    (1) $50 a month for two years.
    (2) $500 down and $60 a month for nine months.
    (3) 30% of the amount Jeff paid for the car.
    (4) $30 a week for 44 weeks.
    (5) $400 down and $80 a month for eight months.

35. John rode his motorcycle 304 miles in $6\frac{1}{2}$ hours on $9\frac{1}{2}$ gallons of gasoline. What average number of miles did he ride on one gallon of gasoline?
    (1) 16
    (2) 19
    (3) 32
    (4) 48
    (5) 64

Items 36–39 refer to the circle graph.

AGE DISTRIBUTION IN THE UNITED STATES (IN YEARS)

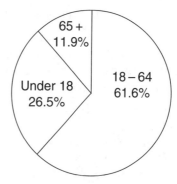

36. The category "under 18" is about what fraction of the total?
    (1) $\frac{1}{3}$
    (2) $\frac{1}{4}$
    (3) $\frac{1}{2}$
    (4) $\frac{2}{3}$
    (5) $\frac{3}{5}$

37. The combined percent of the "over 65" and "under 18" groups is how many percentage points less than the "18–64 yrs" group?
    (1) 49.7%
    (2) 47.0%
    (3) 35.1%
    (4) 23.2%
    (5) 14.6%

38. People from ages 18 to 40 represent about what fraction of the total?
    (1) $\frac{4}{5}$
    (2) $\frac{3}{5}$
    (3) $\frac{1}{3}$
    (4) $\frac{1}{4}$
    (5) Not enough information is given.

39. The population of Capital City is 300,000. If Capital City follows the national average, how many people in Capital City are under 18?
    (1) 184,800
    (2) 92,400
    (3) 79,500
    (4) 35,700
    (5) Not enough information is given.

40. Which of the following expresses the total amount Joe paid for a car if he put $600 down and then paid $80 a month for five years?
    (1) 600 + 60 × 80
    (2) 12 × 600 × 80
    (3) 5 × 500 × 80
    (4) 60(600 + 80)
    (5) 60 × 80 × 600

41. Bianca works two days each week and earns a total of $150. How many weeks does she have to work in order to earn $3000?
    (1) 10
    (2) 12.5
    (3) 17.5
    (4) 20
    (5) Not enough information is given.

Items 42–45 refer to the following situation.

In Capital City, which has a population of 300,000, the estimated number of jobs is 120,000. Elk Electronics plans to build a new factory in Capital City. This will increase the number of jobs by 5%. The new jobs will become available gradually over a three-year period. The first third of the jobs will be open at the end of a year, the next third at the end of two years, and the last third at the end of three years. In addition, Paulson's Plastics plans to open a new shop in one year. Paulson's will increase the current number of jobs by 2%.

42. How many new jobs will be available at Elk Electronics at the end of one year?
    (1) 1200
    (2) 2000
    (3) 2400
    (4) 4000
    (5) 6000

43. By about what percent will the population of Capital City increase two years from now?
    (1) 2%
    (2) 5%
    (3) 7%
    (4) 10%
    (5) Not enough information is given.

44. Assuming no other change, what will be the total number of jobs in Capital City at the end of two years if both Elk Electronics and Paulson's Plastics complete their plans?
    (1) 124,000
    (2) 124,400
    (3) 126,400
    (4) 128,400
    (5) Not enough information is given.

45. Suppose that after one year Elk Electronics sticks to their plan, Paulson's decides not to build a shop, and a bottling plant closes and lays off 700 people. What will be the net change in the current number of jobs in the city caused by these events?
    (1) 2000 more
    (2) 1300 more
    (3) 700 more
    (4) 300 fewer
    (5) Not enough information is given.

46. Nick can paint an average of three rooms a day. Which of the following expresses the number of work days Nick needs to paint the interior of 20 houses if each house has an average of six rooms?

   (1) $3(20 + 6)$
   (2) $\frac{20 + 6}{3}$
   (3) $20 \times 6 \times 3$
   (4) $\frac{3}{20 \times 6}$
   (5) $\frac{20 \times 6}{3}$

47. Janet weighs 160 pounds, which is 80% of what she weighed one year ago. How many pounds has she lost?

   (1) 20
   (2) 25
   (3) 30
   (4) 40
   (5) 60

48. The ratio of wins to losses for Mike's softball team last season was 5:2. The team played a total of 49 games. How many did they win, and how many did they lose?

   (1) won 40, lost 9
   (2) won 35, lost 14
   (3) won 30, lost 19
   (4) won 25, lost 24
   (5) won 33, lost 16

Items 49 and 50 refer to the following chart.

Postal Rates

First Class Letters
1 oz $0.29
Each additional oz 0.23

Certified Mail
(In addition to postage) 1.00

49. Find the cost of sending a first class letter that weighs 4 oz.

   (1) $0.75
   (2) $0.92
   (3) $0.98
   (4) $1.10
   (5) $1.16

50. Find the cost of sending a two-ounce first class letter by certified mail.

   (1) $1.29
   (2) $1.52
   (3) $1.58
   (4) $1.62
   (5) $1.68

Check your answers on page 334. If you have all 50 problems correct, go on to the next chapter. If you answered a question incorrectly, use the following guide to choose which lessons to review.

Review Test Guide

Below is a chart of the items from the Review Test divided into the first five chapters of this book. Find the item number you answered incorrectly on the chart. Next to it you will find the lesson number where you can review similar problems. For example, if you missed item 18, review Chapter 2, Lesson 8.

Chapter 1		Chapter 2		Chapter 3	
Item	Lesson	Item	Lesson	Item	Lesson
3	11	5	13	4	14
7	6	15	10	6	12
9	6	18	8	8	17
10	7	24	11	11	16
13	7	25	5	12	16
14	15	26	12	23	13
16	16	27	12	28	16
17	16	49	12	29	16
19	17	50	12	32	18
20	17			35	19
21	17			48	18
22	18				
33	18				
40	18				
41	13				
46	18				

Chapter 4		Chapter 5	
Item	Lesson	Item	Lesson
1	10	36	2
2	7	37	2
30	11	38	2
31	11	39	2
34	14		
42	10		
43	11		
44	10		
45	10		
47	12		

Chapter 6
ALGEBRA

Algebra is a branch of mathematics that builds on the basic skills of arithmetic. Algebra uses letters to represent numbers, like those in the formulas you have already used in this book.

To find out whether you need to work through this section, try the following preview.

Preview

Solve each problem.

1. Which point on the number line below corresponds to the value 2.25?

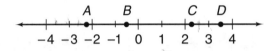

2. $(12) + (-9) + (-15) =$
3. $(-14) - (-18) =$
4. $(8)(-5)(-3) =$
5. $\dfrac{-108}{-12} =$
6. What is the value of $a + c$ when $a = -15$ and $c = -30$?
7. Find the value of xy when $x = 10$ and $y = -0.4$.
8. $(-5y) + (-3y) + (+y) =$
9. $(11w) - (-4w) =$
10. $(3ab)(-4a) =$
11. $\dfrac{-4x^2y^2}{-8xy} =$
12. Find the value of $3m - mn$ if $m = -5$ and $n = -2$.

Check your answers on page 336. If you have all 12 problems correct, do the Level 1 Review on page 200. If you have answered a problem incorrectly, study Level 1 beginning with Lesson 1.

Level 1: Algebra Skills

In this section you will learn to add, subtract, multiply, and divide signed numbers. You will substitute signed numbers into algebraic expressions. You will also learn to perform these four operations with the algebraic expressions called monomials.

Lesson 1: The Number Line

Whole numbers, decimals, and fractions in arithmetic are called **positive numbers**. The value of a positive number is greater than zero. The **number line** below represents all positive numbers.

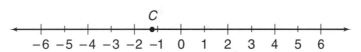

The arrow at the right end of the number line means that the numbers go on and on. Point A is about half-way between 2 and 3. Point A represents the number $2\frac{1}{2}$ or 2.5. Point B is about three-fourths of the way between 4 and 5. Point B represents the number $4\frac{3}{4}$ or 4.75.

The numbers used in algebra are sometimes less than zero. These numbers are called **negative numbers**. The number line below represents both positive and negative numbers.

The arrows mean that the numbers go on and on in both directions. Point C is about one-fourth of the way between -1 and -2. Point C represents the number $-1\frac{1}{4}$ or -1.25. This number is read as "minus one and one-fourth" or "negative one and one-fourth."

In arithmetic the + sign means to add. In algebra the + sign also means a positive (or plus) number. **When a number has no sign, it is understood to be positive.**

In arithmetic the − sign means to subtract. In algebra the − sign also means a negative (or minus) number. A negative number always has a − sign in front of it.

Zero has no sign. Zero is neither positive nor negative.

Example: Which letter on the number line below corresponds to the number $-2\frac{2}{3}$?

Point D is more than half-way between -2 and -3.
Point D corresponds to $-2\frac{2}{3}$.

Lesson 1 Exercise

Use the number line in the last example to tell the letter that represents each of the following values.

1. $+4$
2. -5
3. $+\frac{3}{4}$
4. $-6\frac{1}{3}$
5. $1\frac{2}{3}$
6. $-5\frac{1}{2}$
7. $-\frac{3}{4}$
8. $2\frac{1}{4}$

Check your answers on page 336.

Lesson 2 Adding Signed Numbers

Adding **positive numbers** means moving to the **right** on the number line. To solve the problem $+3 + 2$, start at $+3$ on the number line and move two units to the right. The answer is $+5$ or 5.

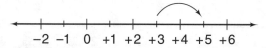

Adding **negative numbers** means moving to the **left** on the number line. To solve the problem $-1 + (-5)$ start at -1 on the number line and move five units to the left. The answer is -6. Notice that -5 is written in parentheses to avoid confusion.

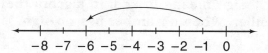

A thermometer is a practical application of the number line. If the temperature on a winter night is $-1°$ and then drops another $5°$, the temperature becomes $-6°$. To drop $5°$ means to become more negative.

The answer to adding a positive number and a negative number depends on the two numbers. To solve the problem +2 + (−6), start at +2 on the number line and move 6 units to the left. The answer is −4.

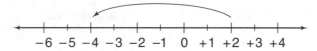

To solve the problem −3 + 7, start at −3 on the number line and move 7 units to the right. The answer is +4 or 4.

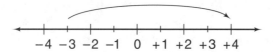

The words **greater** and **less** can be confusing with signed numbers. Any number on the number line is greater than the numbers to its left. For example, +5 is greater than −12. However, we say that −12 has a greater **absolute value** than +5. The absolute value is the distance a number is from zero.

Using this information, you do not have to draw a number line to solve signed number problems. There are rules to follow when adding signed numbers.

When a rule tells you to give an answer the sign of the greater number, give the answer the sign of the number with the greater absolute value.

Use these rules for adding two signed numbers.

> To add two signed numbers:
> 1. If the signs are alike, add and give the answer the sign of the numbers.
> 2. If the signs are different, subtract. Then compare the numbers being added and give the answer the sign of the greater number.

Example 1: −7 + (−9) =

Because the signs are alike, add and make the answer negative. The answer is −16.

$-7 + (-9) = -16$

Example 2: −8 + 13 =

Because the signs are different, subtract. Since 13 is greater than 8, give the answer the same sign as 13. The answer is +5 or 5.

$-8 + 13 = +5 \text{ or } 5$

Example 3: −6 + 2 =

Because the signs are different, subtract. Since 6 is greater than 2, give the answer the same sign as 6. The answer is −4.

$-6 + 2 = -4$

Example 4: (−25) + (+25) =

Because the signs are different, subtract. The answer is 0, which is neither positive nor negative.

$-25 + 25 = 0$

Level 1, Lesson 2: Adding Signed Numbers

To add more than two signed numbers:
1. Add the positive numbers and make the total positive.
2. Add the negative numbers and make the total negative.
3. Subtract the two totals and give the answer the sign of the greater total.

Example 5: +8 + (−9) + 3 =

Step 1. Add the positive numbers, +8 and +3, and make the total positive.   +8 + 3 = +11

Step 2. Because there is only one negative number, the negative total is −9. Subtract the two totals. Because 11 is greater than 9, make the answer positive. The answer is +2.   +11 + (−9) = +2

Example 6: −7 + 2 + 5 + (−6) =

Step 1. Find the total of the positive numbers.   +2 + 5 = +7

Step 2. Find the total of the negative numbers.   −7 + (−6) = −13

Step 3. Subtract the totals. Because 13 is greater than 7, give the answer the same sign as 13. The answer is −6.   −13 + 7 = −6

Example 7: (3) + (−4) + (+9) + (−10) =

Step 1. Add the positive numbers and make the total positive.   +3 + 9 = +12

Step 2. Add the negative numbers and make the total negative.   −4 + (−10) = −14

Step 3. Subtract the totals and make the answer negative because 14 is greater than 12. The answer is −2.   +12 + (−14) = −2

Lesson 2 Exercise

Solve each problem.

1. +6 + (−10) =
2. +15 + (−4) =
3. −8 + (−12) =
4. +7 + (−6) =
5. −13 + 13 =
6. −14 + 18 =
7. (+9) + (−15) =
8. (−8) + (−11) =
9. (−12) + (+12) =
10. (−24) + (+7) =
11. (−3) + (+14) =
12. (−19) + (−19) =
13. +3 + (−9) + 7 =
14. −4 + (−6) + (−3) =
15. +2 + 8 + (−10) =

188   Chapter 6   Algebra

16. (+3) + (+11) + (+5) =
17. (−9) + (−1) + (+4) =
18. +7 + (−4) + (−3) + 2 =
19. −8 + (−1) + (−6) + 10 =
20. (−8) + (+12) + (+16) + (−11) =
21. (+7) + (−12) + (+3) + (+2) =

Check your answers on page 336.

Lesson 3: Subtracting Signed Numbers

To subtract signed numbers you must find the distance between two numbers on the number line. A − sign between parentheses means to subtract. To solve the problem (−5) − (−3), start at −3 on the number line. Then count the number of spaces to −5. From −3 to −5 is two spaces to the left or −2.

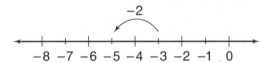

You do not have to use the number line to solve every subtraction problem. The rule is based on the fact that subtraction is the opposite of addition.

> To subtract a signed number:
> 1. Change the subtraction to addition and change the sign of the number being subtracted.
> 2. Then follow the rules for adding signed numbers.

Example 1: (+5) − (−3) =

Step 1. The − sign between the parentheses means that −3 is being subtracted. Change the subtraction to addition and change the sign of −3 to +3.

(+5) − (−3) =
+5 + (+3) = +8

Step 2. Because the signs are alike, add the numbers and make the answer positive. The answer is +8 or 8.

Example 2: (+3) − (−4) =

Step 1. −4 is being subtracted. Change the subtraction to addition and change the sign of −4 to +4.

(+3) − (−4) =
+3 + (+4) = +7

Step 2. Because the signs are the same, add and make the answer positive. The answer is +7 or 7.

Level 1, Lesson 3: Subtracting Signed Numbers

Example 3: $(-1) - (6) =$

Step 1. +6 is being subtracted. Change the subtraction to addition and change the sign of 6 to −6.

$(-1) - (6) =$
$-1 + (-6) = -7$

Step 2. Because the signs are the same, add and make the answer negative. The answer is −7.

Some problems include both addition and subtraction. Change each subtraction to addition and change the sign of every number being subtracted. Then follow the rules for adding signed numbers.

Example 4: $(-5) + (-3) - (-2) =$

Step 1. Only −2 is being subtracted. Change the subtraction to addition and change the sign of −2 to +2.

$(-5) + (-3) - (-2) =$
$-5 + (-3) + (+2) =$
$-8 + 2 = -6$

Step 2. Find the total of the negative numbers −5 and −3.

Step 3. Subtract the two totals and make the answer negative because 8 is greater than 2. The answer is −6.

Lesson 3 Exercise

Solve each problem.

1. $(+8) - (+9) =$
2. $(+7) - (-6) =$
3. $(-9) - (+3) =$
4. $(-3) - (-14) =$
5. $(-10) - (+15) =$
6. $(+6) - (-11) =$
7. $(+20) - (-4) =$
8. $(+2) - (+18) =$ ✓
9. $(-16) - (-5) =$
10. $(-11) - (+15) =$
11. $(+7) - (-8) =$
12. $(-9) - (+9) =$
13. $(+8) - (+7) + (6) =$
14. $(-4) + (-5) - (+3) =$
15. $(-10) - (-7) - (-1) =$
16. $(+11) - (-14) + (-3) =$

Check your answers on page 336.

Lesson 4 Multiplying Signed Numbers

Think about the following four situations. In these examples a + sign refers to money someone owes you and a − sign refers to money that you owe someone else. A + sign also means days in the future and a − sign means days in the past.

If someone pays you $3 a day, then in 5 days you will have $15 **more** than you have now. In algebra this is $(+3)(+5) = +15$.

If you have to pay someone $3 a day, then in 5 days you will have $15 **less** than you have now. In algebra this is $(-3)(+5) = -15$.

If someone has been paying you $3 a day, then 5 days ago you had $15 **less** than you have now. In algebra this is $(+3)(-5) = -15$.

If you have been paying someone $3 a day, then 5 days ago you had $15 **more** than you have now. In algebra this is $(-3)(-5) = +15$.

We can write a simple rule from these examples.

> **When you multiply two signed numbers:**
> 1. The answer is positive if the signs are alike.
> 2. The answer is negative if the signs are different.

Example 1: $(-3)(20) =$

Multiply. The answer is negative because the signs are different. $(-3)(20) = -60$

Example 2: $(-6)(-7) =$

Multiply. The answer is positive because the signs are alike. $(-6)(-7) = +42$

> **When you multiply more than two signed numbers:**
> 1. The answer is positive if there is an even number of negative signs (2, 4, 6, etc.).
> 2. The answer is negative if there is an odd number of negative signs (1, 3, 5, etc.).

Example 3: $(\frac{2}{3})(-\frac{3}{4})(-\frac{4}{5}) =$

Cancel and multiply across. The answer is positive because there are two negative signs.

$$\frac{2}{3} \times -\frac{\cancel{3}}{\cancel{4}} \times -\frac{\cancel{4}}{5} = \frac{2}{5}$$

Example 4: $(-1)(+8)(-3)(-2) =$

Multiply across. The answer is negative because there are three negative signs. $(-1)(+8)(-3)(-2) = -48$

In algebra, multiplication can be shown three ways:
1. A raised dot (·). For example, $9 \cdot 2 = 18$.
2. No sign between a number and a letter. For example, $5a = 5$ times a.
3. No sign before or between parentheses. For example, $8(4) = 32$ or $(-1)(+2) = -2$.

Level 1, Lesson 4: Multiplying Signed Numbers

Lesson 4 Exercise

Solve each problem.

1. $(+6)(-10) =$
2. $(+12)(-1) =$
3. $(-5)(-13) =$
4. $(+8)(+\frac{1}{2}) =$
5. $(-\frac{1}{3})(-15) =$
6. $(-9)(+8) =$
7. $(-\frac{3}{4})(+24) =$
8. $(+20)(-3) =$
9. $(-15)(-2) =$
10. $(+4)(-10)(+8) =$
11. $(+5)(-9)(-\frac{2}{3}) =$
12. $(-\frac{1}{2})(-12)(-5) =$
13. $(-8)(+12)(-1) =$
14. $(-4)(+10)(+\frac{3}{4})(+2) =$
15. $(+5)(-1)(+9)(-6) =$

Check your answers on page 337.

Lesson 5 — Dividing Signed Numbers

The rules for dividing signed numbers are similar to those for multiplying signed numbers.

> When you divide signed numbers:
> 1. The answer is positive if the signs are alike.
> 2. The answer is negative if the signs are different.

Division of signed-number problems frequently look like fractions.

Example 1: $\frac{-36}{-9} =$

Divide. The answer is positive because the signs are alike.

$$\frac{-36}{-9} = +4$$

Example 2: $\frac{56}{-8} =$

Divide. The answer is negative because the signs are different.

$$\frac{56}{-8} = -7$$

In some division problems the answer is a reduced fraction.

Example 3: $\frac{-12}{+16} =$

Reduce. The answer is negative because the signs are different.

$$\frac{-12}{+16} = -\frac{3}{4}$$

Chapter 6 Algebra

A negative fraction can be written two ways. The minus sign can go with the numerator or in front of the entire fraction. The answer to the last example is either $\frac{-3}{4}$ or $-\frac{3}{4}$.

Lesson 5 Exercise

Solve each problem.

1. $\frac{-24}{-8} =$
2. $\frac{-30}{+10} =$
3. $\frac{+5}{-10} =$
4. $\frac{+8}{+2} =$
5. $\frac{-12}{+12} =$
6. $\frac{-15}{-20} =$
7. $\frac{+100}{-20} =$
8. $\frac{-20}{-40} =$
9. $\frac{63}{-9} =$
10. $\frac{-48}{-12} =$
11. $\frac{-80}{+100} =$
12. $\frac{+18}{-6} =$
13. $\frac{150}{-25} =$
14. $\frac{-35}{49} =$
15. $\frac{-20}{-24} =$
16. $\frac{-96}{+12} =$

Check your answers on page 337.

Lesson 6 — Substituting Signed Numbers

Substitution means replacing letters with number values. You learned to substitute when you worked with formulas in earlier chapters.

The rules for combining signed numbers apply to substitution.

Example: Find the value of cd when $c = 12$ and $d = -6$.

Replace c with 12 and d with -6. Remember that two letters written next to each other mean to multiply. Because the signs are different, the answer is negative.

$$cd = (12)(-6) = -72$$

Lesson 6 Exercise

1. Find the value of $a + b$ if $a = -6$ and $b = -4$.
2. Find the value of $m + n + p$ if $m = 3$, $n = 9$, and $p = -7$.
3. Find the value of $x + y$ if $x = 15$ and $y = -15$.
4. Find the value of $r + s + t$ if $r = -1$, $s = -8$, and $t = 7$.
5. Find the value of $a - b$ when $a = -6$ and $b = -7$.
6. Find the value of $m - n$ when $m = 14$ and $n = -4$.

7. Find the value of $w - y$ when $w = -6$ and $y = +7$.
8. Find the value of $c - d$ when $c = 15$ and $d = -9$.
9. Find the value of ab when $a = -7$ and $b = -6$.
10. Find the value of pqr when $p = 5$, $q = -3$, and $r = -10$.
11. Find the value of mn when $m = -9$ and $n = +9$.
12. Find the value of xyz when $x = -1$, $y = -\frac{1}{2}$, and $z = -\frac{1}{4}$.
13. Find the value of $\frac{a}{b}$ when $a = -14$ and $b = -2$.
14. Find the value of $\frac{x}{y}$ when $x = -12$ and $y = 15$.
15. Find the value of $\frac{m}{n}$ when $m = 23$ and $n = -1$.
16. Find the value of $\frac{c}{d}$ when $c = +18$ and $d = -24$.

Check your answers on page 337.

Lesson 7: Adding Monomials

Algebra uses **expressions** to indicate mathematical relationships. An algebraic expression contains a combination of letters and numbers. $x + y$ is an expression that indicates the sum of two numbers. cd is an expression that indicates the product of two numbers.

Algebraic expressions are made up of **terms**. A term contains a number or a letter, or both. They are combined by multiplication or division. There is no addition or subtraction in a term. A **monomial** is an expression that contains only one term. $9m$ is a monomial. $-12pq$ is a monomial. And c^2 is also a monomial.

A **coefficient** is the number in a monomial. The **variable** is the letter. In the expression $9m$, the coefficient is 9 and the variable is m. In the expression $-12pq$, -12 is the coefficient and p and q are the variables. In the expression c^2, the coefficient is 1 even though the 1 is not written, and c is the variable. (Remember that 2 is the exponent or power.)

> To add monomials:
> 1. Be sure the monomials have the same variables.
> 2. Use the rules for adding signed numbers.

Example 1: $2x + 7x =$

Because the signs are alike, add and make the answer positive.

$2x + 7x = 9x$

Example 2: $(8ab) + (-3ab) + (-4ab) =$

Step 1. Add the negative monomials.

Step 2. Subtract and make the answer positive. Notice that you do not have to write the 1.

$(8ab) + (-3ab) + (-4ab) =$
$8ab + (-7ab) = ab$

Lesson 7 Exercise

Solve each problem.

1. $5m + 3m =$
2. $2c + (-6c) =$
3. $8x + (-x) =$
4. $-4p + 5p =$
5. $18xy + (-19xy) =$
6. $-6st + (-5st) =$
7. $9w + (-3w) + 4w =$
8. $-8y + (-3y) + (-y) =$
9. $-d + 5d + (-4d) =$
10. $(-8e) + (-6e) + (+11e) =$
11. $(7n) + (-12n) + (-4n) =$
12. $(-a) + (5a) + (-9a) + (+3a) =$
13. $(-7w) + (-2w) + (-3w) + (11w) =$

Check your answers on page 337.

Lesson 8: Subtracting Monomials

To subtract monomials be sure the variables are the same. Then follow the rules for subtracting signed numbers.

Example 1: $(8x) - (-6x) =$

Step 1. Rewrite the monomials, and change the subtraction sign to addition and change $-6x$ to $+6x$.

$(8x) - (-6x) =$

Step 2. Add and make the answer positive.

$8x + (+6x) = 14x$

Example 2: $(-mn) + (2mn) - (7mn) =$

Step 1. Rewrite the monomials and change the subtraction sign to addition and change $7mn$ to $-7mn$.

$(-mn) + (2mn) - (7mn) =$
$-mn + 2mn + (-7mn) =$
$-8mn + 2mn = -6mn$

Step 2. Add the two negative monomials.

Step 3. Subtract and make the answer negative.

Lesson 8 Exercise

Solve each problem.

1. $(+5p) - (+3p) =$
2. $(+5p) - (-3p) =$
3. $(-5p) - (-3p) =$
4. $(-6a) - (-3a) =$
5. $(+4c) - (-4c) =$
6. $(-7mn) - (+3mn) =$
7. $(-9f) - (-f) + (-2f) =$
8. $(10y) - (3y) - (-7y) =$
9. $(-8cd) + (-2cd) - (3cd) =$
10. $(-12t) - (-3t) + (9t) =$
11. $(11u) - (12u) + (-u) =$
12. $(-9xy) + (-xy) + (-3xy) =$

Check your answers on page 337.

Lesson 9: Multiplying Monomials

When you multiply monomials, the exponents of the variables change. Remember that a number multiplied by itself is the number to the second power. That is, $c \cdot c = c^2$. The variable c is understood to be c^1. Notice that a dot between the letters means to multiply.

To multiply variables, simply add their exponents.

> To multiply monomials:
> 1. Multiply the coefficients according to the rules for multiplying signed numbers.
> 2. Add the exponents of the repeated letters.

Example 1: $a^3 \cdot a^4 =$

Because there are no coefficients to multiply, simply add the exponents for the variables.

$$a^3 \cdot a^4 = a^7$$

When the variables are different, write the letters beside each other in the product.

Example 2: $c \cdot d =$

Write the letters beside each other in the product.

$$c \cdot d = cd$$

When the monomials you are multiplying have more than one variable, add the exponents of each variable separately.

196   Chapter 6 Algebra

Example 3: $(-8m^2n)(-3m^3n^5) =$

Step 1. Multiply the coefficients. Because the signs are alike, the product is positive.

$(-8m^2n)(-3m^3n^5) = +24m^5n^6$

Step 2. Add the exponents of the variable m, $2 + 3 = 5$. Then add the exponents of the variable n, $1 + 5 = 6$.

Lesson 9 Exercise

Solve each problem.

1. $c^2 \cdot c^3 =$
2. $m^4 \cdot m =$
3. $x^3 \cdot x^5 =$
4. $a \cdot a =$
5. $a \cdot b =$
6. $(-a^2b^2)(a^3b^4) =$
7. $(2x)(-3x) =$
8. $(-4y^2)(-2y^5) =$
9. $(-5p)(2p^3) =$
10. $(5yz)(-6yz) =$
11. $(-mn^2)(-4m^2n^3) =$
12. $(-2rs)(-9r^3s) =$

Check your answers on page 338.

Lesson 10 — Dividing Monomials

Think about dividing x^5 by x^2.

$$\frac{x^5}{x^2} = \frac{\cancel{x}\cancel{x}xxx}{\cancel{x}\cancel{x}} = x^3$$

The quotient x^3 is the result of subtracting the exponents of x.

> To divide monomials:
> 1. Follow the rules for dividing signed numbers for the coefficients.
> 2. Subtract the exponents of the same variables.

Example 1: $\frac{-12a^4}{-3a} =$

Because the signs are the same, the quotient is positive. Subtract the exponents of the variable.

$$\frac{-12a^4}{-3a} = 4a^3$$

Sometimes there is no variable to cancel with another.

Level 1, Lesson 10: Dividing Monomials

Example 2: $\frac{-3ab}{6a}$

Because the signs are different, the quotient is negative. The a's cancel and the b remains.

$$\frac{-3ab}{6a} = \frac{-b}{2}$$

Lesson 10 Exercise

Solve each problem.

1. $\frac{m^6}{m^2} =$
2. $\frac{st}{s} =$
3. $\frac{c^5}{c^2} =$
4. $\frac{a^5}{a^4} =$
5. $\frac{x^3 y^2}{xy} =$
6. $\frac{x^6}{x^6} =$
7. $\frac{-12x^2}{3x} =$
8. $\frac{20m^3 n}{-4m} =$
9. $\frac{-36a^5}{-9a^4} =$
10. $\frac{-4m^2 n^2}{16mn} =$
11. $\frac{-18c^3 d^4}{-12c^3 d} =$
12. $\frac{+24x}{+30x} =$

Check your answers on page 338.

Lesson 11: Evaluating Algebraic Expressions

So far you have used only one operation at a time with signed numbers. When there is more than one operation, use the following **order of operations:**

1. Parentheses and division bars
2. Powers and square roots
3. Multiplication and division
4. Addition and subtraction

In the following examples you will see how to use this order of operations. Each of these examples involves **substitution**. Remember that substitution means replacing variables with number values.

Example 1: Find the value of $x + xy$ if $x = -6$ and $y = -4$.

Step 1. Replace each variable with its value from the problem. Notice how the parentheses are used to show multiplication.

$x + xy =$
$-6 + (-6)(-4) =$

Step 2. This problem has addition and multiplication. Because multiplication comes before addition in the order of operations, first multiply -6 and -4.

$-6 + (+24) =$

Step 3. Combine -6 and $+24$ according to the rules for adding signed numbers.

$-6 + 24 = 18$

Example 2: Find the value of $a(a + b)$ if $a = -9$ and $b = 4$.

Step 1. Replace a and b with their values from the problem.

$a(a + b) =$
$-9(-9 + 4) =$

Step 2. The operation $-9 + 4$ is in parentheses. Since parentheses are first in the order of operations, add -9 and $+4$ first.

$-9(-5) = +45$

Step 3. Multiply -9 by -5.

Example 3: Find the value $x(x - y)$ if $x = 2$ and $y = -6$.

Step 1. Replace x and y with their values from the problem. Notice how the extra parentheses separate -6 from the $-$ sign in front.

$x(x - y) =$
$2(2 - (-6)) =$

Step 2. Because the operation $2 - (-6)$ is in parentheses, combine these first according to the rules for subtracting signed numbers.

$2(2 + 6) =$

Step 3. Multiply $+2$ by $+8$.

$2(8) = 16$

Example 4: Find the value of $\frac{m + n}{2}$ if $m = -4$ and $n = -8$.

Step 1. Replace m and n with their values from the problem.

$\frac{m + n}{2}$

Step 2. Because the operation $-4 + (-8)$ is above the division bar, combine them first. The division bar works like parentheses to group -4 and -8 together.

$\frac{-4 + (-8)}{2} =$

Step 3. Divide -12 by 2.

$\frac{-12}{2} = -6$

Lesson 11 Exercise

Solve each problem.

1. Find the value of $ab - c$ if $a = -4$, $b = -3$, and $c = -5$.
2. Find the value of $m(m - n)$ if $m = -2$ and $n = 7$.
3. Find the value of $x^2 y$ if $x = -3$ and $y = -4$.
4. Find the value of $s(s + t) - t$ if $s = -5$ and $t = 1$.
5. Find the value of $e + ef$ if $e = 6$ and $f = -4$.
6. Find the value of $p + pq$ if $p = -6$ and $q = 4$.
7. Find the value of ab^2 if $a = -2$ and $b = -5$.
8. Find the value of $x(x - y)$ if $x = -8$ and $y = 2$.
9. Find the value of $g(g - h)$ if $g = 6$ and $h = -7$.

10. Find the value of $(j - k)^2$ if $j = -3$ and $k = 1$.
11. Find the value of $\frac{a + b}{2}$ if $a = -6$ and $b = -2$.
12. Find the value of $\frac{m - n}{n}$ if $m = 8$ and $n = -4$.

Check your answers on page 338.

Level 1 Review

Solve each problem.

```
   A   B       C   D
 --+-+-+-+-+-+-+-+-+--
  -4 -3 -2 -1  0  1  2  3  4
```

1. Which point on the number line below corresponds to the value $-1\frac{3}{4}$?
2. $(-13) + (8) + (-7) =$
3. $(-9) - (-11) - (+4) =$
4. $(-6)(-\frac{2}{3})(-5) =$
5. $\frac{+24}{-36} =$
6. Find the value of $p + q$ if $p = -20$ and $q = -14$.
7. Find the value of cd if $c = 24$ and $d = -\frac{1}{2}$.
8. Find the value of $\frac{a}{e}$ if $a = 144$ and $e = -12$.
9. $(4m) + (-2m) + (-7m) =$
10. $(-n) - (-6n) + (-4n) =$
11. $(-9x^2y)(-2xy) =$
12. $\frac{12a^2c}{-3a} =$
13. Find the value of $ab - 4b$ if $a = -7$ and $b = -3$.
14. Find the value of c^2d if $c = -5$ and $d = 10$.
15. Find the value of $x(x + y)$ if $x = -4$ and $y = 9$.

Check your answers on page 338. If you have all 15 problems correct, go on to Level 2. If you have answered a problem incorrectly, find the item number on the chart below and review that lesson before you go on.

Review:	If you missed item number:
Lesson 1	1
Lesson 2	2
Lesson 3	3
Lesson 4	4
Lesson 5	5
Lesson 6	6, 7, 8
Lesson 7	9
Lesson 8	10
Lesson 9	11
Lesson 10	12
Lesson 11	13, 14, 15

Level 2 Algebra Applications

In this section you will learn to solve equations with one variable and to solve inequalities. You will also learn to multiply binomials, to factor, and to find the solutions to quadratic equations.

Lesson 12: One-Step Equations

An **equation** is a mathematical statement that two amounts are equal.

The statement $x + 15 = 42$ is an equation. It states that some number called x increased by 15 is equal to 42. The variable x is called the **unknown.** You will often see the word *side* in instructions about equations. The = sign separates an equation into two sides.

To **solve** an equation means to find the value of the unknown that makes the statement true. To solve an equation, you must know about opposite or **inverse** operations.

- Addition is the inverse of subtraction.
- Subtraction is the inverse of addition.
- Multiplication is the inverse of division.
- Division is the inverse of multiplication.

To solve a one-step equation, perform the inverse operation on **both sides** of the equation in order to get a statement that says "Unknown = solution" or "Solution = unknown."

Example 1: Solve the equation $x + 15 = 42$.

In this equation 15 is added to the unknown. The inverse of addition is subtraction. Subtract 15 from both sides of the equation. The solution is $x = 27$.

$$\begin{array}{rr} x + 15 = & 42 \\ -15 & -15 \\ \hline x = & 27 \end{array}$$

To check the solution to an equation, substitute the solution for the unknown. When you do the arithmetic, you should get the same number on both sides.

In the last example replace x with 27. $27 + 15 = 42$. When you do the arithmetic, you get 42 on both sides of the equal sign.

In some equations the unknown is on the right side. In these equations you want a statement that says, "Solution = unknown."

Example 2: Solve and check the equation $18 = y - 6$.

Step 1. In this equation 6 is subtracted from the unknown. Addition is the inverse of subtraction. Add 6 to both sides. The solution is $24 = y$.

$$\begin{array}{r} 18 = y - 6 \\ +\ 6 + 6 \\ \hline 24 = y \end{array}$$

Step 2. To check the solution replace y with 24 in the original equation. You get 18 on both sides. This proves that 24 is correct.

$$18 = 24 - 6$$
$$18 = 18$$

Example 3: Solve and check the equation $9m = 45$.

Step 1. In this equation the unknown is multiplied by 9. Division is the inverse of multiplication. Divide both sides by 9. The solution is $m = 5$.

$$\frac{9m}{9} = \frac{45}{9}$$
$$m = 5$$

Step 2. To check the solution replace m with 5. You get 45 on both sides. This proves that 5 is correct.

$$9 \cdot 5 = 45$$
$$45 = 45$$

Example 4: Solve and check the equation $\frac{n}{8} = 6$.

Step 1. In this equation the unknown is divided by 8. Multiplication is the inverse of division. Multiply both sides by 8. The solution is $n = 48$.

$$\frac{\cancel{8}}{1} \cdot \frac{n}{\cancel{8}} = 6 \cdot 8$$
$$n = 48$$

Step 2. To check the solution replace n with 48. You get 6 on both sides. This proves that 48 is the solution.

$$\frac{48}{8} = 6$$
$$6 = 6$$

Example 5: Solve and check the equation $\frac{2}{3}y = 12$.

Step 1. In this equation the unknown is multiplied by $\frac{2}{3}$. Division is the inverse of multiplication. Remember that you must invert and multiply when you divide by a fraction. Multiply both sides by $\frac{3}{2}$. The solution is $y = 18$.

$$\frac{\cancel{3}}{\cancel{2}} \cdot \frac{\cancel{2}}{\cancel{3}} y = \frac{\cancel{12}}{1} \cdot \frac{3}{\cancel{2}}$$
$$y = 18$$

Step 2. To check the solution replace y with 18. You get 12 on both sides. This proves that 18 is the solution.

$$\frac{2}{3} \cdot \frac{18}{1} = 12$$
$$12 = 12$$

Lesson 12 Exercise

Solve and check each equation.

1. $m + 11 = 30$
2. $8w = 56$
3. $c - 12 = 5$
4. $16 = f - 4$
5. $\frac{c}{4} = 5$
6. $6n = 9$
7. $2 = \frac{y}{9}$
8. $12 = 18p$
9. $14 = a + 3$
10. $g - 9 = 41$
11. $e + 6 = -8$
12. $\frac{n}{12} = 1$
13. $15 = i - 8$
14. $200 = 25r$
15. $\frac{3}{4}s = 24$
16. $10 = \frac{1}{2}w$
17. $21 = d + 16$
18. $10 = \frac{z}{5}$
19. $24f = 12$
20. $p + 14 = 4$
21. $\frac{3}{8}x = 15$

Check your answers on page 338.

Lesson 13 Multistep Equations

To solve some equations you must use more than one inverse operation. Do the **addition and subtraction before multiplication and division.**

Example 1: Solve and check the equation $4a - 3 = 29$.

Step 1. On the left side of the equation there is both multiplication and subtraction. First do the inverse of subtraction by adding 3 to both sides.

$$\begin{array}{r} 4a - 3 = 29 \\ + 3 + 3 \\ \hline 4a = 32 \end{array}$$

Step 2. Since the unknown is multiplied by 4, divide both sides by 4. The solution is $a = 8$.

$$\frac{4a}{4} = \frac{32}{4}$$
$$a = 8$$

Step 3. Replace a with 8 and do the arithmetic. Both sides equal 29. This proves that 8 is the solution.

$$4 \cdot 8 - 3 = 29$$
$$32 - 3 = 29$$
$$29 = 29$$

Example 2: Solve and check the equation $7 = \frac{x}{5} + 1$.

Step 1. On the right side of the equation there is both division and addition. First do the inverse of addition by subtracting 1 from both sides.

$$\begin{array}{r} 7 = \frac{x}{5} + 1 \\ -1 = \phantom{\frac{x}{5}} - 1 \\ \hline 6 = \frac{x}{5} \end{array}$$

Step 2. Since the unknown is divided by 5, multiply both sides by 5. The solution is $30 = x$.

$$5 \cdot 6 = \frac{x}{\cancel{5}} \cdot \cancel{5}$$
$$30 = x$$

Step 3. Replace x with 30 and do the arithmetic. Both sides equal 7. This proves that 30 is the solution.

$$7 = \frac{30}{5} + 1$$
$$7 = 6 + 1$$
$$7 = 7$$

Chapter 6 Algebra

Lesson 13 Exercise

Solve and check each equation.

1. $6m + 5 = 47$
2. $3x - 2 = 28$
3. $\frac{c}{4} + 1 = 8$
4. $17 = 7a + 3$
5. $50 = 9d - 4$
6. $8 = \frac{x}{7} - 2$
7. $2n - 11 = 3$
8. $\frac{3}{4}a + 5 = 17$
9. $\frac{s}{10} - 6 = 3$
10. $5y + 7 = -3$
11. $2 = \frac{w}{9} + 11$
12. $\frac{4}{5}f - 7 = 1$
13. $8z + 3 = 9$
14. $2 = 9p - 4$
15. $-4 = 7t + 3$
16. $10b - 8 = 12$
17. $19 = 4s + 1$
18. $\frac{n}{6} - 5 = 1$

Check your answers on page 339.

Lesson 14: Equations with Separated Unknowns

In some equations the unknown appears in more than one place.

When the unknowns appear on the same side of the equals sign, combine the terms according to the rules for adding monomials.

Example 1: Solve the equation $7m - 3 - 2m = 27$.

Step 1. Combine the terms containing the unknown, m.

Step 2. Add 3 to both sides of the equation.

Step 3. Divide both sides of the equation by 5. The solution is $m = 6$. Your work is more likely to be correct if you keep it carefully lined up, as in the examples and the answers in the key.

$$7m - 3 - 2m = 27$$
$$5m - 3 = 27$$
$$5m - 3 = 27$$
$$+3 \quad +3$$
$$\frac{5m}{5} = \frac{30}{5}$$
$$m = 6$$

When the unknown appears on both sides of the = sign, use inverse operations to combine the unknowns. Remember that you want the unknowns on one side and numbers without unknowns on the other side.

Example 2: Solve the equation $5x + 30 = 8x$.

Step 1. To get a statement that says "Number = unknown," subtract $5x$ from both sides.

Step 2. Divide both sides by 3. The solution is $10 = x$.

$$5x + 30 = 8x$$
$$-5x \quad\quad -5x$$
$$30 = 3x$$
$$\frac{30}{3} = \frac{3x}{3}$$
$$10 = x$$

Level 2, Lesson 14: Equations with Separated Unknowns

To check the equation replace each unknown with the solution. When you do the arithmetic you should get the same number on both sides. For the last example replace each x with 10.

$$5 \cdot 10 + 30 = 8 \cdot 10$$
$$50 + 30 = 80$$
$$80 = 80$$

Lesson 14 Exercise

Solve and check each equation.

1. $9a - 2a = 21$
2. $5m = 18 + 2m$
3. $8r = 15 + 3r$
4. $12 - 5x = 7x$
5. $16 = 13c + 7c$
6. $3p + 7 = 10p$
7. $6y - y = 10$
8. $3t = 9 + 2t$
9. $4 - 2n = 6n$
10. $5x + 4 = 3x + 20$
11. $8w - 5 = 7w + 13$
12. $3p + 12 = 8p - 23$
13. $7c - 3c = c + 27$
14. $9m - 12 = m + 20$
15. $2d - 8 = 7d + 12$

Check your answers on page 340.

Lesson 15 — Equations with Parentheses

To solve an equation with parentheses use the Distributive Property you studied earlier. Multiply each term inside the parentheses by the number outside the parentheses.

Example: Solve the equation $6(x - 2) = 18$.

Step 1. Multiply x by 6 and -2 by 6.

Step 2. Add 12 to both sides.

Step 3. Divide both sides by 6. The solution is $x = 5$.

$$\begin{aligned} 6(x - 2) &= 18 \\ 6x - 12 &= 18 \\ +12 &= +12 \\ \frac{6x}{6} &= \frac{30}{6} \\ x &= 5 \end{aligned}$$

To check, replace the unknown with the solution. For the example replace x with 5. To do the arithmetic you can first subtract 2 from 5 or you can first multiply 6 by both 5 and 2.

$$\begin{array}{ll} 6(5 - 2) = 18 & \text{or} \quad 6(5 - 2) = 18 \\ 6(3) = 18 & \quad\quad 30 - 12 = 18 \\ 18 = 18 & \quad\quad 18 = 18 \end{array}$$

Lesson 15 Exercise

Solve and check each equation.

1. $4(m - 3) = 20$
2. $5(a + 2) = 15$
3. $9 = 2(x - 3)$
4. $3(c + 4) = 2c + 17$
5. $8n - 7 = 6(n - 1)$
6. $9(p + 2) = p + 20$
7. $4(a - 5) = 3(a + 2)$
8. $6(d - 1) = 3(d + 2)$
9. $5(y + 2) = 3(y - 8)$

Check your answers on page 340.

Lesson 16 — Inequalities

An **inequality** is a statement that two amounts are **not** equal. Below are the four symbols that express inequalities.

Symbol		Example
$<$	means less than	$2 < 5$
$>$	means more than.	$9 > 1$
$\leq$	means less than or equal to.	$7 \leq 10$ or $4 \leq 4$
$\geq$	means more than or equal to.	$8 \geq 3$ or $6 \geq 6$

To solve an inequality follow the rules you learned for solving equations.

Example: Solve the inequality $m - 9 > 6$.

Add 9 to both sides of the inequality. The solution is $m > 15$.

$$\begin{array}{r} m - 9 > 6 \\ +9 +9 \\ \hline m > 15 \end{array}$$

The solution $m > 15$ means that any number greater than 15 is a solution to the inequality. Both $15\frac{1}{4}$ and 99 are solutions.

Lesson 16 Exercise

Solve each inequality.

1. $a + 6 > 9$
2. $c - 12 \leq 3$
3. $\frac{n}{2} < 7$
4. $16r \geq 20$
5. $6m - 2 < 22$
6. $\frac{3}{5}x \leq 18$
7. $3p - 4 > p + 6$
8. $9w + 2 \geq w + 10$
9. $3(m - 2) < 9$
10. $\frac{1}{2}y - 3 \geq 1$
11. $4 > 2(n - 9)$
12. $8t - 5 \leq 2t + 1$

Check your answers on page 341.

Lesson 17: Multiplying Binomials

A **binomial** is an algebraic expression with two terms. An example of a binomial is $x + 8$.

Multiplying binomials is an important tool in an area of algebra that you will learn more about later. You already know the rules for multiplying monomials.

To multiply binomials:

1. Put one binomial under the other.
2. Multiply both terms in the top binomial by the term at the right in the bottom binomial.
3. Multiply both terms in the top binomial by the term at the left in the bottom binomial.
4. Add the results of the last two steps.

Study the next examples carefully to see how these problems are set up.

Example 1: $(x + 3)(x + 2) =$

Step 1. Put one binomial under the other.

Step 2. Multiply the top terms by $+2$. $+2 \cdot +3 = +6$ and $+2 \cdot x = +2x$.

Step 3. Multiply the top terms by x. $x \cdot +3 = +3x$ and $x \cdot x = x^2$.

Step 4. Add the results. Notice how $+2x$ and $+3x$ are written under each other.

$$\begin{array}{r} x + 3 \\ x + 2 \\ \hline + 2x + 6 \\ x^2 + 3x \\ \hline x^2 + 5x + 6 \end{array}$$

Example 2: $(x - 5)(x + 4) =$

Step 1. Put one binomial under the other.

Step 2. Multiply $+4$ by -5 and $+4$ by x.

Step 3. Multiply x by -5 and x by x.

Step 4. Add the results.

$$\begin{array}{r} x - 5 \\ x + 4 \\ \hline + 4x - 20 \\ x^2 - 5x \\ \hline x^2 - x - 20 \end{array}$$

Example 3: $(x - 6)(x - 1) =$

Step 1. Put one binomial under the other.

Step 2. Multiply -1 by -6 and -1 by x.

Step 3. Multiply x by -6 and x by x.

Step 4. Add the results.

$$\begin{array}{r} x - 6 \\ x - 1 \\ \hline - x + 6 \\ x^2 - 6x \\ \hline x^2 - 7x + 6 \end{array}$$

Example 4: $(x - 2)(x + 2) =$

Step 1. Put one binomial under the other.

Step 2. Multiply $+2$ by -2 and $+2$ by x.

Step 3. Multiply x by -2 and x by x.

$$\begin{array}{r} x - 2 \\ x + 2 \\ \hline +2x - 4 \\ x^2 - 2x \\ \hline x^2 - 4 \end{array}$$

Step 4. Add the results. Notice that the answer has no middle term since $2x + (-2x) = 0$. The 0 is not written.

Be sure that you understand how to find each term in the examples before you try the exercise.

Lesson 17 Exercise

Multiply each pair of binomials.

1. $(x + 5)(x + 2) =$
2. $(x + 3)(x + 1) =$
3. $(x + 2)(x + 6) =$
4. $(x - 4)(x - 3) =$
5. $(x - 1)(x - 8) =$
6. $(x - 4)(x - 5) =$
7. $(x + 5)(x - 2) =$
8. $(x - 6)(x + 7) =$
9. $(x + 12)(x - 10) =$
10. $(x + 8)(x - 8) =$
11. $(x + 10)(x - 10) =$
12. $(x - 3)(x + 3) =$
13. $(x - 7)(x + 3) =$
14. $(x + 6)(x - 9) =$
15. $(x - 11)(x + 6) =$

Check your answers on page 342.

Lesson 18 Factoring

Factors are numbers that multiply together to give another number. For example, 3 and 2 are factors of 6. Also, 6 and 1 are factors of 6. In this book you will learn to work some of the common types of factoring problems that you may see on the GED Test.

Think about the binomial $6x + 15$. Both terms can be divided evenly by the whole number 3. Therefore, 3 is one of the factors of $6x + 15$.

> To factor a binomial:
> 1. Look for a number or letter that divides evenly into both terms.
> 2. Divide the binomial by the number or letter to find the other factor.

Example 1: Factor the expression $6x + 15$.

Three divides evenly into both $6x$ and 15. Divide 3 into each term. Write 3 outside a set of parentheses. Write the quotient of 3 divided into each term inside the set of parentheses: $6x \div 3 = 2x$; $+15 \div 3 = +5$.

$6x + 15 = 3(2x + 5)$

To check, multiply 3 by each term inside the parentheses. 3 times $2x = 6x$, and 3 times $+5 = +15$.

Example 2: Factor the expression $8y - 4$.

Four divides evenly into each term. Divide 4 into both $8y$ and 4. Write 4 outside a set of parentheses. Write the quotient of 4 divided into each term inside the parentheses.

$8y - 4 = 4(2y - 1)$

To check, multiply 4 by each term inside the parentheses. 4 times $2y = 8y$, and 4 times $-1 = -4$.

In some binomials a letter divides evenly into both terms.

Example 3: Factor the expression $s^2 + 7s$.

The letter s divides evenly into each term. Divide s into both s^2 and $7s$. Write s outside a set of parentheses. Write the quotient of s divided into each term inside the parentheses.

$s^2 + 7s = s(s + 7)$

To check, multiply s by each term inside the parentheses. s times s is s^2, and s times 7 is $+7s$.

Lesson 18 Exercise

Factor each expression.

1. $3x + 9 =$
2. $8w - 12 =$
3. $10c - 5d =$
4. $4x - 16 =$
5. $7a + 21 =$
6. $5y - 20z =$
7. $36f - 12 =$
8. $9m + 21 =$
9. $6w - 9 =$
10. $c^2 + 8c =$
11. $m^2 + 6m =$
12. $x^2 - 5x =$
13. $p^2 + 10p =$
14. $a^2 - a =$
15. $n^2 - 2n =$
16. $e^2 + 7e =$
17. $t^2 - 3t =$
18. $y^2 - y =$

Check your answers on page 342.

Lesson 19

Factoring Quadratic Expressions

One of the most common factoring problems in algebra is factoring a **quadratic** expression. In a quadratic expression the unknown is raised to the second power. The answers to the problems you multiplied in Exercise 17 were all quadratic expressions. The factors for these expressions are two binomials.

To factor a quadratic expression, you must think about each term in the expression. Look carefully at the problem $(x + 9)(x + 5)$.

$$\begin{array}{r} x + 9 \\ x + 5 \\ \hline + 5x + 45 \\ x^2 + 9x \\ \hline x^2 + 14x + 45 \end{array}$$

To factor the expression $x^2 + 14x + 45$, you must understand where each term comes from.

The first term, x^2, is easy. It comes from multiplying x by x.
The third term, 45, is also easy. It comes from multiplying $+5$ by $+9$.
The middle term, $14x$, is more difficult. It comes from adding $5x$ and $9x$.
The most important questions to consider when you factor a quadratic expression are:

"What numbers **multiply** to give the third term?" and

"What terms **add** to give the middle term?"

Think about these two questions as you study the next examples.

Example 1: Factor the expression $x^2 + 10x + 24$.

Think first about $+24$. Several combinations multiply together to make 24; for example, $+12$ times $+2$ or $+8$ times $+3$ or $+6$ times $+4$. Only one of these, however, gives the middle term $10x$: $6x + 4x = +10x$. The factors for $x^2 + 10x + 64$ are $(x + 6)$ and $(x + 4)$.

$x^2 + 10x + 24 =$
$(x + 6)(x + 4)$

When you factor a quadratic expression, **always** check by multiplying the factors together. For the last example,

$$\begin{array}{r} x + 6 \\ x + 4 \\ \hline + 4x + 24 \\ x^2 + 6x \\ \hline x^2 + 10x + 24 \end{array}$$

Level 2, Lesson 19: Factoring Quadratic Equations

Example 2: Factor the expression $x^2 - 7x + 10$.

First think about $+10$. Both 10 times 1 and 5 times 2 equal 10. Look at the middle term. When the middle term is negative and the third term is positive, you are looking for two negative numbers. $-7x$ comes from adding $-5x$ and $-2x$. The $+10$ comes from multiplying -5 by -2. The factors for $x^2 - 7x + 10$ are $(x - 5)(x - 2)$.

$x^2 - 7x + 10 =$
$(x - 5)(x - 2)$

Check the problem by multiplying $x - 5$ by $x - 2$.

Think carefully about the signs in every factoring problem.

Example 3: Factor the expression $x^2 + 2x - 15$.

First think about -15. Both 15 times 1 and 5 times 3 equal 15. Because -15 is negative, there must be one positive number and one negative number. Because $+2x$ is positive, the larger factor must be positive. $(+5x) + (-3x) = +2x$. $+5$ times $-3 = -15$. The factors for $x^2 + 2x - 15$ are $(x + 5)$ and $(x - 3)$. Multiply the factors to check.

$x^2 + 2x - 15 =$
$(x + 5)(x - 3)$

Example 4: Factor the expression $x^2 - 3x - 28$.

First think about -28. Both 14 times 2 and 7 times 4 equal 28. Because -28 is negative, there must be one positive number and one negative number. Because $-3x$ is negative, the larger factor must be negative. $-7x$ and $+4x$ equal $-3x$. -7 times $+4 = -28$. The factors for $x^2 - 3x - 28$ are $(x - 7)$ and $(x + 4)$. Multiply the factors to check.

$x^2 - 3x - 28 =$
$(x - 7)(x + 4)$

Example 5: Factor the expression $x^2 - 25$.

First think about -25. Both 25 times 1 and 5 times 5 equal 25. Because -25 is a negative number, there must be one negative number and one positive number. Because there is no middle term, the factors must be the same. $+5x$ and $-5x$ equal 0. $+5$ times $-5 = -25$. The factors for $x^2 - 25$ are $(x + 5)$ and $(x - 5)$. Multiply the factors to check.

$x^2 - 25 = (x + 5)(x - 5)$

Be sure you understand these examples before you try the exercise.

Lesson 19 Exercise

Factor each expression. Check by multiplying the factors together.

1. $x^2 + 6x + 8 =$
2. $x^2 + 8x + 7 =$
3. $x^2 + 13x + 40 =$
4. $x^2 - 13x + 36 =$
5. $x^2 - 12x + 36 =$
6. $x^2 - 10x + 9 =$
7. $x^2 + 5x - 14 =$
8. $x^2 + 2x - 8 =$
9. $x^2 + x - 56 =$
10. $x^2 - 36 =$
11. $x^2 - 4 =$
12. $x^2 - 81 =$
13. $x^2 - 3x - 10 =$
14. $x^2 - 4x - 96 =$
15. $x^2 - x - 12 =$

Check your answers on page 342. For more practice factoring quadratic expressions, try factoring the solutions to Exercise 17.

Lesson 20: Quadratic Equations

In a **quadratic equation** the unknown is raised to the second power. Most quadratic equations have two correct solutions.

Think about the equation $x^2 = 16$. You know that the square root of 16 is 4. This means that $x = 4$ is a solution. In algebra, however, the square root of 16 can also be -4. When you multiply -4 by itself, you get $+16$ because the signs are alike. The solutions to the equation $x^2 = 16$ are $x = +4$ and $x = -4$.

Not every quadratic equation is this easy to solve. Most quadratic equations are written in the form $x^2 + Ax + B = 0$, where A and B are numbers. This is called the **standard form** of a quadratic equation.

> To solve a quadratic equation:
> 1. Be sure the equation is written in standard form.
> 2. Factor the expression on the left side of the = sign.
> 3. Set each factor equal to zero and solve each new equation.

Example 1: Solve the equation $x^2 + 5x + 6 = 0$.

Step 1. Factor the expression $x^2 + 5x + 6$. The factors that give $+6$ at the right and the $5x$ in the middle are $(x + 3)$ and $(x + 2)$.

$$x^2 + 5x + 6 = 0$$
$$(x + 3)(x + 2) = 0$$

Step 2. Set each factor equal to zero. The solution from one factor is $x = -3$. The solution from the other factor is $x = -2$.

$$x + 3 = -0$$
$$\underline{-3 \quad -3}$$
$$x \quad = -3 \quad \text{and}$$

$$x + 2 = 0$$
$$\underline{-2 \quad -2}$$
$$x \quad = -2$$

Level 2, Lesson 20: Quadratic Equations

To check the solutions to a quadratic equation, substitute each value into the original equation.

When you substitute $x = -3$ into the original equation, you get:

$$(-3)^2 + 5(-3) + 6 = 0$$
$$+9 \; - 15 \; + 6 = 0$$
$$+15 \; - 15 \quad\quad = 0$$

When you substitute $x = -2$ into the original equation, you get:

$$(-2)^2 + 5(-2) + 6 = 0$$
$$+4 \; - 10 \; + 6 = 0$$
$$+10 \; - 10 \quad\quad = 0$$

Example 2: Solve the equation $x^2 - 3x - 28 = 0$.

Step 1. Factor the expression $x^2 - 3x - 28$. $x^2 - 3x - 28 = 0$
The factors that give -28 at the right and $(x - 7)(x + 4) = 0$
$-3x$ in the middle are $(x - 7)$ and $(x + 4)$.

Step 2. Set each factor equal to zero. $x - 7 = 0$ $x + 4 = 0$
The solution from one factor is $x = 7$. $+7 = +7$ $-4 \; -4$
The solution from the other factor is $x = -4$. $x \quad = +7$ and $x \quad = -4$

To check the solutions substitute each value into the original equation.

$$(7)^2 - 3(7) - 28 = 0 \quad \text{and} \quad (-4)^2 - 3(-4) - 28 = 0$$
$$+49 - 21 - 28 = 0 \quad\quad\quad +16 \; + \; 12 - 28 = 0$$
$$+49 - \;\; 49 \quad\quad = 0 \quad\quad\quad\quad\quad + \; 28 - 28 = 0$$

Quadratic equations require careful work. It is easy to make mistakes with the signs. Be sure to check the solutions by substituting your answers into the original equations.

Lesson 20 Exercise

Solve and check each quadratic equation.

1. $x^2 + 7x + 10 = 0$
2. $x^2 + 10x + 9 = 0$
3. $x^2 + 15x + 56 = 0$
4. $x^2 - 7x + 12 = 0$
5. $x^2 - 17x + 60 = 0$
6. $x^2 - 12x + 27 = 0$
7. $x^2 + 2x - 24 = 0$
8. $x^2 + x - 72 = 0$
9. $x^2 + 7x - 30 = 0$
10. $x^2 - 4x - 21 = 0$
11. $x^2 - 5x - 6 = 0$
12. $x^2 - 10x - 24 = 0$
13. $x^2 - 49 = 0$
14. $x^2 - 1 = 0$
15. $x^2 - 144 = 0$

Check your answers on page 342.

Lesson 21: Factoring and Square Roots

A useful application of factoring is simplifying square roots. So far in this book you have found the square root of numbers with exact square roots. Most numbers, however, do not have an exact square root.

Think about the number 20. The square root of 20 is larger than 4, since $4 \times 4 = 16$. The square root of 20 is smaller than 5, since $5 \times 5 = 25$.

> To simplify a square root:
> 1. Find factors that include an exact square root.
> 2. Put the square root of one factor outside the $\sqrt{}$ sign, and leave the other factor inside the $\sqrt{}$ sign.

Example 1: Simplify $\sqrt{20}$.

Step 1. The whole number factors of 20 are 20 and 1, 2 and 10, and 4 and 5. The combination that includes an exact square root is 4 and 5. Put each factor inside a $\sqrt{}$ sign.

$$\sqrt{20} = \sqrt{4} \cdot \sqrt{5} = 2\sqrt{5}$$

Step 2. Put the square root of 4 next to $\sqrt{5}$.

Example 2: Simplify $\sqrt{125}$.

Step 1. The factors of 125 that include an exact square root are 25 and 5. Put each factor inside a $\sqrt{}$ sign.

$$\sqrt{125} = \sqrt{25} \cdot \sqrt{5} = 5\sqrt{5}$$

Step 2. Put the square root of 25 next to $\sqrt{5}$.

To the nearest hundredth the square root of 5 is 2.24. To show that $2\sqrt{5}$ is a sensible answer to $\sqrt{20}$, find (2×2.24) to the second power.

```
    2.24        (4.48)² =    4.48
  ×    2                   × 4.48
    4.48                    35 84
                           1 79 2
                          17 92
                          20.07 04, which is close to 20.
```

To show that $5\sqrt{5}$ is a sensible answer to $\sqrt{125}$, find (5×2.24) to the second power.

$$\begin{array}{r} 2.24 \\ \times5 \\ \hline 11.20 \end{array} \qquad (11.2)^2 = \begin{array}{r} 11.2 \\ \times11.2 \\ \hline 224 \\ 112 \\ 112 \\ \hline 125.44, \text{ which is close to 125.} \end{array}$$

Lesson 21 Exercise

Simplify each square root.

1. $\sqrt{8}$
2. $\sqrt{75}$
3. $\sqrt{18}$
4. $\sqrt{24}$
5. $\sqrt{72}$
6. $\sqrt{54}$
7. $\sqrt{500}$
8. $\sqrt{45}$
9. $\sqrt{96}$
10. $\sqrt{128}$

Check your answers on page 343.

Level 2 Review

Solve.

1. $24 = c - 6$
2. $8m = 12$
3. $10n - 2 = 3$
4. $15 = 9x - 3$
5. $9y - 2y = 35$
6. $4a + 2 + 3a = a + 20$
7. $4(p - 7) = p + 5$
8. $7x - 3 = 5(x + 5)$
9. $6w - 4 \leq 17$
10. $6n - 5 > 4n + 21$
11. Multiply $(a + 9)(a + 1)$.
12. Multiply $(y - 8)(y + 7)$.
13. Factor the expression $m^2 + 12m$.
14. Factor the expression $c^2 - 6c$.
15. Factor the expression $x^2 + 11x + 30$.
16. Factor the expression $x^2 + 6x - 16$.
17. Find the solutions to the equation $x^2 - 100 = 0$.

18. Find the solutions to the equation $x^2 - 6x - 27 = 0$.
19. Simplify $\sqrt{12}$.
20. Simplify $\sqrt{28}$.

Check your answers on page 343. If you have all 20 problems correct, go on to Level 3. If you have answered a problem incorrectly, find the item number on the chart below and review that lesson before you go on.

Review:	If you missed item number:
Lesson 12	1, 2
Lesson 13	3, 4
Lesson 14	5, 6
Lesson 15	7, 8
Lesson 16	9, 10
Lesson 17	11, 12
Lesson 18	13, 14
Lesson 19	15, 16
Lesson 20	17, 18
Lesson 21	19, 20

Level 3 Algebra Problem Solving

In this section you will learn to write algebraic expressions and to write equations from word descriptions. You will also learn to use algebra to solve problems with formulas and to set up and solve multistep algebra word problems.

Lesson 22 Writing Algebraic Expressions

One of the most useful applications of algebra is writing verbal information in algebraic form.

Below are some key words and phrases for each of the basic operations. After each key word or phrase is a verbal example, and after each example is an algebraic expression that represents the example. Study these examples carefully before you try the exercise. In each example a letter stands for the unknown.

Addition
sum   The sum of a number and 8: $x + 8$
total   The total of 12 and a number: $12 + c$
increased by A number increased by 5: $y + 5$
plus   One plus a number: $1 + t$
more than   Three more than a number: $m + 3$

The order in the last examples is not important. For example, $m + 3$ is the same as $3 + m$. In the subtraction examples that follow, the order **is** important.

Subtraction
less than   Ten less than a number: $a - 10$
decreased by   Seven decreased by a number: $7 - b$
subtracted from A number subtracted from 12: $12 - f$
minus   Nine minus a number: $9 - g$

Notice that in the following multiplication examples the coefficient always comes before the unknown.

Multiplication
product   The product of six and a number: $6k$
times   Two times a number: $2m$
a fraction of   One-fourth of a number: $\frac{1}{4}a$

Division
divided by   A number divided by 20: $\frac{d}{20}$

30 divided by a number: $\frac{30}{n}$

Lesson 22 Exercise

Write an algebraic expression for each of the following expressions. Use *x* to stand for the unknown.

1. A number increased by two.
2. Ten decreased by a number.
3. The sum of thirteen and a number.
4. Twelve more than a number.
5. The product of eight and a number.
6. A number divided by four.
7. Nineteen less than a number.
8. Forty plus a number.
9. Six times a number.
10. Three-eighths of a number.
11. Sixteen divided by a number.
12. A number decreased by one.
13. Fourteen minus a number.
14. The total of 100 and a number.
15. Three subtracted from a number.

Check your answers on page 344.

Lesson 23 — Writing Multistep Algebraic Expressions

The examples in the last lesson all showed one operation. Below are examples of multistep operations.

Notice the use of parentheses in some of these examples. A set of parentheses groups numbers together. The words *quantity* and *all* often suggest that some amount should be enclosed in parentheses.

Verbal Expression	Algebraic Expression
A number decreased by one-half the number.	$s - \dfrac{1}{2}s$
The sum of 11 and a number, all multiplied by four.	$4(n + 11)$
One-third of the sum of a number and seven.	$\dfrac{1}{3}(e + 7)$
Eight divided into the difference of a number and two.	$\dfrac{i - 2}{8}$ or $\dfrac{(i - 2)}{8}$
Two less than a number divided by eight.	$\dfrac{i}{8} - 2$

Lesson 23 Exercise

Write an algebraic expression for each of the following verbal expressions. Let x represent each unknown.

1. A number increased by five times the same number.
2. One-third of a number decreased by one-sixth of the same number.
3. The sum of eight and a number all multiplied by six.
4. One more than ten times a number.
5. Nine less than one-half of a number.
6. Twenty decreased by twice a number.
7. Three divided into the sum of a number and seven.
8. The sum of a number and twice the same number all multiplied by four.
9. Three-fourths of the quantity of a number decreased by eleven.
10. Eleven less than three-fourths of a number.

Check your answers on page 344.

Lesson 24 Translating Words into Algebra

You can use algebraic expressions to show the mathematical relationships in practical situations. Changing English words into the numbers and letters of algebra is like translating from one language to another.

In the examples below watch for the words and phrases that tell you what operations to use.

Example 1: Alberto makes x dollars an hour. Write an expression for his hourly wage if he gets a raise of $2 a hour.

A raise of $2 an hour means to add. $x + 2$

Example 2: Beef costs b dollars a pound. How much did Celia pay for four pounds of beef?

Multiply the cost of one pound by four. $4b$

Lesson 24 Exercise

Write an algebraic expression for each of the following situations.

1. Celeste's take-home pay is p dollars a month. She spends one-fourth of her pay for rent. What is her monthly rent?
2. Jack is x years old.
    a. What was his age five years ago?
    b. What will his age be in three years?
3. If Jim drives at an average speed of r miles per hour, how far will he drive in four hours?
4. Let t stand for the total number of workers in the hospital where Mary works. There are 45 women working at the hospital. How many men work there?
5. If Max cuts a length of wire that is l feet long into three equal pieces, how long will each piece be?
6. The Smiths' house was worth x in 1970. Now it is worth twice as much. What is the value of the house now?
7. Let g stand for Larry's gross salary. Larry's employer deducts 22% of Larry's salary for taxes. Write an expression for the amount of the deductions. (Hint: Change the percent to a decimal.)
8. Mike's regular wage is w dollars an hour. For overtime he makes $1\frac{1}{2}$ times his regular wage. What is his overtime wage?
9. If s represents Darlene's weekly pay, how much does she make each hour for a 35 hour week?
10. Colin is x years old. His son is 32 years younger. Write an expression for Colin's son's age.

Check your answers on page 344.

Lesson 25

Writing Equations

In Lessons 22 to 24 you learned to write algebraic expressions from verbal descriptions. To write equations from verbal descriptions, watch for the verb in each description. Words such as *is* and *equals* tell you where to put the = sign.

Example 1: Eight less than three times a number is 19. Write an equation and find the number.

Step 1. Let x represent the number. Put the = sign where the verb *is* appears in the sentence.

$$3x - 8 = 19$$

Step 2. Add 8 to both sides.

Step 3. Divide both sides by 3. The solution is 9.

$$3x - 8 = 19$$
$$+8 \quad +8$$
$$\frac{3x}{3} = \frac{27}{3}$$
$$x = 9$$

Example 2: Twice the difference of a number and five equals twenty. Find the number.

Step 1. Let x represent the number. Put the = sign where the verb *equals* appears.

$$2(x - 5) = 20$$

Step 2. Multiply x and -5 by 2.

Step 3. Add 10 to both sides.

Step 4. Divide both sides by 2. The solution is 15.

$$2x - 10 = 20$$
$$+10 \quad +10$$
$$\frac{2x}{2} = \frac{30}{2}$$
$$x = 15$$

Lesson 25 Exercise

Write and solve an equation for each of the following.

1. A number increased by fifteen is 21.
2. Six times a number is 72.
3. A number divided by 25 is six.
4. Thirty equals a number decreased by 27.
5. Nine more than four times a number equals 33.
6. Six less than seven times a number is 78.
7. Two more than the quotient of a number divided by five is eight.
8. 77 equals three less than 8 times a number.
9. One-half a number decreased by five equals 35.
10. Seven times a number decreased by the same number is 78.

11. Nine times a number decreased by three is the same as twice the same number increased by 25.
12. Five less than ten times a number is nineteen more than twice the same number.
13. Twelve times a number equals 45 more than three times the same number.
14. Five times the quantity of a number decreased by two is thirty.
15. Seven times the quantity of a number decreased by one equals four times the sum of the same number and two.

Check your answers on page 344.

Lesson 26
Using Algebra to Solve Formulas

Algebra is a useful tool for solving formulas. It is especially useful when the information you have is "backwards."

For example, think about the formula $d = rt$ for finding distance. If the information you have includes the distance and the rate, you can use the formula to find the time.

Example 1: Jack walked 10 miles at an average speed of 4 mph. Find the total time that Jack walked.

Step 1. Replace d with 10 and r with 4 in the formula $d = rt$.

$$d = rt$$
$$10 = 4t$$

Step 2. Solve for the time by dividing both sides by 4. The time is $2\frac{1}{2}$ hours.

$$\frac{10}{4} = \frac{4t}{4}$$
$$2\frac{2}{4} = t$$
$$2\frac{1}{2} \text{ hrs} = t$$

The next example uses the cost formula, $c = nr$.

Example 2: Maria spent $7.95 for three pounds of meat. What was the price per pound?

Step 1. Replace c with $7.95 and n with 3 in the formula $c = nr$.

$$c = nr$$
$$\$7.95 = 3r$$

Step 2. Solve for the rate (the price of one pound) by dividing both sides by 3. The price of one pound is $2.65.

$$\frac{\$7.95}{3} = \frac{3r}{3}$$
$$\$2.65 = r$$

The next example uses the formula $i = prt$ for interest.

Example 3: Bill earned $21 in interest on a loan of $350 in one year. Find the interest rate.

Step 1. Replace i with $21, p with $350, and t with 1 in the formula $i = prt$.

$$i = prt$$
$$\$21 = \$350 \cdot r \cdot 1$$

Step 2. To find r (the interest rate) divide both sides by $350. The interest rate is $\frac{3}{50}$.

$$\frac{\$21}{\$350} = \frac{\$350r}{\$350}$$
$$\frac{3}{50} = r$$

Step 3. Change the fraction $\frac{3}{50}$ to a percent by multiplying by 100 and writing a % sign. The interest rate was 6%.

$$\frac{3}{\cancel{50}} \times \frac{\cancelto{2}{100}}{1} = 6\%$$

Lesson 26 Exercise

Use the formulas $d = rt$, $c = nr$, or $i = prt$ to solve each of the following problems.

1. Alfredo drove 210 miles at an average rate of 42 mph. For how many hours did he drive?
2. A plane flew 1194 miles in three hours. Find the average speed of the plane.
3. Isabella hiked 27 miles in six hours. Find her average hiking speed.
4. How long did a train need to go 135 miles if its average speed was 54 mph?
5. Paul drove for 496 miles in 8 hours. Find his average driving speed.
6. Carla spent $4.17 for three pounds of pork. What was the price per pound?
7. Tom spent $22 each for the sweaters he sells in his store. Altogether he spent $528 for sweaters. How many sweaters did he buy?
8. For a dozen baseball gloves Mr. Huston spent $420. Find the average price for each glove.
9. The Rigbys spent $5625 for 4.5 acres of land. Find the price per acre.
10. One ticket for a concert at the Municipal Theatre is $6.50. If the total ticket sales were $3380, how many tickets were sold?
11. Kate earned $51 interest on $850 for one year. Find the rate of interest.
12. On a loan of $350 Carmen had to pay $14 interest at 8% annual interest. Find the length of time of the loan.
13. In one year the Santiagos pay $4800 in interest on a loan at an annual interest rate of 12%. Find the total amount of the loan.
14. Max paid $27 interest in six months on a loan of $600. Find the rate of the loan.
15. Laurie borrowed $600 at 9% annual interest. She paid the loan back in less than a year. The interest she paid was $36. In how many months did she pay back the loan?

Check your answers on page 345.

Lesson 27: Algebra Word Problems

Algebra is a tool for solving complex word problems.
For every algebra word problem choose a letter to stand for the unknown. Then use this letter to write expressions for all the parts of the problem.

Example 1: One number is four times another number. The larger number decreased by two is the same as the smaller number increased by sixteen. Find the two numbers.

Step 1. Let x represent the small number. Let $4x$ represent the large number.

small number = x
large number = $4x$

Step 2. Write an equation from the second sentence in the problem. Notice that the phrase *is the same as* tells you where to put the = sign.

$4x - 2 = x + 16$

Step 3. Subtract x from both sides. Then add 2 to both sides. Then divide both sides by 3.

$$4x - 2 = x + 16$$
$$ -x -x$$
$$3x - 2 = 16$$
$$+2 +2$$
$$\frac{3x}{3} = \frac{18}{3}$$

Step 4. The solution is $x = 6$, which is the small number. Find the value of $4x$, which is the large number. $4 \cdot 6 = 24$.

$x = 6$
$4x = 4 \cdot 6 = 24$

For some problems it helps to make a chart of the information in the problem.

Example 2: Joaquin is 25 years older than his son. In ten years the sum of their ages will be 99. Find their ages now.

Step 1. Make a chart that expresses Joaquin's age and his son's age now and in 10 years.

	age now	age in 10 years
son	x	$x + 10$
Joaquin	$x + 25$	$x + 25 + 10 = x + 35$

Step 2. Write an equation that shows the sum of their ages in 10 years.
Step 3. Combine the unknowns and solve the equation.

$$x + 10 + x + 35 = 99$$
$$2x + 45 = 99$$
$$-45 = -45$$
$$\frac{2x}{2} = \frac{54}{2}$$

Step 4. The unknown x is the son's age. To find Joaquin's age find the value of $x + 25$.

son's age $\quad x = 27$
Joaquin's age $\quad x + 25 = 27 + 25 = 52$

Lesson 27 Exercise

Write an equation for each problem and solve.

1. One number is three times another number. When the larger number is decreased by ten, the result is the same as when the smaller number is increased by eighteen. Find both numbers.
2. One number is seven more than another number. Four less than three times the smaller number is the same as twice the larger number. Find the two numbers.
3. Louise is 26 years older than her daughter. Louise's age is two more than four times her daughter's age. Find both their ages.
4. Douglas is 46 years younger than his grandfather. Three times Douglas's age is eight less than the grandfather's age. Find both their ages.
5. Juan's net pay is five times the amount his employer deducts. Juan's gross weekly pay is $324. Find his weekly take-home pay.
6. There are fifteen more women than men at the Wednesday night exercise class at the Greenport Community Center. Altogether there are 47 people in the class. Find the number of women.
7. Joe is a plumber. He makes $3 an hour more than his assistant, Phil. On a job that they each spent ten hours doing, they made $350. Find the hourly wage for each of them.
8. Altogether Jim, Carmen, and George spent 95 hours campaigning for their friend Ed, who was running for State Assemblyman. Carmen worked five hours more than Jim, and George worked twice as long as Carmen. How many hours did each of them work on the campaign?
9. Chris is three years older than his brother Andy. In five years three times Andy's age will be six more than twice Chris's age. Find their ages now.
10. For every $3 the Martins spend for food they spend $2 on car expenses and $4 on rent. They spend a total of $648 a month on these items. How much do they spend each month on rent?

Check your answers on page 346.

Level 3 Review

Solve.

1. Write an algebraic expression for a number divided by 15.
2. Write an algebraic expression for 20 times a number.
3. Write an algebraic expression for four times the quantity of a number decreased by seven.

4. Write an algebraic expression for the quantity of three less than twice a number, all divided by five.

5. Let p represent Bill's hourly wage now. Write an expression for his wage if he gets one dollar an hour more each hour.

6. Let x represent Sam's age now. Write an expression for Sam's age twelve years ago.

7. Let t represent the total number of students in Ann's math class. There are eight women in the class. Write an expression for the number of men in the class.

8. Let m represent Jean's income for the month. She spends 0.3 of her income for food. Write an expression for the amount she spends on food each month.

9. Five more than six times a number is seventeen. Find the number.

10. Forty-three equals two less than nine times a number. Find the number.

11. Four times the quantity of a number decreased by one is the same as two more than the same number. Find the number.

12. Two times the quantity of a number decreased by three equals the number increased by seven. Find the number.

For problems 13–16, use the formulas $d = rt$ for distance, $c = nr$ for cost, and $i = prt$ for interest.

13. A train traveled 217 miles at an average speed of 62 mph. For how many hours did the train travel?

14. Selma bought $1\frac{1}{2}$ pounds of chicken for $1.26. Find the cost of a pound of chicken.

15. Rick paid back $48 in interest in six months on a loan for which he was charged 12% annual interest. Find the amount of the loan.

16. On a loan of $1200 at 15% annual interest, Shirley had to pay $135 interest. For how many months did she have the loan?

17. One number is five times another. The larger number decreased by nine is twice the quantity of the smaller number increased by three. Find both numbers.

18. At the Spring Street Day Care Center there are 65 children. There are nine more girls than boys. Find the number of girls at the center.

19. Fred and his sons, Tom and Bill, built a recreation room in their basement. Fred worked twice as many hours as Tom, and Bill worked 10 hours more than Tom. Altogether the three of them worked 94 hours. How many hours did Fred work?

20. Now Sarah is twice as old as her sister Rachel. In eight years three times Sarah's age will be ten less than five times Rachel's age. Find Sarah's age now.

Check your answers on page 347. If you have all 20 problems correct, go on to the next chapter. If you answered a problem incorrectly, find the item number on the chart below and review that lesson before you go on.

Review:	If you missed item number:
Lesson 22	1, 2
Lesson 23	3, 4
Lesson 24	5, 6, 7, 8
Lesson 25	9, 10, 11, 12
Lesson 26	13, 14, 15, 16
Lesson 27	17, 18, 19, 20

Chapter 7
GEOMETRY

Geometry is a branch of mathematics that includes identifying and measuring lines, angles, surfaces, and three-dimensional figures.

You have already learned several geometric concepts in this book. You have found the perimeter and area of squares, rectangles, triangles, and circles. You have also found the volume of cubes, rectangular solids, and cylinders.

In this chapter you will become familiar with other geometric concepts. There are many terms in this chapter. Take the time to memorize the terms you do not already know.

Level 1
Geometry Skills

In this section you will learn to identify types of angles. You will also learn the names and characteristics of the most important triangles. To find out whether you need to work through this section, try the following preview.

Preview

Solve each problem.

1. Angle x contains 75°. What kind of angle is it?

2. Angle y contains 137°. What kind of angle is it?

3. How many degrees are there in the complement of a 67° angle?

4. In the picture ∠PQS = 180° and ∠RQS = 72°. Find the measure of ∠PQR.

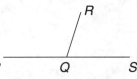

Use the figure to answer questions 5–7.

5. Find the measure of ∠f.
6. Which angle is an alternate interior angle with ∠c?
7. Find the measure of ∠e.

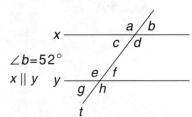

8. In the triangle DEF at the right, ∠D = 56° and ∠E = 62°. What kind of triangle is DEF?
9. In triangle KLM, ∠K = 48° and ∠L = 42°. What kind of triangle is KLM?
10. In triangle WXY, ∠W = 46° and ∠X = 54°. Which side of WXY is longest?

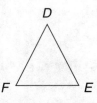

Check your answers on page 348. If you have all 10 problems correct, do the Level 1 Review on page 238. If you have fewer than 10 problems right, study Level 1 beginning with Lesson 1.

230   Chapter 7   Geometry

Lesson 1: Angles

An **angle** is formed by two lines or two **rays** extending from the same point. A **ray** has one end point. The size of an angle depends on how "open" or "closed" the lines are.

The angle called *x* above is larger than the angle called *y* because the lines that form *x* are open more than the lines that form *y*. The point that the lines of an angle extend from is called the **vertex**. The unit of measurement of angles is the degree (°).

An angle is named by the number of degrees it contains. Below is a list of angle names. Notice that a small curve (or sometimes a small square) indicates each angle.

Examples

A **right angle** contains exactly 90°. A box at the vertex always means a right angle. You have already seen right angles in this book. Each corner of a square or a rectangle is a right angle. Also, the height and the base of a triangle meet to form a right angle.

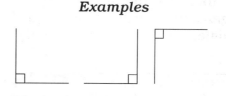

An **acute angle** contains less than 90°. The sides of an acute angle are more closed than the sides of a right angle.

An **obtuse angle** has more than 90° and less than 180°. The sides of an obtuse angle are open more than the sides of a right angle.

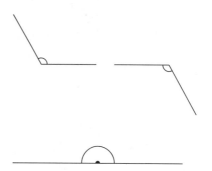

A **straight angle** has exactly 180°. A straight angle looks like a straight line.

A **reflex angle** has more than 180° and less than 360°. You can think of a reflex angle as being bent back upon itself.

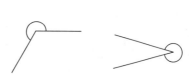

An angle with 360° does not look like an angle. 360° is the angular measurement of a complete circle.

The symbol for an angle is ∠.

There are three ways to refer to an angle. In one method a capital letter near the vertex refers to the angle.

Example 1: What kind of an angle is ∠X in the triangle?

Since there is a small square X, ∠X is a right angle.

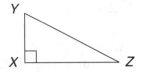

Sometimes three letters are used to refer to an angle. The middle letter is always the vertex.

Example 2: What kind of angle is ∠ABC?

Because the sides that form ∠ABC are closed more than a right angle, ∠ABC is acute.

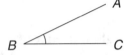

Sometimes a small letter inside the opening of an angle is used to refer to the angle.

Example 3: What kind of angle is ∠m?

Because the sides that form ∠m are open more than a right angle and are closed more than a straight angle, ∠m is an obtuse angle.

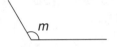

Lesson 1 Exercise

Identify each angle.

1.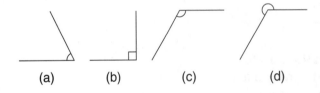

    (a)   (b)   (c)   (d)

2.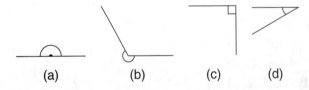

    (a)   (b)   (c)   (d)

Identify each angle.

3.   **a.** 35°   **b.** 60°   **c.** 130°   **d.** 90°
4.   **a.** 250°   **b.** 162°   **c.** 180°   **d.** 85°

Check your answers on page 348.

Lesson 2: Pairs of Angles

Sometimes angles add together to form different types of angles.

Complementary angles are two angles that add up to 90°. Together ∠a and ∠b in the picture make a right angle. ∠a and ∠b are complementary angles. If ∠a = 60°, then ∠b = 90° − 60° = 30°.

Supplementary angles are two angles that add up to 180°. Together ∠c and ∠d in the picture make a straight angle. ∠c and ∠d are supplementary angles. If ∠c = 70°, then ∠d = 180° − 70° = 110°.

When two straight lines intersect, four angles are formed. **Vertical angles** are across from each other when two lines intersect. Vertical angles are equal to each other. In the picture, ∠e and ∠g are vertical angles. ∠f and ∠h are also vertical angles. If ∠e = 50°, then ∠g = 50°. Notice that the angles next to each other are supplementary. If ∠e = 50°, then ∠f = 180° − 50 = 130°. ∠h is also 130°.

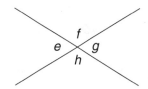

When two straight lines intersect, we say that the angles that are across from each other are **opposite** each other. For example, in the picture above, ∠f and ∠h are opposite each other and vertical. When two angles are next to each other we say that they are **adjacent**. ∠e and ∠f are adjacent and supplementary.

Example 1: Find the complement of a 15° angle.

Subtract 15° from 90°.

$$\begin{array}{r} 90° \\ -\ 15° \\ \hline 75° \end{array}$$

Example 2: Find the supplement of a 15° angle.

Subtract 15° from 180°.

$$\begin{array}{r} 180° \\ -\ 15° \\ \hline 165° \end{array}$$

Lesson 2 Exercise

Solve each problem.

1. What is the complement to an angle that measures 22°?
2. What is the supplement to an angle that measures 22°?

3. In the picture, ∠x = 43°. Find ∠y.

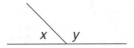

Use the picture to answer questions 4–8.

4. Find the measure of ∠a.
5. Which angle is vertical to ∠d?
6. Which angle is vertical to ∠a?
7. Find the measurement of ∠b.
8. Find the measurement of ∠c.

Use the picture to answer questions 9–12.

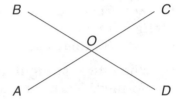

9. What angle is vertical to ∠AOB?
10. What angle is vertical to ∠BOC?
11. If ∠AOB = 58°, what is the measure of ∠AOD?
12. If ∠COD = 72°, what is the measure of ∠AOB?
13. Solve for m in the picture.

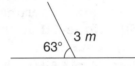

14. One angle is 20° more than another, and the two angles are complementary. Find the measure of each angle.
15. One angle is four times another, and the two angles are supplementary. Find the measure of each angle.

Check your answers on page 348.

Lesson 3
Parallel Lines and Transversals

Two lines that run in the same direction and never cross are called **parallel lines**. $AB \parallel CD$ means that the two lines are parallel.

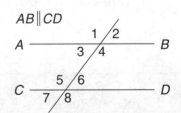

When another line, called a **transversal**, crosses two parallel lines, eight angles are formed. Following are the names of the angles formed when two parallel lines are crossed by a transversal. There are four pairs of **corresponding angles**. For the picture above,

∠1 and ∠5 are corresponding.
∠2 and ∠6 are corresponding.

∠3 and ∠7 are corresponding.
∠4 and ∠8 are corresponding.

Corresponding angles are equal to each other.

Think about each pair of corresponding angles. For example, ∠1 is in the upper left of the top intersection, and ∠5 is in the upper left of the bottom intersection. Notice that each location "corresponds" to the other.

There are two pairs of **alternate interior angles**. Here the word **interior** refers to being between the parallel lines. For the picture above,

∠3 and ∠6 are alternate interior angles and
∠4 and ∠5 are alternate interior angles.

Alternate interior angles are equal to each other. In each pair one angle is at the upper intersection and the other is at the lower intersection, but they are on opposite sides of the transversal.

There are two pairs of **alternate exterior angles**. Here the word **exterior** refers to being outside the parallel lines. For the picture above,

∠1 and ∠8 are alternate exterior angles and
∠2 and ∠7 are alternate exterior angles.

Alternate exterior angles are equal to each other. In each pair one angle is at the upper intersection and the other is at the lower intersection, but they are on opposite sides of the transversal.

When two parallel lines are crossed by a transversal, two angles are either equal to each other or they are supplementary. For example, ∠1 and ∠4 are equal, but ∠1 and ∠2 are supplementary.

Lesson 3 Exercise

Use the picture to answer the following questions.

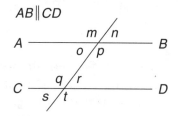

AB∥CD

1. Which angle corresponds to ∠p?
2. Which angle is alternate exterior with ∠n?
3. Which angle is vertical to ∠t?
4. Which angle is alternate interior with ∠q?
5. If ∠m = 118°, what is the measure of ∠t?
6. If ∠n = 49°, what is the measure of ∠q?

Fill in the blanks.

7. ∠m and ∠p are _____ angles.
8. ∠o and ∠r are _____ angles.
9. ∠n and ∠s are _____ angles.
10. ∠p and ∠t are _____ angles.

Check your answers on page 349.

Lesson 4: Triangles

You have already learned to find the perimeter and the area of a triangle. In this lesson you will learn the names of various triangles. You will also learn the relationships among the sides and angles of triangles.

A triangle is a flat figure with three sides. The three angles of a triangle add up to 180°. Each point where two sides meet is called a **vertex**. Below are descriptions of the most common triangles. Notice that the names depend on the relationships between the sides or the angles.

An **equilateral** triangle has **three** equal sides. An equilateral triangle also has three equal angles. Each angle measures 60°. An equilateral triangle can also be called **equiangular**.

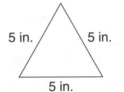
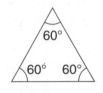

An **isosceles** triangle has **two** equal sides. It also has two equal angles. The equal angles of an isosceles triangle are called the **base angles**. The third angle is called the **vertex angle**.

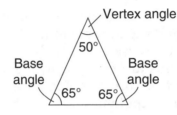
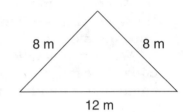

A **scalene** triangle has **no** equal sides and **no** equal angles.

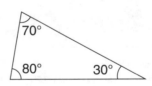
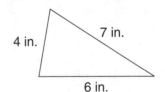

A **right** triangle has one right angle. The side opposite the right angle is called the **hypotenuse**. The other two sides are sometimes called the **legs**.

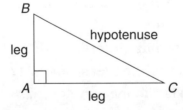
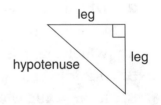

The symbol △ stands for the word *triangle*. Each side of a triangle is usually referred to by the two letters at the endpoints of the side. For example, the hypotenuse in △ABC above is BC. The legs are AB and AC. The angle of a triangle can be identified by one letter. The right angle in △ABC is ∠A.

Sometimes three letters are used to identify an angle of a triangle. With three letters the middle letter is always the vertex of the angle. The right angle in △ABC can be called ∠BAC or ∠CAB.

Example 1: In △XYZ, ∠X = 70° and ∠Y = 40°. Find the measure of ∠Z.

Step 1. Add the measures of ∠X and ∠Y.

$$\begin{array}{r} 70° \\ +\ 40° \\ \hline 110° \end{array}$$

Step 2. Subtract the total of ∠X and ∠Y from 180°.

$$\begin{array}{r} 180° \\ -110° \\ \hline 70° \end{array}$$

Example 2: What kind of triangle is △XYZ in Example 1?

Since two angles are the same (70°), △XYZ is isosceles.

In any triangle **the longest side is opposite the largest angle.** Also, the shortest side is opposite the smallest angle. In triangle ABC notice that side AB is the longest and that it is opposite ∠C, which is the largest angle.

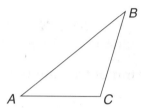

Example 3: In △PQR, ∠P = 45° and ∠Q = 75°. Which side of △PQR is longest?

Step 1. Add the measures of ∠P and ∠Q.

$$\begin{array}{r} 45° \\ +75° \\ \hline 120° \end{array}$$

Step 2. Find the measure of ∠R by subtracting the total of ∠P and ∠Q from 180°.

$$\begin{array}{r} 180° \\ -120° \\ \hline ∠R = \ \ 60° \end{array}$$

PR is longest.

Step 3. The largest angle is ∠Q (75°). The side opposite ∠Q is PR. PR is the longest side.

Lesson 4 Exercise

Solve each problem.

1. In △KLM, ∠K = 32° and ∠L = 74°. What kind of triangle is KLM?
2. In △DEF, ∠D = 35° and ∠E = 55°. What kind of triangle is DEF?
3. In an isosceles triangle each base angle measures 42°. Find the measure of the vertex angle.
4. In another isosceles triangle the vertex angle measures 78°. Find the measure of each base angle.
5. One acute angle of a right triangle is 33°. Find the measure of the other acute angle.
6. In △ABC, ∠A = 30° and ∠B = 60°. Which side of △ABC is longest?

Level 1, Lesson 4: Triangles

7. In the triangle DE = 6 in. and EF = 8 in. The perimeter is 24 in. Which angle is largest?

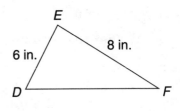

8. In the triangle ∠X = 48° and ∠Y = 66°. Which side is the shortest?

9. In triangle MNO, ∠M = 90°. Which side is the hypotenuse?

10. What kind of triangle is pictured?

Check your answers on page 349.

Level 1 Review

Solve each problem.

1. Angle m is what type of angle?

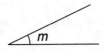

2. Angle n is what type of angle?

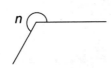

3. Find the measure of the supplement of an angle of 84°.

4. In the picture ∠AOC = 90° and ∠AOB = 36°. Find the measure of ∠BOC.

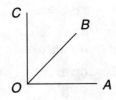

Use this figure to answer problems 5–7.

5. What is the measure of ∠u?
6. Which angle is an alternate exterior angle with ∠s?
7. Find the measurement of ∠z.

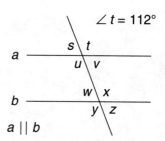

a ∥ b

8. In the triangle ABC ∠A = 33° and ∠B = 57°. Side AB is called the _____.

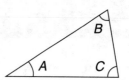

9. In triangle XYZ, ∠X = 49° and ∠Y = 82°. What kind of triangle is XYZ?
10. In △PQR, ∠P = 55° and ∠Q = 47°. Which side of the triangle is longest?

Check your answers on page 349. If you have all 10 problems correct, go on to Level 2. If you answered a problem incorrectly, find the item number on the chart below and review that lesson before you go on.

Review:	If you missed item number:
Lesson 1	1, 2
Lesson 2	3, 4
Lesson 3	5, 6, 7
Lesson 4	8, 9, 10

Level 1: Review

Level 2 Geometry Applications

In this section you will learn about similar triangles and congruent triangles. You will also learn to solve problems with the Pythagorean relationship and to use the rectangular coordinate system.

Lesson 5: Similar Figures

Two geometric figures are **similar** when they have the same shape but are not necessarily the same size.

Two figures are similar if their angles are equal. The triangles at the right are similar because they each have angles of 30°, 60°, and 90°.

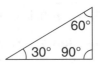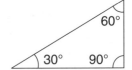

Two triangles are also similar if their sides are proportional. The triangles at the right are similar because their sides are proportional. Each side of the triangle in the figure at the right is twice as long as the corresponding side of the triangle at the left.

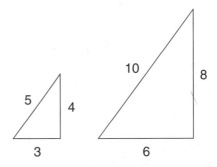

Example 1: In triangle ABC, $\angle A = 77°$ and $\angle B = 58°$. In triangle DEF, $\angle E = 58°$ and $\angle F = 45°$. Are the triangles similar?

Step 1. Find the measure of $\angle C$ by subtracting the total of $\angle A$ and $\angle B$ from 180°.

$$\begin{array}{r} 77° \\ +\ 58° \\ \hline 135° \end{array} \quad \begin{array}{r} 180° \\ -135° \\ \hline 45° = \angle C \end{array}$$

Step 2. Find the measure of $\angle D$ by subtracting the total of $\angle E$ and $\angle F$ from 180°. Since the triangles have the same angles, they are similar.

$$\begin{array}{r} 58° \\ +\ 45° \\ \hline 103° \end{array} \quad \begin{array}{r} 180° \\ -103° \\ \hline 77° = \angle D \end{array}$$

Yes, the triangles are similar.

240   Chapter 7   Geometry

Example 2: Are the rectangles similar?

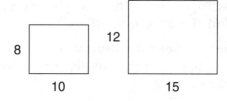

Step 1. Find the ratio of the width to the length of the first rectangle.

$$\frac{w}{l} = \frac{8}{10} = \frac{4}{5}$$

Step 2. Find the ratio of the width to the length of the second rectangle. Since the sides of the rectangles are proportional, they are similar.

$$\frac{w}{l} = \frac{12}{15} = \frac{4}{5}$$

Yes, the rectangles are similar.

Corresponding sides of similar figures are proportional. You can use a proportion to find the length of the side of a figure.

In the diagram for the next example, notice how small curves are used to indicate equal angles.

Example 3: In the triangles, $\angle A = \angle D$, $\angle B = \angle E$, and $\angle C = \angle F$. $AB = 4$ inches, $AC = 6$ inches, and $DE = 18$ inches. Find the length of side DF.

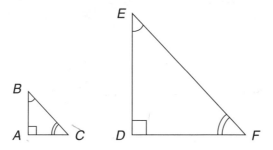

Step 1. Set up the ratio between the height and the base for each triangle and write a proportion stating that the ratios are $=$. Let x stand for the base of $\triangle DEF$.

$$\frac{\text{height}}{\text{base}} = \frac{4}{6} = \frac{18}{x}$$

Step 2. Solve the proportion. The length of DF is 27 inches.

$$4x = 108$$
$$\frac{4x}{4} = \frac{108}{4}$$
$$x = 27 \text{ in.}$$

With some pairs of triangles it is hard to know which sides are corresponding. In the triangles $\angle M$ and $\angle P$ each equal 90°. The sides opposite them are corresponding. This means that side NO corresponds to side QO.

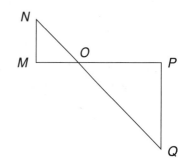

$\angle N = \angle Q$. The sides opposite these angles are corresponding. This means that side MO corresponds to side OP.

$\angle MON = \angle POQ$. The sides opposite these angles are corresponding. This means that side MN corresponds to side PQ.

Example 4: In the last picture on page 241, MN = 10, MO = 16, and PQ = 25. Find the length of side OP.

Step 1. Set up a proportion of the short leg to the long leg of each triangle. Let x stand for the long side of OPQ.

$$\frac{\text{short}}{\text{long}} = \frac{10}{16} = \frac{25}{x}$$

Step 2. Solve the proportion. The length of side OP is 40.

$$10x = 16 \cdot 25$$
$$\frac{10x}{10} = \frac{400}{10}$$
$$x = 40$$

Lesson 5 Exercise

Solve each problem.

1. Is △ABC similar to △DEF?

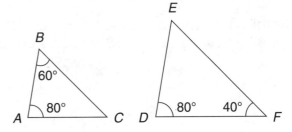

2. In △GHI, ∠G = 65° and ∠H = 55°. In △JKL, ∠J = 60° and ∠K = 50°. Is △GHI similar to △JKL?

3. Is rectangle MNOP similar to rectangle QRST?

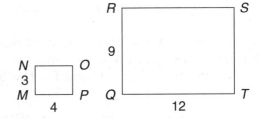

4. Are the two rectangles similar?

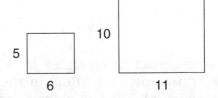

5. A snapshot that is 4 inches wide and 5 inches long was enlarged to be 20 inches wide. How long is the enlargement?

6. ∠M = ∠P and ∠O = ∠R. MO = 30, NO = 18, and PR = 25. Find QR.

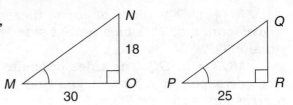

7. A 5-foot vertical pole casts a shadow 3 feet long at the same time that a building casts a shadow 72 feet long. How tall is the building? (Hint: To solve this problem draw a picture to show the height of each object, the shadow of each object, and a line from the top of each object to the ground. The picture should show two similar triangles.)

8. $\angle B$ and $\angle D$ are each 90°, $\angle A = \angle E$, $AB = 10$ ft, $BC = 3$ ft, and $CD = 24$ ft. DE is the distance across a river. Find the length of DE.

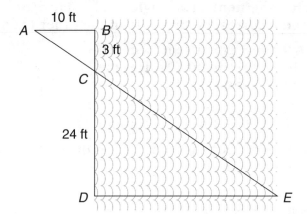

9. Every window in a new building is 6 feet high and 3 feet wide. The shape of the doors is proportional to the shape of the windows. Every door in the new building is 7 feet high. Find the width of each door.

10. Find the length of PQ.

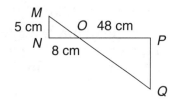

Check your answers on page 349.

Lesson 6

Congruent Figures

Two geometric figures are **congruent** when they have the same shape and the same size. In congruent figures all corresponding parts are equal. Corresponding angles are equal, and corresponding sides are equal.

Three sets of conditions guarantee that two triangles are congruent.

1. The **SAS** or **side, angle, side** requirement:

   Two triangles are congruent if two sides and an included angle of one triangle are equal to two sides and a corresponding angle of the other triangle.

Level 2, Lesson 6: Congruent Figures   243

These two triangles are congruent because they satisfy the SAS requirement. Notice how the small marks on the sides indicate equal sides.

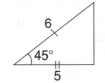

2. The **ASA** or **angle, side, angle** requirement:

Two triangles are congruent if two angles and a side of one triangle are equal to two angles and a corresponding side of the other triangle.

These two triangles are congruent because they satisfy the ASA requirement. Notice the small curves that indicate equal angles.

3. The **SSS** or **side, side, side** requirement:

Two triangles are congruent if the three sides of one triangle are equal to the three sides of the other triangle.

These two triangles are congruent because they satisfy the SSS requirement. Notice the small marks that indicate equal sides.

Example 1: Are the two triangles congruent?

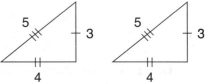

Yes. Two sides and an angle of one triangle are equal to two sides and a corresponding angle of the other triangle. The triangles satisfy the SAS requirement.

Example 2: Are the two triangles congruent?

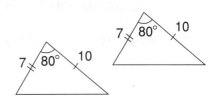

No. Two angles of one triangle are equal to two angles of the other triangle, but the equal sides do not correspond. One 9-inch side is opposite the 45° angle. The other 9-inch side is not. These triangles do not satisfy any of the requirements for congruence.

Example 3: Along with the conditions shown which of the following is sufficient to guarantee that the triangles are congruent?

Given: ∠M = ∠P and MN = PQ

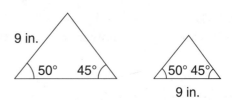

(1) MN = MO
(2) MO = PR
(3) MN = PR
(4) MO = PQ

Choice (2) is enough to satisfy the SAS requirement.
Choice (1) is about the triangle at the left only.
Choice (3) is about sides that do not correspond.
Choice (4) is also about sides that do not correspond.

Lesson 6 Exercise

For each pair of triangles in 1–6, decide if the two triangles are congruent. If they are congruent, tell which requirement (SAS, ASA, or SSS) the triangles satisfy. If they are not congruent, tell which requirement the triangles fail to satisfy.

1.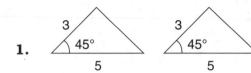

2.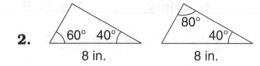

3.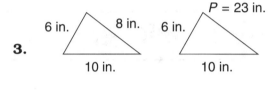

4.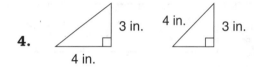

5.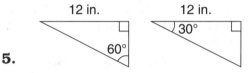

6.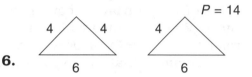

7. Of the following conditions, which, together with the information given, guarantees that the triangles are congruent?

    (1) $\angle C = \angle E$ (3) $\angle B = \angle D$
    (2) $\angle A = \angle E$ (4) $\angle B = \angle E$

    Given: $\angle C = \angle F$
    $AC = DF$

    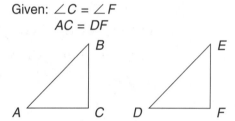

8. Of the following conditions, which, along with the information given, guarantees that the triangles are congruent?

    (1) $GI = JL$ (3) $HI = JL$
    (2) $GH = JL$ (4) $GI = JK$

    Given: $GH = JK$
    $HI = KL$

    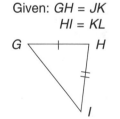 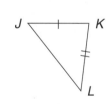

9. Which of the conditions below, along with the given information, makes the triangles congruent?
    A. $JL = KL$
    B. $JL = MO$
    C. $KL = MN$

    (1) A only   (4) A or B
    (2) B only   (5) B or C
    (3) C only

    Given: $JK = MN$
    $KL = NO$

    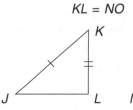 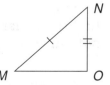

Level 2, Lesson 6: Congruent Figures   245

10. Which of the conditions below, along with the information given, ensures that the triangles at the right are congruent?
   A. $PR = SU$
   B. $\angle P = \angle U$
   C. $\angle Q = \angle T$

   Given: $PQ = ST$
   $\angle P = \angle S$

   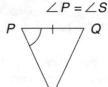

   (1) A only
   (2) B only
   (3) C only
   (4) A or B
   (5) A or C

Check your answers on page 350.

Lesson 7

The Pythagorean Relationship

In the sixth century B.C. a mathematician named Pythagoras discovered that the sides of a right triangle have a special relationship. He found that the square of the hypotenuse equals the sum of the squares of the other two sides. We call this relationship the **Pythagorean** relationship or theorem.

Remember that the hypotenuse is the side across from the right angle.

> The formula for the Pythagorean relationship is:
> $c^2 = a^2 + b^2$, where c = the hypotenuse and a and b are the legs of a right triangle.

Example 1: What is the length of the hypotenuse of the triangle?

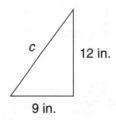

Step 1. Replace a with 9 and b with 12 in the formula for the Pythagorean relationship.

$c^2 = a^2 + b^2$
$c^2 = 9^2 + 12^2$

Step 2. Solve the formula for c. Notice that the formula gives c to the second power. To solve for c find the square root of 225. The length of the hypotenuse is 15 inches.

$c^2 = 81 + 144$
$c^2 = 225$
$c = \sqrt{225}$
$c = 15$ in.

Remember that c always stands for the hypotenuse. In some problems you must find the length of one of the legs.

Example 2: Find the length of the side labeled a.

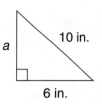

Step 1. Replace c with 10 and b with 6 in the formula for the Pythagorean relationship.

$$c^2 = a^2 + b^2$$
$$10^2 = a^2 + 6^2$$

Step 2. Solve for a. First subtract 36 from both sides. Then find the square root of 64. Side a is 8 inches long.

$$100 = a^2 + 36$$
$$\underline{-\ 36 = \ -\ 36}$$
$$64 = a^2$$
$$\sqrt{64} = a$$
$$8 \text{ in.} = a$$

Lesson 7 Exercise

Solve each problem.

1. Find the length of the hypotenuse.

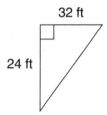

2. The legs of a right triangle measure 36 inches and 48 inches respectively. Find the length of the hypotenuse.

3. In the right triangle what is the length of side AB?

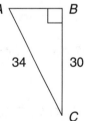

4. The hypotenuse of a right triangle measures 13 feet and one leg measures 5 feet. Find the length of the other leg.

5. The diagram shows the plan of a rectangular garden. What is the diagonal distance shown by the dotted line?

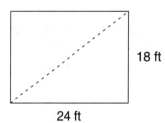

6. Find the length of line AC.

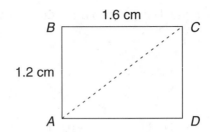

7. The isosceles triangle has a height of 24 inches. Sides XY and YZ are each 26 inches long. What is the length of the base?

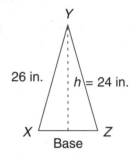

8. Find the length of the side labeled b.

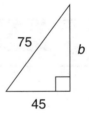

9. Figure MNOP is a rectangle. Side MP is 15 feet and the diagonal distance MO is 17 feet. Find the length of OP.

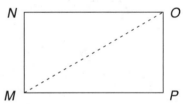

10. Geraldine drove 20 miles east and then 15 miles south. What was the shortest distance from her starting point to her ending point? (Hint: Draw a diagram to show her path.)

Check your answers on page 350.

Lesson 8

Rectangular Coordinates

A graph called the **rectangular coordinate system** is a tool for picturing algebraic and geometric relationships. To understand rectangular coordinates you need skills in both algebra and geometry.

The rectangular coordinate system is a plane (a flat surface) divided by two perpendicular lines. The horizontal line is called the **x-axis**. The vertical line is called the **y-axis**. Each axis looks like the number line you studied earlier. On the x-axis positive numbers are to the right of zero and negative numbers are to the left. On the y-axis positive numbers are above zero and negative numbers are below. The lines intersect (cross) at zero on each axis. This point is called the **origin**.

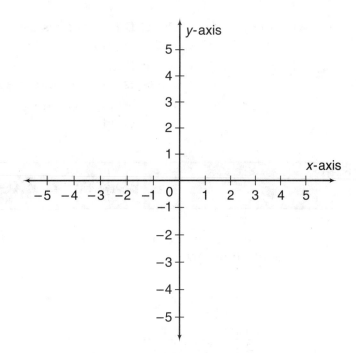

The location of every point on the plane can be described by two numbers called the **coordinates** of a point. The numbers, written inside parentheses, tell how far a point is from each axis.

The first number, called the **x-coordinate**, tells the distance to the right or left of the vertical axis. A positive x-coordinate is to the right of the vertical axis and a negative x-coordinate is to the left.

The second number, called the **y-coordinate** tells the distance above or below the horizontal axis. A positive y-coordinate is above the horizontal axis, and a negative y-coordinate is below. A comma (,) separates the coordinates.

Examples: What are the coordinates for each lettered point on the rectangular coordinate system?

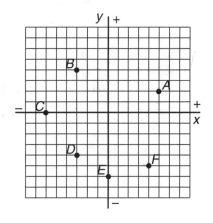

A = (+5, +2)   Point A is 5 spaces right of the y-axis and 2 spaces above the x-axis.

B = (−3, +4)   Point B is 3 spaces left of the y-axis and 4 spaces above the x-axis.

C = (−6, 0)   Point C is 6 spaces left of the y-axis and directly on the x-axis. Notice that the y-coordinate is zero.

Level 2, Lesson 8: Rectangular Coordinates

$D = (-3, -4)$ Point D is 3 spaces left of the y-axis and 4 spaces below the x-axis.

$E = (0, -6)$ Point E is directly on the y-axis and 6 spaces below the x-axis. Notice that the x-coordinate is zero.

$F = (+4, -5)$ Point F is 4 spaces right of the y-axis and 5 spaces below the x-axis.

Lesson 8 Exercise

1. Write the coordinates of each lettered point on the rectangular coordinate system below.

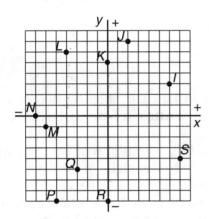

I =   N =
J =   P =
K =   Q =
L =   R =
M =   S =

2. Put the points that correspond to the following coordinates on the rectangular coordinate system below.

A = (+6, +5)   E = (−4, −3)
B = (+1, +8)   F = (−2, −7)
C = (−3, +6)   G = (+4, −6)
D = (−5, 0)   H = (+5, 0)

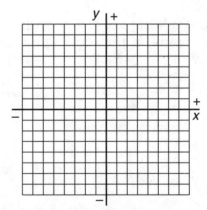

Check your answers on page 351.

Lesson 9: Finding the Distance Between Points

The distance between two points on the rectangular coordinate system is measured in numbers with no units. Each space corresponds to a whole number. In some problems you can simply count the spaces to find the distance between two points.

Example 1: What is the distance between points M and N?

M is 2 spaces left of the y-axis.
N is 4 spaces right of the y-axis.
Distance MN is $2 + 4 = 6$.

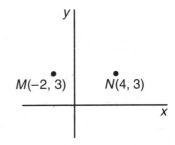

Example 2: What is the distance between points P and Q?

P is 5 spaces above the x-axis.
Q is 3 spaces below the x-axis.
Distance PQ is $5 + 3 = 8$.

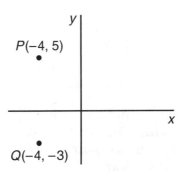

When two points in a plane do not lie on a horizontal or vertical line, the distance (d) between the points is: $d = \sqrt{(x_2 - x_1)^2 + (y_2 - y_1)^2}$ where (x_1, y_1) and (x_2, y_2) are the coordinates of two points.

Notice the small numbers in the expressions (x_1, y_1) and (x_2, y_2). These numbers, called **subscripts**, simply distinguish between different x-coordinates and different y-coordinates.

In the formula the distance d is the hypotenuse of a right triangle. The expression $(x_2 - x_1)$ is the length of the horizontal leg, and the expression $(y_2 - y_1)$ is the length of the vertical leg.

Example 3: What is the distance between points C and D?

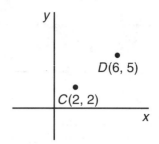

Step 1. Replace x_2 with 6, x_1 with 2, y_2 with 5, and y_1 with 2 in the formula for the distance between two points in a plane.

$C = (x_1, y_1) = (2, 2)$ and
$D = (x_2, y_2) = (6, 5)$
$d = \sqrt{(x_2 - x_1)^2 + (y_2 - y_1)^2}$
$d = \sqrt{(6 - 2)^2 + (5 - 2)^2}$

Step 2. Solve the formula for d. The distance between points C and D is 5.

$d = \sqrt{(4)^2 + (3)^2}$
$d = \sqrt{16 + 9}$
$d = \sqrt{25}$
$d = 5$

Example 4: What is the distance between points K and L?

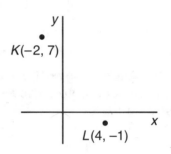

Step 1. Replace x_2 with 4, x_1 with -2, y_2 with -1, and y_1 with 7 in the formula for the distance between two points in a plane. Notice how the extra set of parentheses around -2 and 7 separate these numbers from the minus signs in the formula.

$K = (x_1, y_1) = (-2, 7)$ and
$L = (x_2, y_2) = (4, -1)$
$d = \sqrt{(x_2 - x_1)^2 + (y_2 - y_1)^2}$
$d = \sqrt{(4 - (-2))^2 + (-1 - (7))^2}$
$d = \sqrt{(4 + 2)^2 + (-1 - 7)^2}$

Step 2. Solve the formula for d. The distance between points K and L is 10.

$d = \sqrt{(6)^2 + (-8)^2}$
$d = \sqrt{36 + 64}$
$d = \sqrt{100}$
$d = 10$

You may be asked on the GED Test to find the coordinates of the midpoint of a line that connects two points. The formula for the midpoint M is:

$$M = \left(\frac{x_1 + x_2}{2}, \frac{y_1 + y_2}{2}\right),$$ where (x_1, y_1) and (x_2, y_2) are the coordinates of two points in a plane.

This formula tells you to find the average of the *x*-coordinates and the average of the *y*-coordinates for the two points.

Example 5: What are the coordinates of the midpoint M which lies half-way between points R and S on the graph?

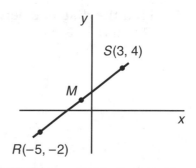

Step 1. Replace x_1 with -5, x_2 with 3, y_1 with -2 and y_1 with 4 in the formula for the coordinates of the midpoint between two points in a plane.

$R = (x_1, y_1) = (-5, -2)$
$S = (x_2, y_2) = (3, 4)$
$$M = \left(\frac{x_1 + x_2}{2}, \frac{y_1 + y_2}{2}\right)$$
$$M = \left(\frac{-5 + 3}{2}, \frac{-2 + 4}{2}\right)$$

Step 2. Simplify the expression. The coordinates for the midpoint M between R and S are $(-1, +1)$.

$$M = \left(\frac{-2}{2}, \frac{+2}{2}\right)$$
$$M = (-1, +1)$$

Lesson 9 Exercise

Solve each problem.

Use the illustration for problems 1–3.

1. What is the distance between points A and B?
2. What is the distance between points B and C?
3. What is the distance between points C and D?

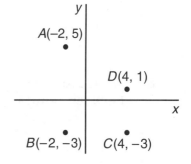

4. Find the distance between points E and F.

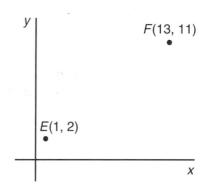

Level 2, Lesson 9: Finding the Distance Between Points 253

5. Find the distance between points G and H.

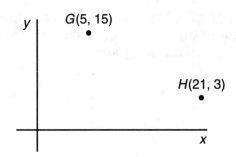

6. What is the distance between I and J?

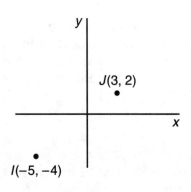

7. Find the distance between points K and L.

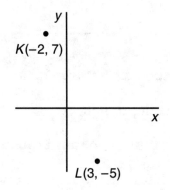

8. What are the coordinates of the midpoint between points N and P?

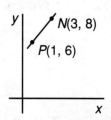

9. Find the coordinates of the midpoint between points Q and R.

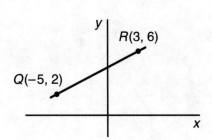

10. What are the coordinates of the midpoint between points S and T?

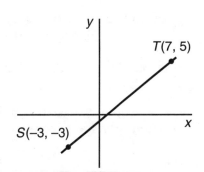

11. Find the coordinates of the midpoint between U and V.

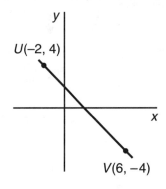

Check your answers on page 351.

Lesson 10

Graphs of Linear Equations

So far you have solved equations that had either one solution (for example, $3x + 2 = 14$ where $x = 4$) or two solutions (for example, $x^2 = 25$ where $x = +5$ or -5).

Think about the equation $y = x - 3$. For every value you can think of for x there is a corresponding value for y. For example,

When $x = 7$, $y = 7 - 3 = 4$.
When $x = 4$, $y = 4 - 3 = 1$.
When $x = -2$, $y = -2 - 3 = -5$.

Each value of x and the corresponding value of y give the coordinates of a point on the rectangular coordinate system. The three points from the examples are $(7, 4)$, $(4, 1)$, and $(-2, -5)$. The picture at the right shows these three points on the rectangular coordinate system.

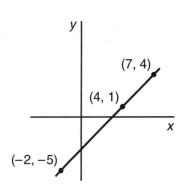

The three points lie on a straight line. The line is called the graph of the equation $y = x - 3$.

Level 2, Lesson 10: Graphs of Linear Equations 255

An equation whose graph is a straight line is called a **linear equation**. The coordinates of every point on the line are solutions of the equation.

Example 1: Draw the graph of the equation $y = 3x + 2$. Find values of y when $x = 2, -1,$ and -3.

Step 1. Replace x with 2, -1, and -3 to find the corresponding values of y.

$y = 3(2) + 2 = 6 + 2 = 8$
$y = 3(-1) + 2 = -3 + 2 = -1$
$y = 3(-3) + 2 = -9 + 2 = -7$

Step 2. Make a chart of each x-coordinate and the corresponding y-coordinate.

x	y
2	8
-1	-1
-3	-7

Step 3. Put each point on a coordinate system and connect the points with a line.

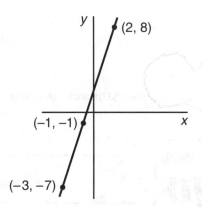

You can find whether a point lies on the graph of an equation by using substitution. Substitute the x-coordinate into the equation. The result should equal the y-coordinate.

Example 2: Does the point (2, 4) lie on the graph of the equation $y = -x + 6$?

Step 1. The x-coordinate is 2. Replace x with 2 in the equation $y = -x + 6$.

$y = -x + 6$
$y = -(2) + 6$

Step 2. Solve for y. Since $y = 4$, the point (2, 4) does lie on the graph.

$y = -2 + 6$
$y = 4$
Point (2, 4) is on the graph.

Example 3: Is point $(-3, 5)$ on the graph of the equation $y = -x + 6$?

Step 1. The x-coordinate is -3. Replace x with -3 in $y = -x + 6$.

$y = -x + 6$
$y = -(-3) + 6$

Step 2. Solve for y. Since $y = +9$, the point $(-3, 5)$ is **not** on the graph.

$y = +3 + 6$
$y = +9$
Point $(-3, 5)$ is not on the graph.

Lesson 10 Exercise

For problems 1–4 find a value of y for each value of x. Fill in the chart with the corresponding values of x and y. Then put each point on the graph and connect the points with a line.

1. $y = x + 4$
   Let $x = 3, -2,$ and -5.

x	y

   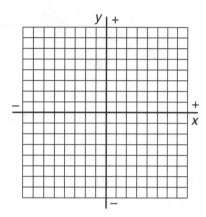

2. $y = \frac{x}{2} + 1$
   Let $x = 8, 4,$ and -6.

   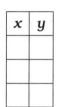

   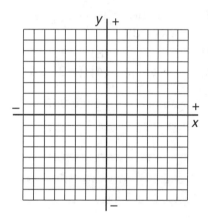

3. $y = -3x + 4$
   Let $x = 3, 1,$ and -2.

   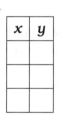

   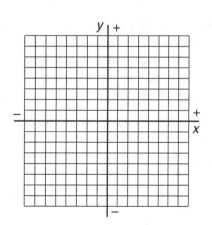

Level 2, Lesson 10: Graphs of Linear Equations

4. $y = -2x - 3$
   Let $x = 2, -3,$ and -4.

x	y

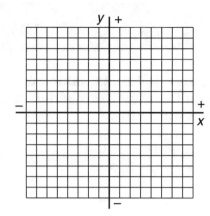

5. Is (1, 8) a solution of the equation $y = 5x + 3$?
6. Is (2, −4) a solution of the equation $y = -3x + 1$?
7. Does point (3, 5) lie on the graph of the equation $y = -x + 6$?
8. Does point (−8, −8) lie on the graph of the equation $y = \frac{3}{4}x - 2$?
9. Which of the following points is on the graph of the equation $y = x - 5$?
   (1) (4, 0)   (3) (2, −3)
   (2) (3, −4) (4) (−1, −5)
10. Which of the following points lies on the graph of the equation $y = -2x - 3$?
    (1) (3, −9)   (3) (−1, +1)
    (2) (2, −4)   (4) (−3, +4)

Check your answers on page 352.

Lesson 11

Graphs and Quadratic Equations

In the last lesson you learned to recognize linear equations. The graphs of these equations are straight lines. Graphs of equations with the unknown raised to the second power are not straight lines.

Think about the equation $y = x^2 + 3$.

When $x = 2, y = (2)^2 + 3 = 4 + 3 = 7$.
When $x = 1, y = (1)^2 + 3 = 1 + 3 = 4$.
When $x = 0, y = (0)^2 + 3 = 0 + 3 = 3$.
When $x = -1, y = (-1)^2 + 3 = 1 + 3 = 4$.
When $x = -2, y = (-2)^2 + 3 = 4 + 3 = 7$.

Each value of x and the corresponding value of y give the coordinates of a point on the rectangular coordinate system. The five points from the examples are (2, 7), (1, 4) (0, 3), (−1, 4) and (−2, 7). The picture at the right shows these five points on the rectangular coordinate system.

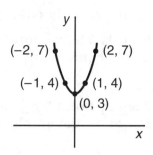

The five points lie on a curve called a **parabola**. An equation whose graph is a parabola is called a **quadratic equation**.

Example 1: Draw a graph of the equation $y = x^2 + x$. Find values of y when $x = 2, 1, 0, −1, −2,$ and $−3$.

<u>Step 1.</u> Replace x with 2, 1, 0, −1, −2, and −3 to find the corresponding values of y.

$y = (2)^2 + 2 = 4 + 2 = 6$
$y = (1)^2 + 1 = 1 + 1 = 2$
$y = (0)^2 + 0 = 0 + 0 = 0$
$y = (−1)^2 + (−1) = 1 − 1 = 0$
$y = (−2)^2 + (−2) = 4 − 2 = 2$
$y = (−3)^2 + (−3) = 9 − 3 = 6$

<u>Step 2.</u> Make a chart of each x-coordinate and the corresponding y-coordinate.

x	y
2	6
1	2
0	0
−1	0
−2	2
−3	6

<u>Step 3.</u> Put each point on the coordinate system and connect the points with a curved line.

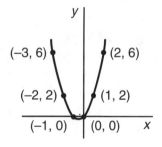

To find whether a point lies on the graph of an equation, replace x in the equation with the x-coordinate of the point. The value should equal the y-coordinate.

Example 2: Does the point (3, 12) lie on the graph of $y = x^2 + 3$?

<u>Step 1.</u> The x-coordinate is 3. Replace x with 3 in the equation $y = x^2 + 3$.

$y = x^2 + 3$
$y = (3)^2 + 3$

<u>Step 2.</u> Solve for y. Since $y = 12$, the point (3, 12) does lie on the graph.

$y = 9 + 3 = 12$
Point (3, 12) is on the graph.

Level 2, Lesson 11: Graphs and Quadratic Equations

Example 3: Is point $(-2, 15)$ on the graph of the equation $y = x^2 - 4x + 5$?

Step 1. The *x*-coordinate is -2. Replace *x* with -2 in the equation $y = x^2 - 4x + 5$.

$y = x^2 - 4x + 5$
$y = (-2)^2 - 4(-2) + 5$

Step 2. Solve for *y*. Since $y = 17$, the point $(-2, 15)$ is **not** on the graph.

$y = 4 + 8 + 5 = 17$
Point $(-2, 15)$ is not on the graph.

Lesson 11 Exercise

For problems 1–4 find a value of *y* for each value of *x*. Fill in the chart with the corresponding values of *x* and *y*. Then put each point on the graph and connect the points with a curved line.

1. $y = x^2 + 2$
   Let $x = 2, 1, 0, -1,$ and -2

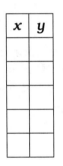
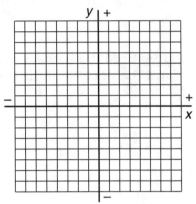

2. $y = x^2 - 2x$
   Let $x = 4, 3, 2, 1, 0, -1,$ and -2.

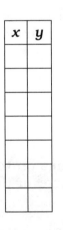
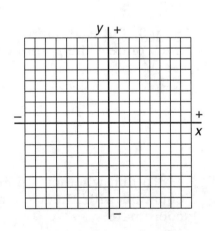

260

3. $y = x^2 + x - 2$
   Let $x = 2, 1, 0, -1, -2,$ and -3.

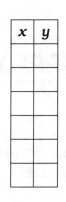

4. $y = x^2 + x - 4$
   Let $x = 2, 1, 0, -1, -2,$ and -3.

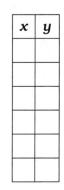
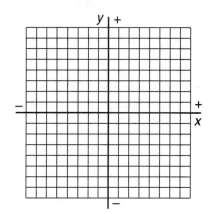

5. Is the point (5, 4) on the graph of the equation $y = x^2 - 4x$?
6. Does point (4, 23) lie on the graph of the equation $y = x^2 + x + 3$?
7. Is point (−3, 11) on the graph of the equation $y = x^2 - x - 1$?
8. Does point (−2, −4) lie on the graph of the equation $y = x^2 + 3x - 2$?
9. Which of the following points is on the graph of $y = x^2 + 2x$?
   (1) (4, 20)   (3) (1, 2)
   (2) (3, 15)   (4) (−2, 1)
10. Which of the following lies on the graph of the equation $y = x^2 - 5x + 3$?
    (1) (3, 3)   (3) (−1, 8)
    (2) (2, −4)   (4) (−2, 17)

Check your answers on page 353.

Lesson 12: Slope and Intercepts

Slope and **intercept** are used to describe the graph of a linear equation. Slope tells how much a line slants or leans. When a line rises from left to right, the slope is **positive**. The graph for the equation $y = x + 4$ has positive slope.

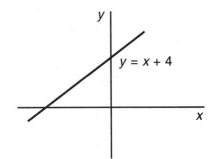

When a line falls from left to right, the slope is **negative**. The graph for the equation $y = -x + 2$ has negative slope.

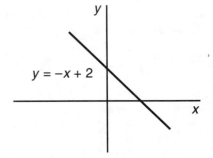

A horizontal line has **zero** slope. The graph of the equation $y = 3$ has zero slope.

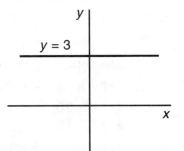

A vertical line has what is called **undefined** slope. The slope of the equation $x = 5$ has undefined slope.

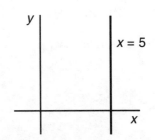

The formula for the slope (m) of a line is

$$m = \frac{y_2 - y_1}{x_2 - x_1}$$

where (x_1, y_1) and (x_2, y_2) are the coordinates of two points in a plane.

Example 1: What is the slope of the line that passes through points A and B?

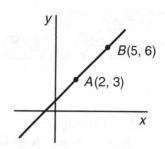

Step 1. Replace y_2 with 6, y_1 with 3, x_2 with 5, and x_1 with 2 in the formula for the slope.
Step 2. Solve for m. The slope is +1.

$A = (x_1, y_1) = (2, 3)$ and
$B = (x_2, y_2) = (5, 6)$

$$m = \frac{y_2 - y_1}{x_2 - x_1} = \frac{6 - 3}{5 - 2} = \frac{3}{3} = +1$$

An **intercept** is the point where two lines cross. For an equation on the rectangular coordinate system, an intercept tells where the line of the equation crosses the x-axis or the y-axis.

The y-intercept tells where the line of an equation crosses the y-axis. The x-coordinate of a y-intercept is always zero. To find a y-intercept substitute 0 for x in an equation.

Example 2: What are the coordinates of the y-intercept for the equation $y = -2x + 3$?

Step 1. Replace x with 0 in $y = -2x + 3$.

$y = -2x + 3$
$y = -2(0) + 3$

Step 2. Solve for y. The value of y when x is zero is +3. The coordinates of the y-intercept are (0, 3).

$y = 0 + 3 = 3$
y-intercept = (0, 3)

The x-intercept tells where the line of an equation crosses the x-axis. The y-coordinate of an x-intercept is always zero. To find an x-intercept substitute 0 for y in an equation.

Example 3: What are the coordinates of the x-intercept for the equation $y = 2x - 4$?

Step 1. Replace y with 0 in $y = 2x - 4$.

Step 2. Solve for x. The value of x when y is zero is +2. The coordinates of the x-intercept are (2, 0).

$y = 2x - 4$
$0 = 2x - 4$
$+4 \qquad +4$
$\dfrac{4}{2} = \dfrac{2x}{2}$
$2 = x$
x-intercept = (2, 0)

Level 2, Lesson 12: Slope and Intercepts

Lesson 12 Exercise

Solve each problem.

1. What is the slope of a line that passes through points C and D?

   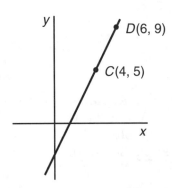

2. Find the slope of the line that passes through points E and F.

   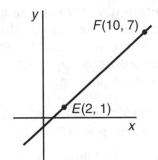

3. What is the slope of a line that passes through points G and H?

   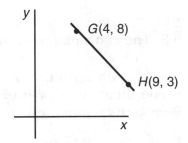

4. Find the slope of the line that passes through points I and J.

   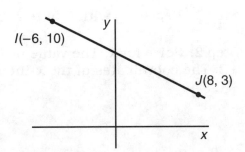

264 Chapter 7 Geometry

5. Which line has a positive slope?

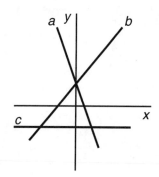

In problems 6–8 find the coordinates of the y-intercept.

6. $y = 2x - 3$
7. $y = -2x + 1$
8. $y = 5x - 7$

Find the coordinates of the x-intercept.

9. $y = 3x - 9$
10. $y = 8x + 4$

Check your answers on page 354.

Level 2 Review

Solve each problem.

1. A photograph that is 3 inches wide and 5 inches long was enlarged to be 15 inches wide. How long was the enlargement?
2. The illustration shows a diagram of two sails that are similar triangles. Find the height of the larger sail.

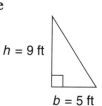
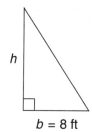

3. $MN = 12$, $NO = 20$, and $PQ = 15$. Find distance OP.

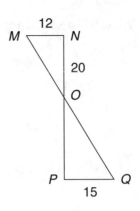

4. For the two triangles, $DE = GH$ and $\angle E = \angle H$. Along with these conditions, which of the following is sufficient to make the two triangles congruent?

   (1) $\angle D = \angle I$
   (2) $\angle D = \angle G$
   (3) $\angle F = \angle G$
   (4) $\angle E = \angle I$

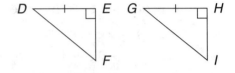

5. For the triangles, $RS = UV$ and $\angle R = \angle U$. Along with these conditions, which of the following is sufficient to ensure that the triangles are congruent?

   (1) $RS = UW$
   (2) $RT = UV$
   (3) $UV = UW$
   (4) $RT = UW$

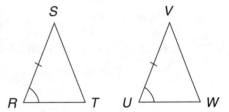

6. Find the diagonal distance MO for the rectangle.

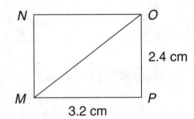

7. Find the length of the side labeled b.

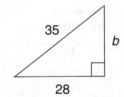

Use the illustration to answer questions 8 and 9.

8. Which of the following represents the coordinates for point A?
   (1) (3, 3)
   (2) (5, 5)
   (3) (5, 2)
   (4) (2, 2)

9. Which of the following represents the coordinates for point B?
   (1) (−3, 4)
   (2) (4, 3)
   (3) (4, −3)
   (4) (−3, −4)

Use the illustration for problems 10 and 11.

10. What is the distance between E and F?

11. The coordinates of the midpoint M that lies half-way between two points in a plane is $M = \left(\dfrac{x_1 + x_2}{2}, \dfrac{y_1 + y_2}{2}\right)$ where (x_1, y_1) and (x_2, y_2) are the coordinates of the two points. Find the coordinates of the midpoint between D and E.

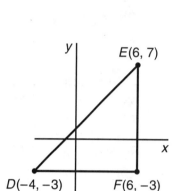

12. Which of the following points lies on the graph of the equation $y = x + 2$?
    (1) (0, 5)
    (2) (1, 3)
    (3) (2, 7)
    (4) (3, 8)

13. Of the points shown by the following coordinates, which lies on the graph of the equation $y = 3x - 4$?
    (1) (−3, −10)
    (2) (−1, −5)
    (3) (2, 3)
    (4) (4, 8)

14. Which of the following points lies on the graph of the equation $y = x^2 - 3x$?
    (1) (−3, 18)
    (2) (−2, 9)
    (3) (1, −1)
    (4) (4, 8)

Level 2: Review   267

15. Of the following points, which lies on the graph of the equation $y = x^2 + 2x - 1$?
    (1) (5, 36)
    (2) (3, 14)
    (3) (−2, −2)
    (4) (−4, 6)

16. Which of the following tells the slope of a straight line that connects points A and B?
    (1) $-\frac{2}{3}$
    (2) $+\frac{2}{3}$
    (3) $-\frac{3}{2}$
    (4) $+\frac{3}{2}$

17. Which of the following are the coordinates of the y-intercept of the equation $y = 5x + 3$?
    (1) (0, 5)
    (2) (0, 3)
    (3) (−3, 0)
    (4) (5, 0)

18. Which of the following are the coordinates of the x-intercept of the equation $y = -2x + 7$?
    (1) (−7, 0)
    (2) $(0, -\frac{7}{2})$
    (3) (2, 0)
    (4) $(\frac{7}{2}, 0)$

Check your answers on page 354. If you have all 18 problems correct, go on to Level 3. If you answered a question incorrectly, find the item number on the chart below and review that lesson before you go on.

Review:	If you missed item number:
Lesson 5	1, 2, 3
Lesson 6	4, 5
Lesson 7	6, 7
Lesson 8	8, 9
Lesson 9	10, 11
Lesson 10	12, 13
Lesson 11	14, 15
Lesson 12	16, 17, 18

Level 3 Geometry Problem Solving

In this section you will learn to use algebra skills to solve geometry word problems.

Lesson 13 Using Algebra to Solve Formulas

In the algebra chapter you learned to solve distance, cost, and interest problems when the information you had was "backwards."

You can use the same skills to solve many perimeter, area, and volume problems.

Example 1: The area of the rectangle is 144 sq in. Find the length.

$w = 9$ in.

l

Step 1. Replace A with 144 and w with 9 in the formula $A = lw$.

$A = lw$
$144 = l \cdot 9$

Step 2. Solve for the length by dividing both sides by 9. The length is 16 inches.

$$\frac{144}{9} = \frac{l \cdot 9}{9}$$
$16 \text{ in.} = l$

Example 2: Find the measure of a side of a square that would have the same area as a rectangle that is 25 feet long and 16 feet wide.

Step 1. Find the area of the rectangle.

$A = lw$
$A = 25 \cdot 16$
$A = 400 \text{ ft}^2$

Step 2. To find the side of the square, replace A with 400 in the formula $A = s^2$.

$A = s^2$
$400 = s^2$

Step 3. Solve for the side by finding the square root of 400. The side is 20 feet.

$\sqrt{400} = s$
$20 \text{ ft} = s$

Level 3, Lesson 13: Using Algebra to Solve Formulas

Lesson 13 Exercise

Solve each problem. Use any formulas from pages 17 and 18 that you need.

1. The area of the rectangle is 132 square feet. Find the width.

   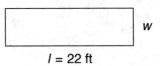

2. The distance around the garden in Mike's yard is 124 feet. The garden is a rectangle with a width of 16 feet. Find the length.

3. The perimeter of the triangle is 55 inches. Find the length of the missing side.

   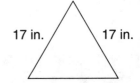

4. A circle has a circumference of 15.7 meters. Find the diameter of the circle.

5. What is the length of one side of a square that has an area of 81 square yards?

6. The perimeter of the figure is 150 feet. Find the length of one side.

7. The area of the triangle is 108 square feet. What is the height?

   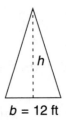

8. The volume of the concrete in a patio is 180 cubic feet. The patio is 10 feet wide and $\frac{1}{2}$ foot deep. What is the length?

9. The rectangle has a perimeter of 46 cm. Find the width of the rectangle.

   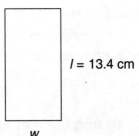

10. A triangle has an area of 84 square inches. The height of the triangle is 14 inches. What is the length of the base?

11. The square and the rectangle at the right have the same perimeter. Find the width of the rectangle.

 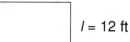

$l = 12$ ft

$s = 16$ ft

12. Find the length of a side of a square that has the same area as a rectangle that is 25 inches long and 9 inches wide.

Check your answers on page 355.

Lesson 14: Algebraic Expressions in Formulas

In the last lesson you substituted numbers into formulas and solved for the missing numbers. In some problems there is also an algebraic relationship between the parts of a formula, such as the length and the width in the formula for the area of a rectangle. Study the next examples carefully.

Example 1: A rectangle has a perimeter of 60 feet. The length of the rectangle is twice the width. Find both the length and the width.

Step 1. Let x represent the width and let $2x$ represent the length.

$w = x$ and $l = 2x$

Step 2. Replace P with 60, l with $2x$, and w with x in the formula $P = 2l + 2w$.

$P = 2l + 2w$
$60 = 2(2x) + 2x$

Step 3. Solve for x.

$60 = 4x + 2x$
$$\frac{60}{6} = \frac{6x}{6}$$
$10 = x$

Step 4. The solution is $x = 10$, which is the width. Find the value of $2x$, which is the length. The length is 20 ft and the width is 10 ft.

$w = 10$ ft
$l = 2 \cdot 10 = 20$ ft

Example 2: The volume of the rectangular container is 160 cubic feet. The height of the container is twice the width. Find the width and the height of the container.

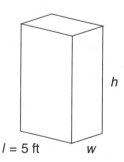

$l = 5$ ft w

Step 1. Let x represent the width and let $2x$ represent the height.

$w = x$ and $h = 2x$

Step 2. Replace V with 160, l with 5, w with x, and h with $2x$ in the formula $V = lwh$.

$V = lwh$
$160 = 5 \cdot x \cdot 2x$

Step 3. Solve the equation for x. (Notice that we are interested only in the positive value of x.)

$$\frac{160}{10} = \frac{10x^2}{10}$$
$$16 = x^2$$
$$\sqrt{16} = x$$
$$4 = x$$

Step 4. The solution is x = 4, the width. Find the value of 2x, the height. The width is 4 ft and the height is 8 ft.

$$w = 4 \text{ ft}$$
$$h = 2 \cdot 4 = 8 \text{ ft}$$

Lesson 14 Exercise

Solve each problem. Use any formulas on pages 17 and 18 that you need.

1. The distance around the kitchen in Miriam's house is 48 feet. The length of the kitchen is 6 feet more than the width. Find the length and the width.
2. The length of a rectangle is three times the width. The perimeter is 64 feet. Find the length and the width.
3. In the triangle side b is 2 inches more than side a, and side c is 4 inches more than side a. The perimeter is 36 inches. Find the length of each side.

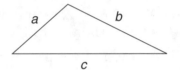

4. In the isosceles triangle each equal side is twice as long as the base. The perimeter is 45 meters. Find the length of each long side.

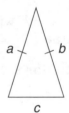

5. The length of a rectangle is twice the width. The area of the rectangle is 98 square yards. Find the length and the width.
6. In another rectangle the width is half the length. The area is 50 square inches. Find the length and the width.
7. The ratio of the length to the width of the living room in Ann's house is 4:3. The area of the floor of the room is 300 square feet. Find the length and the width.
8. The ratio of the base to the height of the triangle is 2:3. The area of the triangle is 108 square inches. Find the base and the height.

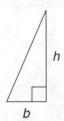

9. The height of a triangle is one-third of the base, and the area is 24 square meters. Find the height and the base.

10. The volume of the rectangular container is 1440 cubic inches. The height is eight times the length. Find the length and the height.

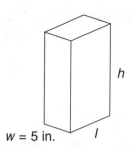

11. Another rectangular container has a volume of 3000 cubic feet and a length of 10 feet. The ratio of the width to the height of the container is 3 : 4. Find the width and the height.

12. The volume of the rectangular container is 64 cubic inches. The width of the container is half the length. Find the length and the width.

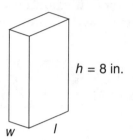

Check your answers on page 356.

Lesson 15: Factoring and the Pythagorean Relationship

In the algebra chapter you learned to simplify square roots by factoring. You can use this skill to express the answers to some right triangle problems.

Example 1: Find the hypotenuse of the right triangle.

Step 1. Replace a with 5 and b with 5 in the formula for the Pythagorean relationship.

$$c^2 = a^2 + b^2$$
$$c^2 = 5^2 + 5^2$$
$$c^2 = 25 + 25$$
$$c^2 = 50$$

Step 2. Solve the equation for c.

$$c = \sqrt{50}$$

Step 3. The factors of 50 that include an exact square are 25 and 2. Put each factor inside a $\sqrt{}$ sign. Then write the square root of 25 next to $\sqrt{2}$.

$$c = \sqrt{25} \cdot \sqrt{2}$$
$$c = 5\sqrt{2}$$

Sometimes there is no factor with a perfect square.

Level 3, Lesson 15: Factoring and the Pythagorean Relationship

Example 2: Find the hypotenuse of the triangle.

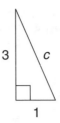

Step 1. Replace a with 1 and b with 3 in the formula for the Pythagorean relationship.

$c^2 = a^2 + b^2$
$c^2 = 1^2 + 3^2$

Step 2. Solve the equation for c. Since no factor of 10 has an exact square root, $\sqrt{10}$ is the simplest form for the answer.

$c^2 = 1 + 9$
$c^2 = 10$
$c = \sqrt{10}$

Lesson 15 Exercise

Solve each problem.

1. Express in simplest form the length of the hypotenuse of the triangle.

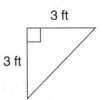

2. What is the hypotenuse of the triangle? Express the answer in simplest form.

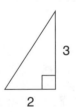

3. Find the diagonal distance AC of the square.

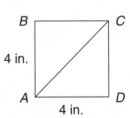

For problems 4 and 5, first make a sketch of the distance the person travels in the problem. Then solve.

4. Pete bicycled 3 miles west and then 6 miles south. Find the shortest distance from his starting point to his end point. Express the answer in simplest form.

5. Claire walked 3 miles north, then 8 miles to the west, and finally 5 miles to the north. Find the shortest distance from her starting point to her end point. Express the answer in simplest form.

Check your answers on page 357.

274   Chapter 7   Geometry

Level 3 Review

Solve each problem.

1. The perimeter of the triangle is 48 inches. Find the missing side.

   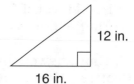

2. The distance around Silvia's rectangular vegetable garden is 40 feet. The garden is 9 feet wide. Find the length.

3. A rectangle has a width of 5 feet and an area of 95 square feet. What is the length of the rectangle?

4. The volume of a rectangular container is 420 cubic feet. The length of the container is 7 feet and the height is 10 feet. What is the width?

5. Find the length of one side of a square that has the same area as the rectangle.

   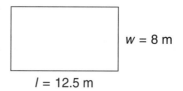

6. A rectangle has an area of 75 square feet. The width is one-third of the length. Find the width and the length.

7. The length of a rectangle is 8 inches more than the width. The perimeter is 100 inches. Find the width and the length.

8. The ratio of the length to the width of the rectangle is $3:2$. Find the length and width.

   $A = 294 \text{ ft}^2$

9. The height of a triangle is twice the base. The area of the triangle is 64 square meters. Find the base and the height.

10. A rectangular container has a volume of 216 cubic inches. The height of the container is 12 inches. The length is twice the width. Find the length and the width.

11. Express in simplest form the hypotenuse of the triangle.

12. Jeff walked 4 miles east and then 8 miles north. Find the shortest distance from his starting point to his end point. Express your answer in simplest form.

13. What is the diagonal distance *AC* on the square?

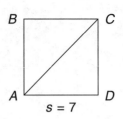

14. Ellen bicycled 2 miles south and then 6 miles east. Express in simplest form the shortest distance from her starting point to her end point.

Check your answers on page 358. If you have all 14 problems correct, go on to the Posttest. If you have answered a question incorrectly, find the item on the chart below and review that lesson before you go on.

Review:	If you missed item number:
Lesson 13	1, 2, 3, 4, 5
Lesson 14	6, 7, 8, 9, 10
Lesson 15	11, 12, 13, 14

POSTTEST

Like the actual Mathematics Test, the items on the Posttest appear in mixed order. The test provides you with two kinds of practice necessary for the GED: practice on the items themselves and practice on switching from one type of mathematics to another.

While the Posttest has 50 multiple-choice items (the actual test has 56), it is similarly challenging. By taking the Posttest, you can gain valuable test-taking experience and will know what to expect when you sit down to take the actual GED Mathematics Test.

Try to complete the test in one sitting with as little distraction as possible. Write your answers on a sheet of paper or use an answer sheet. If a question gives you trouble, take an educated guess and then move on.

Compare your answers to those in the answer key on page 358. Whether you answer an item correctly or not, look over the solution in the answer key. This will help reinforce your ability with mathematics and develop your test-taking skills.

After scoring your test with the answer section, use the Posttest Guide to direct you to parts of the book where you can review areas in which you need additional work.

Directions: Choose the one best answer for each item. Use any of the formulas on pages 17 and 18 that you need.

1. Of the 120 employees at the High Rock Mill, 80% drive to work. How many of the employees at the mill drive to work?
   (1) 60
   (2) 72
   (3) 80
   (4) 90
   (5) 96

2. Margaret tutors five students—Mary, age 56; Kevin, 22; Ramona, 41; James, 29; and Laura, 32. Find the mean (average) age of the students.
   (1) 23
   (2) 29
   (3) 32
   (4) 36
   (5) 41

3. George takes home $380 a week and saves $60 each week. How many weeks will he need to save $750?
   (1) 6.25
   (2) 10
   (3) 12.5
   (4) 15
   (5) Not enough information is given.

4. Which of the following tells Fernando's net income for the year if his gross monthly pay is $1800 and his employer deducts $400 each month for taxes and social security?
   (1) $12(\$1800 - \$400)$
   (2) $\frac{\$1800 + \$400}{12}$
   (3) $12 \times \$1800 \times \$400$
   (4) $52 \times \$1800 \times \$400$
   (5) $52(\$1800 - \$400)$

277

5. Of 1200 people interviewed, only 360 said that the present administration should stay in office. What is the ratio of the number of people in favor of the administration to the total number interviewed?
   (1) 3 : 6
   (2) 3 : 10
   (3) 3 : 12
   (4) 1 : 36
   (5) 1 : 120

6. Which of the following represents the total distance a hiker traveled if he walked for two hours at 4.5 mph and then for four hours at 3.5 mph?
   (1) $\frac{2 \times 4.5}{4 \times 3.5}$
   (2) $2 \times 4.5 \times 4 \times 3.5$
   (3) $4(4.5) + 2(3.5)$
   (4) $(2 + 4) + (4.5 + 3.5)$
   (5) $2(4.5) + 4(3.5)$

Items 7 to 9 refer to the following situation.

Tom works in the shipping department of Elk Electronics where he makes $1750 a month. His boss offered him a choice of two promotions. With choice A, Tom would become the head of the shipping department in the plant where he works and would get a raise of 8%. With choice B, Tom would head the shipping department in another plant and would get a raise of 10%. Tom estimates that his travel expenses to the other plant would be $40 a month.

7. How much would Tom make in a month if he accepts choice A?
   (1) $1758
   (2) $1830
   (3) $1890
   (4) $1905
   (5) $1925

8. With choice B, Tom's monthly salary will be how much more than his current salary?
   (1) $ 75
   (2) $100
   (3) $125
   (4) $150
   (5) $175

9. If Tom chooses job A, how much will he have at the end of a year in the company's pension plan?
   (1) $ 850
   (2) $1100
   (3) $1350
   (4) $1485
   (5) Not enough information is given.

10. In this picture, ∠COE = 90° and ∠DOE = 23°. Find ∠COD.
    (1) 23°
    (2) 57°
    (3) 67°
    (4) 77°
    (5) 157°

11. The figure is a 5-inch scale. Find the distance between points C and D.

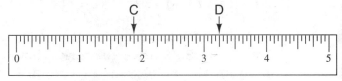

   (1) $1\frac{3}{8}$ in.
   (2) $1\frac{5}{8}$ in.
   (3) $1\frac{7}{8}$ in.
   (4) $4\frac{1}{8}$ in.
   (5) $4\frac{7}{8}$ in.

12. Find the volume in cubic meters of the cylinder.
    (1) 31.4
    (2) 39.3
    (3) 62.8
    (4) 78.5
    (5) 157.0

13. On a test Yoshiko answered 48 problems correctly and 12 problems incorrectly. What percent of the problems did she answer correctly?
    (1) 75%
    (2) 80%
    (3) 85%
    (4) 90%
    (5) 95%

14. An 8-foot tall vertical pole casts a shadow 3 feet long at the same time that a tree casts a shadow 48 feet long. Find the height of the tree.
    (1) 16 ft
    (2) 32 ft
    (3) 64 ft
    (4) 96 ft
    (5) 128 ft

15. If x represents a number, which of the following expresses a number raised to the third power?
    (1) $\frac{x}{3}$
    (2) $3x$
    (3) x^3
    (4) 3^x
    (5) $\sqrt{x}$

16. A patio in the shape of a square measures 15 feet on each side. What is the area of the patio in square feet?
    (1) 1500
    (2) 225
    (3) 90
    (4) 60
    (5) 30

17. In an isosceles triangle, each base angle contains 72°. Find the measurement of the vertex angle.
    (1) 36°
    (2) 54°
    (3) 64°
    (4) 72°
    (5) 108°

18. Suppose n represents the number of men in Paul's night school class. The number of women in the class is seven less than twice the number of men. Which expression represents the number of women?
    (1) $n + 7$
    (2) $2n - 7$
    (3) $2n + 7$
    (4) $7n - 2$
    (5) $7n + 2$

19. What is the ratio of 21 inches to one yard?
    (1) $1:7$
    (2) $7:1$
    (3) $5:12$
    (4) $1:21$
    (5) $7:12$

20. In this rectangle the length is twice the width and the area is 800. Find the width.
    (1) 20
    (2) 40
    (3) 80
    (4) 120
    (5) 160

    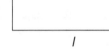

21. The two triangles are similar. Find the height of the triangle on the right.
    (1) 12
    (2) 18
    (3) 60
    (4) 75
    (5) 150

    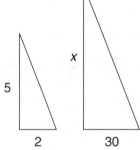

22. Solve for a in $8a - 2 = 6$.
    (1) $\frac{1}{2}$
    (2) 1
    (3) 2
    (4) 4
    (5) 8

23. Which of the following is equal to the expression $18m + 24$?
    (1) $6(3m + 4)$
    (2) $18m(24)$
    (3) $6(m + 4)$
    (4) $18(m + 3)$
    (5) $6(3m + 8)$

24. These two figures have equal areas. Find the measurement of the side of the square.

    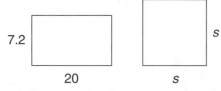

    (1) 10
    (2) 12
    (3) 14.4
    (4) 27.2
    (5) 36

25. Which point on the number line corresponds to $\frac{5}{2}$?

    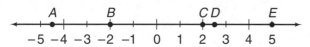

    (1) A
    (2) B
    (3) C
    (4) D
    (5) E

Items 26 to 28 refer to the following table.

Sales of Records, Cassettes, and CDs (in millions)

	1980	1985	1989
Long-playing records (LPs)	487.1	287.7	71.2
Prerecorded cassettes	110.2	339.1	446.3
Compact disks (CDs)	(X)	22.6	207.2

Source: Statistical Abstract of the U.S., 1991.

26. How many more LP records were sold in 1980 than in 1989?
    (1) 199.4 million
    (2) 200.6 million
    (3) 216.5 million
    (4) 415.9 million
    (5) 416.1 million

27. Which of the following describes the number of cassettes sold in 1989 compared to the number sold in 1980?
    (1) more than 4 times the number sold in 1980
    (2) about twice the number sold in 1980
    (3) nearly 3 times the number sold in 1980
    (4) about the same as the number sold in 1980
    (5) about half the number sold in 1980

28. Which of the following describes the number of CDs sold in 1989 compared to the number of LPs sold that year?
    (1) CD sales were nearly 5 times the sales of LPs
    (2) CD sales were nearly 4 times the sales of LPs
    (3) CD sales were nearly 3 times the sales of LPs
    (4) CD sales were about the same as sales of LPs
    (5) CD sales were about half of the sales of LPs

29. Solve for t in $5(t + 1) < 20$.
    (1) $t < 2\frac{1}{2}$
    (2) $t < 3$
    (3) $t < 4$
    (4) $t < 5$
    (5) $t < 7\frac{1}{2}$

30. In the diagram $\angle x = 121°$. Find the measurement of $\angle w$.
    (1) 121°
    (2) 118°
    (3) 62°
    (4) 59°
    (5) 31°

31. What is the area of a triangle with a base of $2\frac{3}{4}$ inches and a height of $1\frac{1}{2}$ inches?

    (1) $4\frac{3}{4}$ in.^{2}
    (2) $3\frac{1}{8}$ in.^{2}
    (3) $2\frac{7}{8}$ in.^{2}
    (4) $2\frac{1}{16}$ in.^{2}
    (5) $1\frac{15}{16}$ in.^{2}

Items 32 and 33 refer to the following information.

A grocery store donated 60 cans of tomato soup, 48 cans of vegetable soup, and 36 cans of chicken soup to the Community Day Care Center. The cans are unlabeled.

32. What is the probability that the first can the cook opens will be tomato soup?

    (1) $\frac{5}{12}$
    (2) $\frac{7}{12}$
    (3) $\frac{1}{60}$
    (4) $\frac{1}{3}$
    (5) $\frac{1}{4}$

33. Of the first 12 cans of soup that the cook opens, 3 are tomato, 3 are vegetable, and 6 are chicken. What is the probability that the next can the cook opens will be chicken soup?

    (1) $\frac{1}{30}$
    (2) $\frac{1}{6}$
    (3) $\frac{5}{22}$
    (4) $\frac{17}{22}$
    (5) $\frac{5}{6}$

34. Choose the equation that expresses the following: Twice the quantity of a number decreased by five is 12.

    (1) $2x - 5 = 12$
    (2) $2x + 5 = 12$
    (3) $5(x - 2) = 12$
    (4) $2(x + 5) = 12$
    (5) $2(x - 5) = 12$

Items 35 to 37 refer to the following graph.

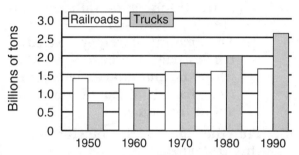

Intercity Freight Carried by Trucks and Railroads

35. In which year shown in the graph was the amount of freight carried by truck first more than the amount carried by railroad?

    (1) 1950
    (2) 1960
    (3) 1970
    (4) 1980
    (5) 1990

36. In which year shown in the graph did the amount of freight carried by truck first reach 2 billion tons?

    (1) 1950
    (2) 1960
    (3) 1970
    (4) 1980
    (5) 1990

37. According to the graph, which of the following is true?
    (1) The amount of freight carried by truck is usually about half the amount carried by railroad.
    (2) While the amount of freight carried by truck has declined, the amount carried by railroad has increased.
    (3) The amount of freight carried by both truck and railroad has steadily decreased.
    (4) The amount of freight carried by both truck and railroad has steadily increased.
    (5) The amount of freight carried by truck has increased steadily while the amount carried by railroad has declined or remained about the same.

38. Find the length of the missing side of this triangle.
    (1) 25
    (2) 20
    (3) 15
    (4) 14
    (5) 12

39. Of the registered voters in Greenport, the ratio of those who voted in the last election to those who did not vote was 3:2. 5,400 people voted. How many registered voters are there in Greenport?
    (1) 3,240
    (2) 3,600
    (3) 5,400
    (4) 9,000
    (5) 12,000

40. For these two triangles $DE = GH$ and $DF = GI$. Which of the following, along with the information given, is enough to guarantee that the triangles are congruent?

    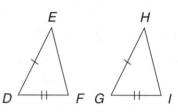

    A. $EF = HI$
    B. $\angle D = \angle E$
    C. $\angle G = \angle I$
    (1) A only
    (2) B only
    (3) C only
    (4) both A and C
    (5) both B and C

41. What is the slope of the line that passes through points F and G?

    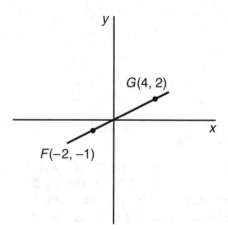

    (1) -1
    (2) -2
    (3) $+2$
    (4) $-\frac{1}{2}$
    (5) $+\frac{1}{2}$

42. A machine can label 15 envelopes per minute. How many minutes will the machine need to label 3000 envelopes?
    (1) 3000×15
    (2) $\frac{15}{3000}$
    (3) $\frac{3000}{15}$
    (4) $\frac{60}{3000}$
    (5) $\frac{3000}{60}$

43. Which of the following is equal to $x^2 - 100$?
    (1) $(x + 25)(x - 4)$
    (2) $(x + 50)(x - 50)$
    (3) $(x + 10)^2$
    (4) $(x - 20)(x + 5)$
    (5) $(x + 10)(x - 10)$

44. Which of the following represents the solution for x in the proportion $x:5 = 9:2$?
    (1) $\frac{9 \cdot 5}{2}$
    (2) $9 \cdot 5 \cdot 2$
    (3) $\frac{2}{9 \cdot 5}$
    (4) $9 + 5 + 2$
    (5) $9 + 2 - 5$

45. Solve for y in $y^2 + y - 12 = 0$.
    (1) $y = +12$ and -1
    (2) $y = -4$ and $+3$
    (3) $y = +6$ and -2
    (4) $y = -12$ and -1
    (5) $y = -6$ and $+2$

46. For every dollar the Chung family spends in a month on car payments, they spend two dollars on food and three dollars on mortgage payments. The Chungs spend a total of $696 a month for these three items. How much do they spend each month on mortgage payments?
    (1) $174
    (2) $232
    (3) $261
    (4) $348
    (5) $464

47. For a concert at the Civic Auditorium, 75% of the seats were filled. There were 1500 people at the concert. How many seats are there in the auditorium?
    (1) 1125
    (2) 1200
    (3) 1500
    (4) 2000
    (5) 2500

Items 48 to 50 refer to this graph.

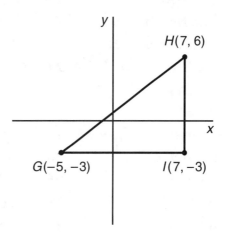

48. What is the distance between points H and I?
    (1) 12
    (2) 9
    (3) 7
    (4) 3
    (5) 2

49. What is the distance between points G and H?
    (1) 8
    (2) 9
    (3) 12
    (4) 13
    (5) 15

50. The coordinates of the midpoint M between two points in a plane are given by the expression $M = (\frac{x_1 + x_2}{2}, \frac{y_1 + y_2}{2})$ where (x_1, y_1) and (x_2, y_2) are two points in the plane. What are the coordinates of the midpoint between G and H?

(1) $(1, 2)$
(2) $(2, 3)$
(3) $(1, 1\frac{1}{2})$
(4) $(2\frac{1}{2}, 1)$
(5) $(1\frac{1}{2}, 1)$

Check your answers on page 358.

Post-Test Guide

1. Percents/6
2. Whole Numbers/6
3. Whole Numbers/13
4. Whole Numbers/18
5. Fractions/14
6. Decimals/13
7. Percents/14
8. Percents/14
9. Percents/14
10. Geometry/2
11. Fractions/11
12. Decimals/10
13. Percents/11
14. Fractions/15
15. Whole Numbers/7
16. Whole Numbers/10
17. Geometry/4
18. Algebra/23
19. Fractions/14
20. Algebra/26
21. Geometry/5
22. Algebra/13
23. Algebra/18
24. Algebra/26
25. Algebra/1
26. Decimals/12
27. Decimals/12
28. Decimals/12
29. Algebra/16
30. Geometry/2
31. Fractions/13
32. Fractions/16
33. Fractions/16
34. Algebra/25
35. Graphs/3
36. Graphs/3
37. Graphs/3
38. Geometry/7
39. Fractions/18
40. Geometry/6
41. Geometry/12
42. Fractions/17
43. Algebra/19
44. Fractions/15
45. Algebra/20
46. Fractions/18
47. Percents/8
48. Geometry/9
49. Geometry/9
50. Geometry/9

SIMULATED TEST

This test is much like the Mathematics Test. The number of items and their degree of difficulty are the same as on the real test. The time limit and the mixed order of the items are also the same. By taking the Simulated Test, you will gain valuable test-taking experience and get a better idea of how well prepared you are to take the actual test.

Using the Simulated Test to Your Best Advantage

You should take this test under the same conditions in which you will take the real test.
- When you take the GED Test, you will have 90 minutes to complete it. Although this will probably be more than enough time, set aside at least 90 minutes to take this test so you can work without interruption.
- Do not talk to anyone or consult any books as you take the test. If you have a question, ask your instructor.
- If you are not sure of an answer, make an educated guess. Guessing a correct answer will improve your score, and, as on the real GED Test, guessing a wrong answer will not affect your score.

As you take the Simulated Test, write your answers on a clean sheet of paper or use an answer sheet. When time is up, you may wish to circle the item that you answered last and then continue with the test. This way, when you score your test, you can see how much of a factor time was in your performance.

Using the Answer Key

Use the Answers and Solutions (page 361) to check your answers, and mark each item you answered correctly. Regardless of whether you answered an item correctly, you should look over each problem solution. This will reinforce your test-taking skills and your understanding of the material.

How to Use Your Score

If you answered 45 items or more correctly, you will have done 80 percent work or better. This shows that you should do well on the actual Mathematics Test. If you got slightly fewer than 45 items correct, you probably need to do some light reviewing. If your score was far below the 80 percent mark, you should spend additional time reviewing the lessons that will strengthen your weak areas. The Simulated Test Guide at the end of the test will help you identify your strong and weak areas.

Directions: Choose the one best answer for each item. Use any of the formulas on pages 17 and 18 that you need.

When you finish, check your answers. Following the test is a guide to the lessons in the book where the skills tested in each problem are presented.

1. Shirley works three days a week and earns $180 a week. How many weeks does she have to work in order to earn $4500?
   - (1) 40
   - (2) 25
   - (3) 20
   - (4) 15
   - (5) Not enough information is given.

2. On the walls of the Neighborhood Community Center, the ratio of green paint to white paint used in a mixture is 3:4. A dealer told the volunteers who are painting the center that they will need 56 gallons of paint altogether to finish the job. How many gallons of green paint will they need?
   - (1) 12
   - (2) 15
   - (3) 18
   - (4) 20
   - (5) 24

3. Find the sales tax on three $15 shirts if the tax rate is 6%.
   - (1) $0.90
   - (2) $1.35
   - (3) $1.80
   - (4) $2.70
   - (5) Not enough information is given.

4. Let c represent the number of cases of cola Bill sells in a week in his store. The number of cases of orange drink he sells in a week is 20 less than half the number of cases of cola. Which expression tells the number of cases of orange drink he sells in a week?
   - (1) $\frac{c+2}{20}$
   - (2) $\frac{c-20}{2}$
   - (3) $\frac{1}{2}c - 20$
   - (4) $\frac{c+20}{2}$
   - (5) $\frac{1}{2}c + 20$

5. The rainfall in Capital City during June was 3.6 in. the first week, 2.45 in. the second week, 4.63 in. the third week, and 3.84 in. the fourth week. Find the mean (average) weekly rainfall for that period.
   - (1) 3.48 in.
   - (2) 3.54 in.
   - (3) 3.63 in.
   - (4) 3.68 in.
   - (5) 3.72 in.

Items 6 and 7 refer to the following information.

John received a shipment of jackets to sell in his store. There were 15 small jackets, 25 medium-sized ones, and 20 large ones. The sizes were mixed together.

6. What is the probability that the first jacket John took from the box was large?
   - (1) $\frac{1}{3}$
   - (2) $\frac{1}{2}$
   - (3) $\frac{1}{6}$
   - (4) $\frac{1}{20}$
   - (5) $\frac{1}{30}$

7. In fact, the first two jackets John took from the box were medium and the next three were large. What is the probability that the fifth jacket he takes from the box will be small?
   - (1) $\frac{1}{3}$
   - (2) $\frac{1}{4}$
   - (3) $\frac{3}{11}$
   - (4) $\frac{1}{15}$
   - (5) $\frac{15}{16}$

286 Simulated Test

8. Jeff makes $6 an hour for the first seven hours of the work day and then $9 an hour for each additional hour. Which of the following tells the amount he makes in a ten-hour work day?
   (1) 7($9) + 3($6)
   (2) 7($6) + 3($9)
   (3) 7 × $9 × 3 × $6
   (4) $10 × 7
   (5) 10 × $9

Items 9 to 12 refer to the following situation.

Jose wants to buy a video recorder. The model he likes is for sale at Sav-a-Lot for $389. Sales tax is 5%, and the delivery charge is $10. Sav-a-Lot offers a time payment plan of $100 down and $16 a month for 24 months.

The same model of video recorder is selling for $439 at Al's Appliances in a nearby state where the sales tax is 6% and the delivery charge is $8. The time payment plan at Al's is $60 down and $36 a month for 12 months.

9. What is the total cost of the recorder at Sav-a-Lot with the time payment plan?
   (1) $384
   (2) $444
   (3) $484
   (4) $524
   (5) Not enough information is given.

10. The total cost of the recorder at Al's with the time payment plan is how much more than Al's list price for the recorder?
    (1) $ 53
    (2) $ 60
    (3) $ 73
    (4) $100
    (5) Not enough information is given.

11. Find the total price of the recorder and a three-year guarantee at Al's.
    (1) $449
    (2) $459
    (3) $476
    (4) $496
    (5) Not enough information is given.

12. At Sav-a-Lot what is the total price of the recorder including tax and delivery if you do not choose the time payment plan?
    (1) $418.45
    (2) $460.95
    (3) $468.95
    (4) $578.45
    (5) Not enough information is given.

13. Geraldine drove for two hours at 35 mph and then for three hours at 55 mph. Which expression tells the total distance she drove?
    (1) 5 × 90
    (2) 2(55) + 3(35)
    (3) 2 × 35 × 3 × 55
    (4) 2(35) + 3(55)
    (5) (35 + 55) × (3 + 2)

Items 14 and 15 refer to the following figure.

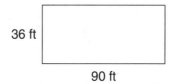

14. This figure shows the plan of a rectangular lot. Find the number of yards of fencing needed to enclose the lot.
    (1) 84
    (2) 126
    (3) 168
    (4) 252
    (5) 324

15. A packet of grass seed covers 300 square feet of ground. How many packets are needed to cover the lot with grass?
    (1) 3.2
    (2) 10.8
    (3) 12.4
    (4) 21.6
    (5) 32.4

16. Find, to the nearest tenth of a cubic meter, the volume of this cube.

    (1) 2.4
    (2) 3.4
    (3) 4.5
    (4) 6.0
    (5) 33.6

    s = 1.5 m

17. The Nieves family have paid off $16,000 of their $40,000 mortgage. What is the ratio of the amount they have paid to the amount they still owe?

    (1) 1 : 2
    (2) 2 : 5
    (3) 3 : 5
    (4) 2 : 3
    (5) 3 : 4

18. This triangle has an area of 150 square inches. The height is one-third of the base. Find the measure of the base in inches.

    (1) 10
    (2) 15
    (3) 25
    (4) 30
    (5) 50

    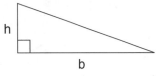

19. Fred is an electrician. His assistant Gordon makes $5 an hour less than Fred. Together they made $760 on a job they both worked on for 40 hours. Find the hourly rates for each man.

    (1) Fred gets $12, and Gordon gets $7.
    (2) Fred gets $13, and Gordon gets $8.
    (3) Fred gets $14, and Gordon gets $9.
    (4) Fred gets $15, and Gordon gets $10.
    (5) Fred gets $16, and Gordon gets $11.

20. Simplify the expression $10^4 - 5^2$.

    (1) 15
    (2) 30
    (3) 975
    (4) 9,975
    (5) 10,025

21. Which expression below is equal to the quantity of eight more than three times a number, all divided by six?

    (1) $\frac{3x + 8}{6}$
    (2) $\frac{3}{6}x + 8$
    (3) $\frac{6x + 8}{3}$
    (4) $\frac{8x + 3}{6}$
    (5) $\frac{8x - 3}{6}$

22. In an isosceles triangle, the vertex angle measures 55°. What is the measure of each base angle?

    (1) 55°
    (2) 62.5°
    (3) 70°
    (4) 110°
    (5) 140°

Items 23 and 24 refer to the table below.

Number (in thousands) of Newspapers and Periodicals

	1985	1986	1987	1988	1989	1990
Newspapers	9.1	9.1	9.0	10.1	10.5	11.5
Periodicals	11.1	11.3	11.6	11.2	11.6	11.1

Source: Statistical Abstract of the U.S., 1991.

23. For which year shown in the table was the number of periodicals first less than the number of newspapers?

    (1) 1986
    (2) 1987
    (3) 1988
    (4) 1989
    (5) 1990

24. Which of the following best describes the pattern shown in the table?
    (1) The number of newspapers generally increased, and the number of periodicals remained about the same.
    (2) The number of newspapers generally decreased, and the number of periodicals increased.
    (3) The number of newspapers and the number of periodicals both increased steadily.
    (4) The number of newspapers and the number of periodicals both decreased steadily.
    (5) The number of newspapers and the number of periodicals remained the same.

25. A machine produces 100 motor parts per hour and runs for eight hours a day. Which of the following tells the number of days needed to produce 12,000 motor parts?
    (1) $8 \times 100 \times 12{,}000$
    (2) $\frac{12{,}000}{8 \times 100}$
    (3) $\frac{100 \times 12{,}000}{8}$
    (4) $\frac{8 + 100}{12{,}000}$
    (5) $\frac{100}{8 \times 12{,}000}$

26. Solve for n in $2n - 7 \geq 5$.
    (1) $n \geq 1$
    (2) $n \geq 2$
    (3) $n \geq 4$
    (4) $n \geq 6$
    (5) $n \geq 12$

27. In the diagram, $k \parallel l$ and $\angle b = 106°$. Find $\angle g$.
    (1) 16°
    (2) 26°
    (3) 74°
    (4) 84°
    (5) 106°

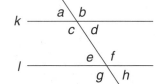

28. Find the length of TS in the picture.
    (1) 18
    (2) 21
    (3) 35
    (4) 84
    (5) 98

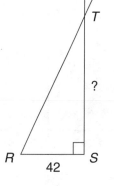

29. Which of the following is equal to $(-5m^3)(-2m)$?
    (1) $-10m^2$
    (2) $+10m^4$
    (3) $+7m^4$
    (4) $-7m^2$
    (5) $+3m^2$

30. Solve for z in $8z - 1 = 6z + 4$
    (1) $\frac{1}{2}$
    (2) 1
    (3) 2
    (4) $1\frac{1}{2}$
    (5) $2\frac{1}{2}$

31. Simplify the expression $\frac{15c^2d}{-25cd}$.
    (1) $\frac{-5d}{3}$
    (2) $\frac{3c^3d^2}{5}$
    (3) $\frac{-3cd}{5}$
    (4) $\frac{-3c}{5}$
    (5) $\frac{3c^2d}{5}$

32. In the figure, $a = 48$ and $b = 36$. Find the length of c.
    (1) 36
    (2) 40
    (3) 50
    (4) 56
    (5) 60

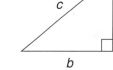

33. What are the coordinates of the point labeled A?

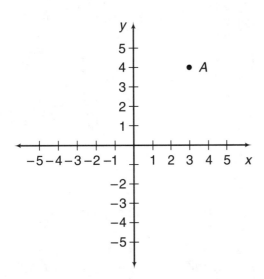

(1) (3, 3)
(2) (3, 4)
(3) (4, 3)
(4) (4, 4)
(5) (−3, −3)

34. Marlene has to pay 18% yearly interest on her charge card. If she has an outstanding balance of $900, what is her monthly interest charge?

(1) $180.00
(2) $162.00
(3) $135.00
(4) $ 16.20
(5) $ 13.50

35. Mike is riding the train from his hometown to the state capital. The train stopped in two hours at a distance 96 miles from Mike's hometown. At this point Mike estimated that the train had gone 80% of the distance to the capital. How many more miles is it to the capital?

(1) 100
(2) 96
(3) 48
(4) 24
(5) 12

36. Jeanne uses $2\frac{1}{4}$ yards of material to make a pair of curtains. How many curtains can she make from $15\frac{3}{4}$ yards of material?

(1) 7
(2) 8
(3) 9
(4) 10
(5) Not enough information is given.

37. Max's employer puts $32.40 a week of Max's pay into a savings plan. This represents 6% of Max's pay. Find his weekly pay.

(1) $194.40
(2) $324.00
(3) $540.00
(4) $572.40
(5) $600.00

38. The scale on a map is 1 inch = 40 miles. How far apart are two towns that are $3\frac{1}{4}$ inches apart on the map?

(1) 150 miles
(2) 130 miles
(3) 120 miles
(4) 110 miles
(5) 100 miles

Items 39 to 41 refer to the following graph.

Median Sales Price, Single Family Houses

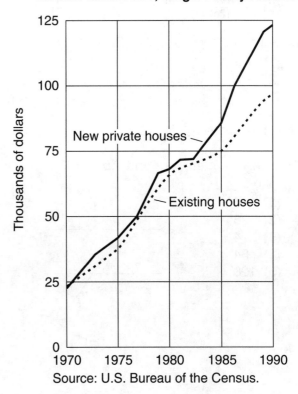

Source: U.S. Bureau of the Census.

39. For what year were the median prices of new private houses and existing houses the same?
    (1) 1970
    (2) 1975
    (3) 1980
    (4) 1985
    (5) 1990

40. Which of the following best tells the difference between the median price of new private houses and the median price of existing houses in 1990?
    (1) $ 5,000
    (2) $10,000
    (3) $15,000
    (4) $20,000
    (5) $25,000

41. Which of the following best describes the trend shown in the graph?
    (1) The price of new private houses and the price of existing houses increased about the same amount.
    (2) The price of new private houses decreased while the price of existing houses increased.
    (3) The price of new private houses and the price of existing houses decreased about the same amount.
    (4) The price of new private houses increased more than the price of existing houses.
    (5) The price of new private houses increased and the price of existing houses stayed about the same.

42. Which of the following represents the area in square meters of a rectangle that is 9 meters long and 3.25 meters wide?
    (1) $2(9) + 2(3.25)$
    (2) $\frac{1}{2}(9 \times 3.25)$
    (3) 9×3.25
    (4) $9^2 + 3.25^2$
    (5) $\frac{9 \times 3.25}{3}$

43. Which of the following equals $x^2 + 2x - 48$?
    (1) $(x + 24)(x - 24)$
    (2) $(x + 24)(x - 2)$
    (3) $(x - 16)(x + 3)$
    (4) $(x + 8)(x - 6)$
    (5) $(x + 4)(x - 12)$

44. Find, in square feet, the area of a parallelogram with a base of 10 feet and a height of $6\frac{1}{2}$ feet.
    (1) 50
    (2) 55
    (3) 60
    (4) 65
    (5) 70

45. Which of the following equals $a^2 - 12a$?
    (1) $a(a - 12)$
    (2) $a^2(-12a)$
    (3) $a(a - 4)$
    (4) $(a - 4)(a + 4)$
    (5) $(a - 6)(a + 2)$

46. Find the value of a in the expression $a = 4b(c - 3)$ when $b = 5$ and $c = 9$.
    (1) 24
    (2) 72
    (3) 108
    (4) 120
    (5) 240

47. Which of the following represents the price of 8 gallons of gasoline at a cost of $1.19 per gallon?
    (1) $8 \times \$1.19$
    (2) $8 + \$1.19$
    (3) $\frac{\$1.19}{8}$
    (4) $\frac{8}{\$1.19}$
    (5) $\$1.19 - 8$

Simulated Test

Items 48 and 49 refer to the following graph.

Children's Rescue Fund Budget

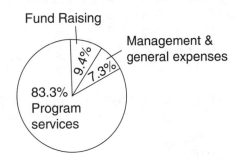

48. This graph shows how the funds for a children's rescue organization are spent. The amount spent for program services is approximately how many times the amount spent on fund raising?

(1) 2 times
(2) 3 times
(3) 5 times
(4) 6 times
(5) 9 times

49. If the yearly budget for the organization is $6,000,000, how much is spent in a year on management and general expenses?

(1) $8,760,000
(2) $ 876,000
(3) $ 438,000
(4) $ 219,000
(5) $ 109,500

50. Which of the following is the same as 2.6×10^5?

(1) 13,000,000
(2) 2,600,000
(3) 1,300,000
(4) 260,000
(5) 130,000

51. What is the sale price of a coat that originally sold for $80 and is now on sale for 15% off?

(1) $80 + 0.15($80)
(2) $80 − 0.15($80)
(3) $80 − 0.15
(4) $80 + 0.15
(5) 0.15($80 + $80)

52. What are the coordinates of the y-intercept of the graph of the equation $y = +3x - 4$?

(1) (0, 3)
(2) (0, −3)
(3) (−3, 0)
(4) (−4, 0)
(5) (0, −4)

53. Solve for c in $c^2 + 3c - 18 = 0$.

(1) $c = +6$ and -3.
(2) $c = -6$ and $+3$
(3) $c = +9$ and -2
(4) $c = +18$ and -1
(5) $c = -18$ and $+1$

54. If $x = -15$, what is the value of $\frac{2}{3}x$?

(1) −30
(2) $-22\frac{1}{2}$
(3) $+22\frac{1}{2}$
(4) +10
(5) −10

Items 55 and 56 refer to the following figure.

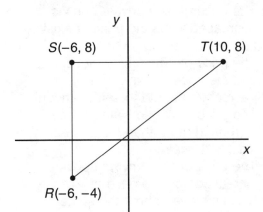

55. What is the distance from S to T?

(1) 6
(2) 8
(3) 10
(4) 16
(5) 18

56. What is the distance from R to T?

(1) 10
(2) 12
(3) 14
(4) 18
(5) 20

Check your answers on page 361.

Simulated Test Guide

1. Whole Numbers/13	20. Whole Numbers/7	39. Graphs/5
2. Fractions/18	21. Algebra/23	40. Graphs/5
3. Percents/10	22. Geometry/4	41. Graphs/5
4. Algebra/23	23. Decimals/12	42. Decimals/11
5. Decimals/13	24. Decimals/12	43. Algebra/19
6. Fractions/16	25. Whole Numbers/18	44. Fractions/13
7. Fractions/16	26. Algebra/16	45. Algebra/18
8. Whole Numbers/18	27. Geometry/3	46. Algebra/6
9. Percents/14	28. Geometry/5	47. Whole Numbers/18
10. Percents/14	29. Algebra/9	48. Graphs/2
11. Percents/14	30. Algebra/14	49. Graphs/2
12. Percents/14	31. Algebra/10	50. Decimals/9
13. Whole Numbers/18	32. Geometry/7	51. Percents/10
14. Fractions/10	33. Geometry/8	52. Geometry/12
15. Decimals/13	34. Percents/9	53. Algebra/20
16. Decimals/11	35. Percents/12	54. Algebra/11
17. Fractions/14	36. Fractions/6	55. Geometry/9
18. Algebra/26	37. Percents/8	56. Geometry/9
19. Algebra/27	38. Fractions/17	

ANSWERS AND SOLUTIONS

Chapter 1 WHOLE NUMBERS

Level 1 Preview, pg. 8

1. **(2) 8**
2. **(4) 4**
3. **(3) 7000**
4. **(1) 4**
5. $\quad\begin{array}{r}9000\\-\ 496\\\hline \mathbf{8504}\end{array}$
6. $\quad\begin{array}{r}473\\\times\ \ 90\\\hline \mathbf{42{,}570}\end{array}$
7. $\quad\begin{array}{r}\mathbf{509}\\15\overline{)7635}\\\underline{75}\\13\\\underline{\ 0}\\135\\\underline{135}\end{array}$
8. $\quad\begin{array}{r}49\\207\\5653\\+\ \ 28\\\hline \mathbf{5937}\end{array}$
9. $\quad\begin{array}{r}\mathbf{2{,}806}\\6\overline{)16{,}836}\\\underline{12}\\48\\\underline{48}\\03\\\underline{\ 0}\\36\\\underline{36}\end{array}$
10. **(1)** $\begin{array}{r}30{,}050\\-\ 5{,}207\\\hline \mathbf{24{,}843}\end{array}$
11. **(2)** $\begin{array}{r}12{,}906\\385\\+\ 4{,}059\\\hline \mathbf{17{,}350}\end{array}$
12. **(3)** $\begin{array}{r}\mathbf{470\ r\ 9}\\21\overline{)9879}\\\underline{84}\\147\\\underline{147}\\09\\\underline{\ 0}\\9\end{array}$

Lesson 1 Exercise, pg. 10

1. **89   16**
2. **980   600**
3. **15,000   90,000**
4. **900**
5. **6,000**
6. **40,000**
7. **500,000**
8. **300**
9. **50,000**
10. **2,000,000**

Lesson 2 Exercise, pg. 12

1. **(c)** $\begin{array}{r}9015\\493\\+\ \ 76\\\hline \mathbf{9584}\end{array}$
2. **(b)** $\begin{array}{r}22{,}500\\-\ 6{,}087\\\hline \mathbf{16{,}413}\end{array}$
3. **(a)** $\begin{array}{r}5030\\-\ 763\\\hline \mathbf{4267}\end{array}$
4. **(b)** $\begin{array}{r}704\\86\\+10{,}471\\\hline \mathbf{11{,}261}\end{array}$
5. $\quad\begin{array}{r}78\\4062\\+\ 529\\\hline \mathbf{4669}\end{array}$
6. $\quad\begin{array}{r}42{,}003\\-\ 8{,}346\\\hline \mathbf{33{,}657}\end{array}$
7. $\quad\begin{array}{r}18{,}206\\-11{,}954\\\hline \mathbf{6{,}252}\end{array}$
8. $\quad\begin{array}{r}30{,}005\\-19{,}472\\\hline \mathbf{10{,}533}\end{array}$
9. $\quad\begin{array}{r}428\\61\\593\\+\ \ 7\\\hline \mathbf{1{,}089}\end{array}$
10. $\quad\begin{array}{r}9{,}048\\-\ 336\\\hline \mathbf{8{,}712}\end{array}$

Lesson 3 Exercise, pg. 15

1. (c)
```
      803
   12)9636
      96
      ‾‾
       03
        0
       ‾‾
       36
       36
       ‾‾
        0
```

2. (a)
```
      704
    ×  18
    ‾‾‾‾‾
    5 632
    7 04
    ‾‾‾‾‾‾
   12,672
```

3. (c)
```
       536
    ×  800
    ‾‾‾‾‾‾‾
   428,800
```

4. (b)
```
      907
    8)7256
      72
      ‾‾
       05
        0
       ‾‾
       56
       56
       ‾‾
        0
```

5.
```
       78
    46)3588
       322
       ‾‾‾
        368
        368
        ‾‾‾
          0
```

6.
```
      506
    ×  38
    ‾‾‾‾‾‾
    4 048
   15 18
   ‾‾‾‾‾‾
   19,228
```

7.
```
        817
     ×  409
     ‾‾‾‾‾‾‾
      7 353
    326 80
    ‾‾‾‾‾‾‾
    334,153
```

8.
```
       80 r 5
    17)1365
       136
       ‾‾‾
         05
          0
         ‾‾
          5
```

9.
```
      230
    ×  34
    ‾‾‾‾‾
      920
      690
    ‾‾‾‾‾
    7820
```

10.
```
          14
     316)4424
         316
         ‾‾‾
        1264
        1264
        ‾‾‾‾
           0
```

11.
```
         39
     78)3042
        234
        ‾‾‾
         702
         702
         ‾‾‾
           0
```

12.
```
          4 r 300
     523)2392
         2092
         ‾‾‾‾
          300
```

Level 1 Review, pg. 16

1. (2) 7
2. (2) 8
3. (3) 500
4. (4) 20,000
5.
```
      890
       23
     4017
    + 605
    ‾‾‾‾‾
     5535
```

6.
```
         920 r 3
    12)11,043
       10 8
       ‾‾‾
          24
          24
          ‾‾
           03
            00
            ‾‾
             3
```

7.
```
     12,050
    − 9,947
    ‾‾‾‾‾‾‾
      2,103
```

8.
```
       308
    ×   76
    ‾‾‾‾‾‾‾
     1 848
    21 56
    ‾‾‾‾‾‾
    23,408
```

9.
```
       3,096
    8)24,768
      24
      ‾‾
       07
        0
       ‾‾
       76
       72
       ‾‾
        48
        48
        ‾‾
         0
```

10. (2)
```
       675
    ×  208
    ‾‾‾‾‾‾‾
     5 400
    135 00
    ‾‾‾‾‾‾‾
   140,400
```

11. (1)
```
     306,471
    − 28,295
    ‾‾‾‾‾‾‾‾
     278,176
```

12. (2)
```
          804
     32)25,728
        25 6
        ‾‾‾‾
           12
            0
           ‾‾
          128
          128
          ‾‾‾
            0
```

Lesson 4 Exercise, pg. 19

1. 80   160   3200   2430
2. 800   1300   6600   400
3. 3000   42,000   29,000   150,000
4. 800,000   300,000   600,000
   3,500,000
5. 6,000,000   12,000,000
   32,000,000   189,000,000

Lesson 5 Exercise, pg. 20

1. $d = rt$
   $d = 4 \times 3 =$ **12 miles**
2. $d = rt$
   $d = 65 \times 4 =$ **260 miles**
3. $d = rt$
   $d = 475 \times 5 =$ **2375 miles**
4. $d = rt$
   $d = 15 \times 3 =$ **45 miles**
5. $c = nr$
   $c = 3 \times \$18 =$ **$54**
6. $c = nr$
   $c = 5 \times \$3.60 =$ **$18.00**
7. $c = nr$
   $c = 12 \times \$6 =$ **$72**
8. $c = nr$
   $c = 30 \times \$65 =$ **$1950**

Lesson 6 Exercise, pg. 22

1. $ 14   **$18**
     22   3)$54
   + 18
   $ 54

2. In order: $14 $18 $22
   $18 is the median.

3.   185   **127 pounds**
      138   4)508
       97
   +   88
      508

4.   65   **82**
       86   5)410
       79
       92
   +   88
      410

5.   284   **238 miles**
      191   5)1190
      297
      162
   + 256
     1190

6.   11   **11 gallons**
      9   3)33
   + 13
     33

7.   213   **249 people**
     191   4)996
     289
   + 303
     996

8. In order: 191 213 289 303
   Find the mean of 213 and 289.
     213   **251 people**
   + 289   2)502
     502

9. In order: $1.16 $1.25 $1.29 $1.39 $1.49
   The median is **$1.29.**

10. $24,800   **$14,375**
      22,500   4)$57,500
       6,000
   +   4,200
     $57,500

Lesson 7 Exercise, pg. 24

1. $2^4 = 2 \times 2 \times 2 \times 2 =$ **16**
   $9^2 = 9 \times 9 =$ **81**
   $3^3 = 3 \times 3 \times 3 =$ **27**
   $8^1 =$ **8**

2. $13^2 = 13 \times 13 =$ **169**
   $50^2 = 50 \times 50 =$ **2500**
   $6^3 = 6 \times 6 \times 6 =$ **216**
   $12^0 =$ **1**

3. $2^5 = 2 \times 2 \times 2 \times 2 \times 2 =$ **32**
   $1^5 =$ **1**
   $16^2 = 16 \times 16 =$ **256**
   $40^2 = 40 \times 40 =$ **1600**

4. $10^3 = 10 \times 10 \times 10 =$ **1000**
   $18^0 =$ **1**
   $25^2 = 25 \times 25 =$ **625**
   $4^4 = 4 \times 4 \times 4 \times 4 =$ **256**

5. $5^2 - 2^3 =$
   $5 \times 5 - (2 \times 2 \times 2) =$
   $25 - 8 =$ **17**
   $8^2 + 3^3 =$
   $8 \times 8 + 3 \times 3 \times 3 =$
   $64 + 27 =$ **91**
   $10^2 - 4^2 + 5^2 =$
   $10 \times 10 - 4 \times 4 + 5 \times 5 =$
   $100 - 16 + 25 =$ **109**

6. $4^3 + 6^1 - 2^4 =$
   $4 \times 4 \times 4 + 6 - (2 \times 2 \times 2 \times 2) =$
   $64 + 6 - 16 =$ **54**
   $12^2 - 5^0 - 3^2 =$
   $12 \times 12 - 1 - (3 \times 3) =$
   $144 - 1 - 9 =$ **134**
   $10^3 - 10^2 =$
   $10 \times 10 \times 10 - (10 \times 10) =$
   $1000 - 100 =$ **900**

Lesson 8 Exercise, pg. 26

1. $\sqrt{289} = \mathbf{17}$
   Guess 20.

   ```
        14              14          17
   20)289        +      20       2)34
        20              34
        89
        80
         9
   ```

   $\sqrt{784} = \mathbf{28}$
   Guess 30.

   ```
        26              26          28
   30)784        +      30       2)56
        60              56
       184
       180
         4
   ```

   $\sqrt{1444} = \mathbf{38}$
   Guess 40.

   ```
         36             36          38
   40)1444        +     40       2)76
        120             76
        244
        240
          4
   ```

   $\sqrt{484} = \mathbf{22}$
   Guess 20.

   ```
         24             24          22
   20)484         +     20       2)44
        40             44
        84
        80
         4
   ```

2. $\sqrt{1521} = \mathbf{39}$
   Guess 40.

   ```
         38             38          39
   40)1521        +     40       2)78
        120            78
        321
        320
          1
   ```

   $\sqrt{1849} = \mathbf{43}$
   Guess 40.

   ```
         46             46          43
   40)1849        +     40       2)86
        160            86
        249
        240
          9
   ```

$\sqrt{529} = \mathbf{23}$
Guess 20.

```
      26              26          23
20)529         +      20       2)46
     40              46
    129
    120
      9
```

$\sqrt{2704} = \mathbf{52}$
Guess 50.

```
       54             54          52
50)2704        +      50       2)104
    250             104
    204
    200
      4
```

3. $\sqrt{4096} = \mathbf{64}$
   Guess 60.

   ```
          68            68          64
   60)4096        +     60       2)128
        360           128
        496
        480
         16
   ```

   $\sqrt{8836} = \mathbf{94}$
   Guess 90.

   ```
          98            98          94
   90)8836        +     90       2)188
        810           188
        736
        720
         16
   ```

   $\sqrt{6084} = \mathbf{78}$
   Guess 80.

   ```
          76            76          78
   80)6084        +     80       2)156
        560           156
        484
        480
          4
   ```

   $\sqrt{7056} = \mathbf{84}$
   Guess 80.

   ```
          88            88          84
   80)7056        +     80       2)168
        640           168
        656
        640
         16
   ```

Answers and Solutions

Lesson 9 Exercise, pg. 28

1. $P = 4s$
   $P = 4 \times 11$
   $P = $ **44 ft**

2. $P = 2l + 2w$
   $P = 2 \times 9 + 2 \times 7$
   $P = 18 + 14$
   $P = $ **32 m**

3. $P = 2l + 2w$
   $P = 2 \times 26 + 2 \times 13$
   $P = 52 + 26$
   $P = $ **78 ft**

4. $P = 4s$
   $P = 4 \times 20$
   $P = $ **80 in.**

5. $P = a + b + c$
   $P = 14 + 14 + 14$
   $P = $ **42 in.**

6. $P = 4s$
   $P = 4 \times 18$
   $P = $ **72 yd**

7. $P = 2l + 2w$
   $P = 2 \times 10 + 2 \times 2$
   $P = 20 + 4$
   $P = $ **24**

8. $P = a + b + c$
   $P = 12 + 16 + 20$
   $P = $ **48**

9. $P = a + b + c$
   $P = 7 + 8 + 9$
   $P = $ **24 cm**

10. $P = 2l + 2w$
    $P = 2 \times 21 + 2 \times 15$
    $P = 42 + 30$
    $P = $ **72 in.**

11. $P = 4s$
    $P = 4 \times 21$
    $P = $ **84**

12. $P = a + b + c$
    $P = 19 + 19 + 10$
    $P = $ **48 in.**

Lesson 10 Exercise, pg. 30

1. $A = lw$
   $A = 16 \times 9$
   $A = $ **144 in.²**

2. $A = s^2$
   $A = 8^2$
   $A = $ **64 sq ft**

3. $A = lw$
   $A = 20 \times 14$
   $A = $ **280**

4. $A = lw$
   $A = 15 \times 3$
   $A = $ **45 yd²**

5. $A = s^2$
   $A = 15^2$
   $A = $ **225 in.²**

6. $A = s^2$
   $A = 11^2$
   $A = $ **121 sq ft**

7. $A = lw$
   $A = 13 \times 3$
   $A = $ **39**

8. $A = s^2$
   $A = 16^2$
   $A = $ **256 m²**

Lesson 11 Exercise, pg. 32

1. $V = s^3$
   $V = 6^3$
   $V = 6 \times 6 \times 6$
   $V = $ **216 cu ft**

2. $V = lwh$
   $V = 12 \times 8 \times 5$
   $V = $ **480 cu in.**

3. $V = lwh$
   $V = 20 \times 6 \times 5$
   $V = $ **600 ft³**

4. $V = s^3$
   $V = 8^3$
   $V = 8 \times 8 \times 8$
   $V = $ **512 cm³**

5. $V = s^3$
   $V = 1^3$
   $V = $ **1 cu yd**

6. $V = lwh$
   $V = 100 \times 24 \times 8$
   $V = $ **19,200 cu ft**

7. $V = lwh$
   $V = 50 \times 4 \times 1$
   $V = $ **200 ft³**

8. $V = s^3$
   $V = 3^3$
   $V = 3 \times 3 \times 3$
   $V = $ **27 ft³**

Lesson 12 Exercise, pg. 35

1. (3) 6×9
2. (2) $20 + 14$
3. (2) $(3 \times 12) \times 8$
4. (3) $15 + (4 + 10)$
5. (1) $7 \times 8 - 7 \times 1$
6. (4) $30 + 20$
7. (3) $2(19 + 7)$
8. (4) $50 \times 3 + 50 \times 4$
9. (2) $4(9 - 2)$
10. (1) $3 \times 10 - 3 \times 1$

Level 2 Review, pg. 36

1. **2,400,000**
2. **500,000**
3. $c = nr$
   $c = 12 \times \$1.20 = $ **\$14.40**
4. $d = rt$
   $d = 435 \times 7$
   $d = $ **3045 miles**
5. 12 **23 pounds**
   36    $4\overline{)92}$
   19
   +25
   92
6. In order: 12 19 25 36
   Find the mean of 19 and 25.
   19   **22 pounds**
   +25 $2\overline{)44}$
   44
7. \$2480   **\$2020**
   1630 $3\overline{)\$6060}$
   +1950
   \$6060
8. In order: \$1950 \$2155 \$3080 \$6075 \$8470
   The median is **\$3080**.

9. $15^2 - 10^2 + 25^1 =$
   $15 \times 15 - (10 \times 10) + 25 =$
   $225 - 100 + 25 = $ **150**
10. $20^2 - 3^3 + 10^0 =$
    $20 \times 20 - (3 \times 3 \times 3) + 1 =$
    $400 - 27 + 1 = $ **374**
11. $\sqrt{3481} = $ **59**
    Guess 60.
    ```
         58                           59
    60)3481        58              2)118
       300        + 60               
       ---        ---
       481        118
       480
       ---
         1
    ```
12. $\sqrt{7744} = $ **88**
    Guess 90.
    ```
         86                           88
    90)7744        86              2)176
       720        + 90
       ---        ---
       544        176
       540
       ---
         4
    ```
13. $P = a + b + c$
    $P = 8 + 11 + 14$
    $P = $ **33 m**
14. $P = 2l + 2w$
    $P = 2 \times 35 + 2 \times 21$
    $P = 70 + 42 = $ **112 meters**
15. $A = lw$
    $A = 36 \times 20$
    $A = $ **720 sq yd**
16. $A = lw$
    $A = 40 \times 24$
    $A = $ **960**
17. $V = lwh$
    $V = 25 \times 12 \times 8$
    $V = $ **2400 cu ft**
18. $V = s^3$
    $V = 12^3$
    $V = 12 \times 12 \times 12$
    $V = $ **1728 in.^{3}**
19. **(2) 9(15 + 20)**
20. **(3) (7 × 20) + (7 × 1)**

Lesson 13 Exercise, pg. 39

1. a. **subtraction**
   b. $ 593,650
      −108,212

      $485,438
2. a. **division** b. **4**
   c.   **$ 6,984**
        4)$27,936
3. a. **subtraction**
   b. $ 18,500
      − 3,518

      $14,982
4. a. **combined**
   b. $ 16,456
       11,294
      + 3,367

      $31,117
5. a. **3** b. **division**
   c.   **$21**
        3)$63
6. a. **1** b. **230** c. **multiplication**
   d. $1,850
        × 230
        ------
        55 500
        370 0

        $425,500
7. a. **more** b. **addition**
   c. 14,273
      + 9,467

      23,740
8. a. **division**
   b. **17 inches**
       6)102
9. a. **volume**
   b. $V = lwh$
      $V = 80 \times 20 \times 6$
      $V = $ **9600 cu ft**
10. a. **subtraction**
    b. 1990
       −1859
       -----
       131 years
11. a. **perimeter**
    b. $P = 4s$
       $P = 4 \times 6$
       $P = $ **24 miles**
12. a. **altogether**
    b. $ 462
        436
        194
        323
       +245

       $1660
13. a. **5** b. **1** c. **division**
    d.   **$ 59**
         5)$295
14. a. **area**
    b. $A = lw$
       $A = 30 \times 15$
       $A = $ **450 sq yd**

Lesson 14 Exercise, pg. 42

The numbers used in the estimated solutions are rounded.

1. **(2) a little less than $20,000**
   Estimate: Exact answer:
   $20,000   **$19,658**
   3)$60,000   3)$58,974

Answers and Solutions   299

2. **(3) a little less than $60,000**
   Estimate: Exact answer:
   $30,000 $ 29,260
   20,000   18,420
   +10,000   + 9,455
   $ 60,000 $57,135

3. **(2) a little more than $25**
   Estimate: Exact answer:
   $ 25 $ 28
   4)$100   4)$112

4. **(1) about 30 inches**
   Estimate: Exact answer:
   30 inches   29 inches
   3)90   3)87

5. **(2) around $60,000**
   Estimate: Exact answer:
   $ 2000 $ 2,160
   × 30   × 30
   $60,000 $64,800

6. **(2) about 1400**
   Estimate: Exact answer:
   300   328
   200   217
   400   421
   200   186
   +300   +313
   1400   1465

7. **(1) between $60 and $70**
   Estimate: Exact answer:
   $ 60 $ 65.50
   5)$300   5)$327.50

8. **(3) between 1200 and 1400**
   Estimate: Exact answer:
   $A = lw$ $A = lw$
   $A = 60 × 20$ $A = 62 × 21$
   $A = $ **1200 sq ft** $A = $ **1302 sq ft**

Lesson 15 Exercise, pg. 44

1. $ 52 $ 780
    ×15   −650
    260         $130
    52
    $780

2. 3 × 12 = 36
   4 × 12 = 48
   2 × 12 = +24
            108 cans

3. $235 $475
   +240   × 4
   $475 $1900

4. $290 $1300
   +400   − 690
   $690 $ 610

5. $ 25 $12 $ 8 $500
   × 20   × 8   × 6   96
   $500 $96 $ 48   + 48
                        $644

6. $6.50 $9.75 $227.50
   × 35   × 8   + 78.00
   32 50 $78.00 $305.50
   195 0
   $227.50

7. $260 $290 $3120 $270
   × 12   × 6   +1740   18)$4860
   520        $1740 $4860   36
   260   126
   $3120   126
                                     00

8. $d = rt = 55 × 4 = $ 220 miles
   $d = rt = 30 × 2 = $ +60
                      280 miles

9. $60,000 $ 35,000
   +45,000   3)$105,000
   $105,000

10. $1,500 $ 1,150
    − 350   × 12
    $1,150   2 300
               11 50
               $13,800

11. LR area: $A = lw$
             $A = 15 × 12$
             $A = 180$ sq ft
    DR area: $A = lw$
             $A = 10 × 8$
             $A = 80$ sq ft
    Total: 180 + 80 = 260 sq ft
    Cost: 260 × $8 = **$2080**

12. total area: $A = lw$
                      $A = 30 × 18$
                      $A = 540$ in.^{2}
    unshaded area: $A = lw$
                      $A = 12 × 10$
                      $A = 120$ in.^{2}
    Subtract to find shaded
    area: 540 − 120 = **420 in.^{2}**

Lesson 16 Exercise, pg. 46

1. **New York**

2. New York $1504
   Florida   − 856
              $ 648

3. The two middle values are $1068 and $1440.
   Find the mean of these two values.
   $1068   **$1254**
   + 1440   2)$2508
   $2508

4. U.S. avg. $1221
   Texas   − 898
              $ 323

5. $1,221
   ×   300,000
   $366,300,000 to the nearest
   $10 million = **$370,000,000**

6. expenditures $856
   taxes   − 694
                 $162

7. $902
      × 15,000
      4 510 000
         9 02
   $13,530,000 to the nearest
   hundred thousand = **$13,500,000**

8. New York $1164
   Florida   − 694
               $470

Lesson 17 Exercise, pg. 47

1. $ 950 $1500
   + 550   ×   12
   $1500   3 000
                 15 00
                 $18,000

2. $ 950 $1500 $5,700
   ×   6   ×   6   + 9,000
   $5700 $9000 $14,700

3.   **$1,225**
   12)$14,700
      12
       27
       24
        30
        24
         60
         60
          0

4. $1500 $1,750
   + 250   ×   12
   $1750   3 500
              17 50
              $21,000

5.   8
      35
    + 10
   53 employees

6. $ 216,000
     756,000
   + 144,000
   $1,116,000

7.   **$ 93,000**
   12)$1,116,000
      1 08
         36
         36
          0 000

8.   **$ 21,600**
   35)$756,000
      70
       56
       35
       21 0
       21 0
          000

9. $ 14,400 $ 1,200
   10)$144,000   12)$14,400
      10   12
      44   24
      40   24
       40   000
       40
        0
        0

Lesson 18 Exercise, pg. 49

1. (3) $\frac{24,720}{12}$ 2. (1) 40 × 29

3. (4) (2 × 20) + (3 × 60)   4. (2) $\frac{80 + 95 + 74}{3}$

5. (1) (4 × 7.99) + 1.92

6. (3) (8 × 350) + (6 × 425)

7. (3) 12(1800 − 360)

8. (1) $\frac{2500 + 4850 + 4200}{5}$

9. (2) 12(20 − 2)

10. (4) (2 × 100) + (2 × 30)

Level 3 Review, pg. 50

1.   **16 pounds**
   48)768
      48
      288
      288

2.   **$13,500**
   5)$67,500

3.   **19 miles**
   23)437
      23
      207
      207
        0

4. (2) **about $1900**
   Estimation: $1300
                + 600
                $1900

5. $315 $ 450 $12,360
   + 135   ×   12   − 5,400
   $ 450 $ 900 $ 6,960
             450
            $ 5400

6. 760   1250   410   4560
   × 6   × 3   × 3   3750
   4560   3750   1230   +1230
                              9540

7. 4 years = 4 × 12 = 48 months
            48 × $350 = $16,800
   2 years = 2 × 12 = 24 months
            24 × $250 = $6,000
   Total: $16,800
            5,000
          + 6,000
           $27,800

8. $\begin{array}{r} 313 \\ -277 \\ \hline 36 \end{array}$

9. $\begin{array}{r} 298 \\ \times\ \$30 \\ \hline \$8940 \end{array}$

10. Connecticut $\begin{array}{r} \$456 \\ -\ 297 \\ \hline \$159 \end{array}$
    United States

11. $\begin{array}{r} \$18.00 \\ +\ 12.50 \\ \hline \$30.50 \end{array}$ $\begin{array}{r} \$30.50 \\ \times\quad 4 \\ \hline \$122.00 \end{array}$

12. $\begin{array}{r} \$12.00 \\ +\ 10.25 \\ \hline \$22.25 \end{array}$ $\begin{array}{r} \$22.25 \\ \times\quad 4 \\ \hline \$89.00 \end{array}$

13. $\begin{array}{r} 100 \\ \times\ .20 \\ \hline \$20.00 \end{array}$ $\begin{array}{r} \$\ 89.00 \\ +\ 20.00 \\ \hline \$109.00 \end{array}$

14. $\begin{array}{r} 200 \\ \times\ .20 \\ \hline \$40.00 \end{array}$ $\begin{array}{r} \$\ 89.00 \\ +\ 40.00 \\ \hline \$129.00 \end{array}$

15. $A = lw$
    $A = 15 \times 10$
    $A = \mathbf{150\ sq\ yd}$

16. $\begin{array}{r} 150 \\ \times\ 25 \\ \hline 750 \\ 300\ \\ \hline \$3750 \end{array}$

17. $\begin{array}{r} 150 \\ \times\ 20 \\ \hline \$3000 \end{array}$ $\begin{array}{r} \$3000 \\ +\ 500 \\ \hline \$3500 \end{array}$

18. $\begin{array}{r} \$3750 \\ -\ 3500 \\ \hline \$\ 250 \end{array}$

19. **(4)** $\dfrac{33 + 25 + 47 + 19}{4}$

20. **(4)** $3 \times 12 + 4 \times 15$

Chapter 2 DECIMALS

Level 1 Preview, pg. 54

1. **(3) 0.708**
2. **(1) 5.003**
3. **(4) forty-five thousandths**
4. **18.023**
5. $\begin{array}{r} 0.409 \\ 0.28 \\ +\ 0.7 \\ \hline \mathbf{1.389} \end{array}$
6. $\begin{array}{r} 0.820 \\ -0.197 \\ \hline \mathbf{0.623} \end{array}$

7. $\begin{array}{r} 4.5 \\ \times 0.26 \\ \hline 270 \\ 90\ \\ \hline 1.170 = \mathbf{1.17} \end{array}$

8. $\begin{array}{r} \mathbf{0.026} \\ 18\overline{)0.468} \\ 36\ \ \ \\ \hline 108 \\ 108 \\ \hline 0 \end{array}$

9. $\begin{array}{r} \mathbf{4.5} \\ 2.4\overline{)108.0} \\ 96\ \ \ \\ \hline 12\ 0 \\ 12\ 0 \\ \hline 0 \end{array}$

10. **0.067 is greater.**

11. **0.2 is greater.**

12. **0.013, 0.031, 0.31, 0.4**

Lesson 1 Exercise, pg. 56

1. (a) **two** (b) **three** (c) **three**
   (d) **five** (e) **two** (f) **one**
2. (a) **20.067** (b) **.409** (c) **28.7**
   (d) **1.208** (e) **3.6** (f) **4.5**
3. (a) 0.⑧9 (b) 3.④
   (c) 10.③07 (d) 5.③681
4. (a) 0.1②5 (b) 4.2⑥79
   (c) 5.8⓪9 (d) 0.7①6
5. (a) 4.05⑥ (b) 0.01②3
   (c) 28.39①7 (d) 0.98⑦2
6. **(d) 4.02**
7. **(a) 0.9**
8. **(d) 4.02**
9. **(b) 0.879**
10. **(b) 4.009**

Lesson 2 Exercise, pg. 57

1. c
2. f
3. e
4. j
5. d
6. h
7. b
8. i
9. a
10. g

11. **.4 or 0.4**
12. **8.09**
13. **.036 or 0.036**
14. **14.003**
15. **.0519 or 0.0519**
16. **72.0006**
17. **1.000005**
18. **.00032 or 0.00032**
19. **4.018**
20. **.418 or 0.418**

Lesson 3 Exercise, pg. 59

1.   0.360
     0.500
   + 0.607

     1.467

2.   0.380
     0.619
   + 0.200

     1.199

3.   0.3
     0.9
   + 0.7

     1.9

4.   0.006
     0.050
   + 0.800

     0.856

5.   2.50
     18.00
   + 1.07

     21.57

6.   0.506
     3.100
   + 9.000

    12.606

7.   38.000
      4.078
   + 0.0195

    42.0975

8.   9.100
     0.870
   + 0.143

    10.113

9.   6.0
    − 2.5

     3.5

10.   8.00
    − 0.19

     7.81

11.   0.300
    − 0.258

     0.042

12.   5.900
    − 2.114

     3.786

13.   0.015
    − 0.009

     0.006

14.   1.0000
    − 0.0865

     0.9135

15.   9.00
    − 0.32

     8.68

16.   0.60
    − 0.24

     0.36

Lesson 4 Exercise, pg. 62

1.   3.5
    ×   7

     24.5

2.   29
    × 0.04

     1.16

3.   0.06
    × 0.5

     0.030 = **0.03**

4.   0.47
    ×   16
    ------
      2 82
      4 7

     7.52

5.   0.185
    ×   0.4

     0.0740 = **0.074**

6.   .59
    × .004

     0.00236

7.   2.09
    ×   30

     62.70 = **62.7**

8.   0.0065
    ×   0.6

     0.00390 = **0.0039**

9.   215
    × 0.04

     8.60 = **8.6**

10.   **2.7**
    6)16.2

11.   **0.48**
    9)4.32

12.   **8.065**
    3)24.195

13.   **1.9**
    25)47.5
       25
       ---
       22 5
       22 5

         0

14.   **0.018**
    64)1.152
       64

       512
       512

         0

15.   **6.6**
    31)204.6
       186
       ---
       18 6
       18 6

          0

16.   **400**
    0.32)128.00
         128
         ---
           0 00

17.   **700**
    0.08)56.00

18.   **90**
    1.2)108.0
        108
        ---
          0 0

19.   **300**
    0.008)2.400

20.   **43 5**
    0.6)261.0

21.   **12**
    0.026)0.312
          26
          --
          52
          52
          --
           0

22.   **.7**
    0.9)0.6 3

23.   **4.8**
    0.03)0.14 4

24.   **6**
    0.052)0.312
          312

            0

Lesson 5 Exercise, pg. 63

1. a. 0.056 = 0.056
      0.05 = 0.050
      0.056 is greater.

   b. 0.19 = 0.19
      0.2 = 0.20
      0.2 is greater.

   c. 1.08 = 1.080
      1.082 = 1.082
      1.082 is greater.

   d. 0.075 = 0.075
      0.57 = 0.570
      0.57 is greater.

2. a. 0.021 = 0.021, 0.012 = 0.012,
      0.21 = 0.210, 0.201 = 0.201
      From least to greatest:
      0.012, 0.021, 0.201, 0.21

   b. 0.045 = 0.045, 0.54 = 0.540, 0.5 = 0.500,
      0.005 = 0.005
      From least to greatest:
      0.005, 0.045, 0.5, 0.54

   c. 3.2 = 3.20, 2.33 = 2.33, 3.22 = 3.22,
      3.3 = 3.30
      From least to greatest:
      2.33, 3.2, 3.22, 3.3

   d. 1.008 = 1.008, 0.8 = 0.800, 1.09 = 1.090,
      0.9 = 0.900
      From least to greatest:
      0.8, 0.9, 1.008, 1.09

3. a. 0.38 = 0.380, 0.8 = 0.800, 0.083 = 0.083,
      0.308 = 0.308
      From greatest to least:
      0.8, 0.38, 0.308, 0.083

   b. 5.0 = 5.000, 0.5 = 0.500, 5.05 = 5.050,
      0.055 = 0.055
      From greatest to least:
      5.05, 5.0, 0.5, 0.055

   c. 0.9 = 0.900, 0.09 = 0.090, 0.999 = 0.999,
      9.0 = 9.000
      From greatest to least:
      9.0, 0.999, 0.9, 0.09

   d. 2.075 = 2.075, 2.75 = 2.750, 2.7 = 2.700,
      2.5 = 2.500
      From greatest to least:
      2.75, 2.7, 2.5, 2.075

Answers and Solutions

4. 0.705 = 0.705, 0.75 = 0.750, 0.075 = 0.075
   0.75 meter is greatest.

5. 1.2 = 1.200, 1.099 = 1.099, 1.209 = 1.209
   1.209 kg is heaviest.

Level 1 Review, pg. 63

1. (1) **1.005**

2. (4) **0.989**

3. (3) **seven and six hundredths**

4. **60.0012**

5.   0.385
      0.600
   + 0.090
      1.075

6.   0.0580
   − 0.0496
      0.0084

7.   8.06
   ×   29
      72 54
    161 2
    233.74

8.   12.8
   ×   .3 5
      64 0
     3 84
    4.48 0 = **4.48**

9.   **0.032**
   76)2.432
      2 28
       152
       152
         0

10.   **9.4**
    0.07)0.65 8

11. **0.085**

12. **0.4, 0.44, 1.4, 4.1**

Lesson 6 Exercise, pg. 66

1. **0.6   0.4   0.1   5.2   0.3**

2. **0.53   0.48   2.02   8.30   0.91**

3. **0.139   0.058   1.781   0.105   6.433**

4. **4 pounds**

5. **9 kilometers**

Lesson 7 Exercise, pg. 67

1. 2.5 × 100 = **250 cm**

2. 3 × 1000 = **3000 g**

3. 4.8 × 1000 = **4800 ml**

4. 6.5 × 1000 = **6500 m**

5.   **1.25 km**
   1000)1250.00

6.   **0.8 l**
   10)8.0

7.   **0.385 kg**
   1000)385.000

8.   **1.95 m**
   100)195.00

9. (a) **1.9 liters**   (b) **4.7 kilograms**
   (c) **0.1 meter**

10. (a) **5 kilometers**   (b) **28 grams**   (c) **10 liters**

Lesson 8 Exercise, pg. 69

1. **2.4 cm**

2. **3.9 cm**

3. **4.4 cm**

4. **7.1 cm**

5. **8.5 cm**

6.   7.1
   − 3.9
     3.2 cm

7.   4.4
   − 2.4
     2.0 cm

8.   8.5
   − 7.1
     1.4 cm

Lesson 9 Exercise, pg. 70

1. $(0.5)^2$ = 0.5 × 0.5 = **0.25**
   $(0.02)^3$ = 0.02 × 0.02 × 0.02 = **0.000008**
   $(0.4)^2$ = 0.4 × 0.4 = **0.16**
   $(0.12)^2$ = 0.12 × 0.12 = **0.0144**

2. $(0.07)^2$ = 0.07 × 0.07 = **0.0049**
   $(0.009)^2$ = 0.009 × 0.009 = **0.000081**
   $(0.1)^4$ = 0.1 × 0.1 × 0.1 × 0.1 = **0.0001**
   $(1.5)^2$ = 1.5 × 1.5 = **2.25**

3. $\sqrt{0.16}$ = **0.4** $\sqrt{0.81}$ = **0.9**
   $\sqrt{0.0036}$ = **0.06** $\sqrt{0.0001}$ = **0.01**

4. $\sqrt{0.0004}$ = **0.02** $\sqrt{0.0121}$ = **0.11**
   $\sqrt{0.000009}$ = **0.003** $\sqrt{0.0625}$ = **0.25**

Lesson 10 Exercise, pg. 72

1. a. $r = \frac{d}{2}$
   $r = \frac{8}{2}$
   r = **4 in.**
   b. $C = \pi d$
   $C = 3.14 \times 8$
   C = **25.12 in.**
   c. $A = \pi r^2$
   $A = 3.14 \times 4^2$
   $A = 3.14 \times 16$
   A = **50.24 sq in.**

2. **a.** $d = 2r$
   $d = 2 \times 30$
   $d = $ **60 ft**
   b. $C = \pi d$
   $C = 3.14 \times 60$
   $C = $ **188.4 ft**
   c. $A = \pi r^2$
   $A = 3.14 \times 30^2$
   $A = 3.14 \times 900$
   $A = $ **2826 ft²**

3. **a.** $d = 2r$
   $d = 2 \times 1.2$
   $d = $ **2.4 m**
   b. $C = \pi d$
   $C = 3.14 \times 2.4$
   $C = $ **7.536 to the nearest tenth = 7.5 m**
   c. $A = \pi r^2$
   $A = 3.14 \times (1.2)^2$
   $A = 3.14 \times 1.44$
   $A = $ **4.5216 to the nearest tenth = 4.5 m²**

4. **a.** $r = \frac{d}{2}$
   $r = \frac{40}{2}$
   $r = $ **20 in.**
   b. $C = \pi d$
   $C = 3.14 \times 40$
   $C = $ **125.6 in.**
   c. $A = \pi r^2$
   $A = 3.14 \times 20^2$
   $A = 3.14 \times 400$
   $A = $ **1256 in.²**

5. **a.** $V = \pi r^2 h$
   $V = 3.14 \times 5^2 \times 20$
   $V = 3.14 \times 25 \times 20$
   $V = $ **1570 ft³**
   b. $V = \pi r^2 h$
   $V = 3.14 \times (0.2)^2 \times 1.5$
   $V = 3.14 \times 0.04 \times 1.5$
   $V = $ **0.1884 m³**
   c. $V = \pi r^2 h$
   $V = 3.14 \times 30^2 \times 1$
   $V = 3.14 \times 900 \times 1$
   $V = $ **2826 cu ft**
   d. $V = \pi r^2 h$
   $V = 3.14 \times 20^2 \times 0.5$
   $V = 3.14 \times 400 \times 0.5$
   $V = $ **628 cu m**

Lesson 11 Exercise, pg. 73

1. $c = nr$
   $c = 6.5 \times \$8.99$
   $c = \$58.435$ to the nearest cent = **\$58.44**

2. $d = rt$
   $d = 55 \times 6.25$
   $d = 343.75$ to the nearest mile = **344 miles**

3.
   ```
     8.25          5.85 lb
     4.5         4)23.40
     7.65
   + 3.
    23.40
   ```

4. The middle values are 4.5 and 7.65. The median is
   ```
     4.5          6.075
   + 7.65      2)12.150
    12.15
   ```

5. $c = nr$
   $c = 45 \times \$0.036$
   $c = $ **\$1.62**

6. $P = a + b + c$
   $P = 1.65 + 2.4 + 1.38$
   $P = $ **5.43 m**

7. **a.** $P = 2l + 2w$
   $P = 2(8.5) + 2(4.2)$
   $P = 17 + 8.4 = $ **25.4 m**
   b. $A = lw$
   $A = 8.5 \times 4.2$
   $A = $ **35.7 m²**

8. **a.** $P = 4s$
   $P = 4 \times 3.4$
   $P = $ **13.6 cm**
   b. $A = s^2$
   $A = (3.4)^2$
   $A = 3.4 \times 3.4$
   $A = $ **11.56 cm²**

9. $V = s^3$
   $V = (2.2)^3$
   $V = 2.2 \times 2.2 \times 2.2$
   $V = $ **10.648 m³**

10. $V = lwh$
    $V = 20 \times 4.5 \times 3.6$
    $V = $ **324 cm³**

Level 2 Review, pg. 74

1. **6.30**
2. **0.284**
3. $2.4 \times 1000 = $ **2400 grams**
4.
   ```
           0.655 m
   1000)655.000
   ```
5. **6.8 cm**
6. **2.7 cm**
7.
   ```
     6.8
   - 2.7
     4.1 cm
   ```
8. $(0.5)^3 = 0.5 \times 0.5 \times 0.5 = $ **0.125**
9. $\sqrt{0.0049} = $ **0.07**
10. $C = \pi d$
    $C = 3.14 \times 12$
    $C = $ **37.68 in.**
11. $d = 2r$
    $d = 2 \times 0.4$
    $d = $ **0.8 m**
    $C = \pi d$
    $C = 3.14 \times 0.8$ m
    $C = 2.512$ to the nearest tenth = **2.5 m**

12. $A = \pi r^2$
    $A = 3.14 \times 8^2$
    $A = 3.14 \times 64$
    $A = 200.96$ to the nearest sq. in. = **201 in.²**
13. $V = \pi r^2 h$
    $V = 3.14 \times 3^2 \times 5$
    $V = 3.14 \times 9 \times 5$
    $V = $ **141.3 ft³**
14. $c = nr$
    $c = 20 \times \$.109$
    $c = \$2.180 = $ **$2.18**
15. $d = rt$
    $d = 4 \times 0.25$
    $d = 1.00 = $ **1 mile**
16. $P = a + b + c$
    $P = 5.4 + 4.9 + 6.2$
    $P = $ **16.5 m**
17. $d = rt$
    $d = 60.5 \times 3.5$
    $d = 211.75$
    to the nearest mile = **212 miles**
18. $c = nr$
    $c = 3.6 \times \$3.48$
    $c = \$12.528$
    to the nearest cent = **$12.53**
19. $V = lwh$
    $V = 10 \times 3.4 \times 2.5$
    $V = $ **85 m³**
20.
    ```
        2.7           3.46  to the nearest
        3.5        3)10.38  kilogram = 3 kg
        4.18
     + ____
       10.38
    ```

Lesson 12 Exercise, pg. 78

1.
   ```
      9.3
    − 5.7
    ─────
      3.6
   ```
2.
   ```
      8.9
    − 7.9
    ─────
      1.0
   ```
3.
   ```
            15              15
      1000)15,000          × 5
                        ────────
                        75 beds
   ```
4.
   ```
            280             280
      1000)280,000         × 6
                        ────────
                        1680 beds
   ```
5. **(4) The rate gradually decreased.**
6. **1980 and 1989**
7.
   ```
      14.1           47.7
      11.6         − 37.2
    + 11.5         ─────────────
    ──────         10.5 million
      37.2
   ```
8.
   ```
     55.7
   − 30.4
   ────────────────
   25.3 million
   ```

9.
   ```
     11.6
   − 10.5
   ─────────────
    1.1 million
   ```
10. **Professional basketball.** 16.6 million is more than twice 7.6 million.

Lesson 13 Exercise, pg. 80

1. ←Horst • Jana→
   $d = rt = 55 \times 0.5 = 27.5$ mi
   $d = rt = 45 \times 0.5 = 22.5$ mi
   ```
      27.5
    + 22.5
    ────────
    50.0 mi
   ```
2. 40 mph →
   30 mph →
   $d = rt = 40 \times 2.5 = 100$ mi
   $d = rt = 30 \times 2.5 = 75$ mi
   ```
      100
    −  75
    ──────
    25 mi
   ```
3. (1) $12 \times 1.25 + 10 \times 1.25$
   ←Miriam • Marcia→
   $d = rt = 12 \times 1.25$
   $d = rt = 10 \times 1.25$
4.
   ```
             23.8   to the nearest whole mile
       8.4)200.0 0         = 24 miles
           168
           ─────
            32 0
            25 2
            ─────
             6 80
             6 72
   ```
5. express →
   local →
   Express travels from 10:30 to 1:00 or 2.5 hours.
   $d = rt = 50 \times 2.5 = 125$ mi
   Local travels from 11:30 to 1:00 or 1.5 hours.
   $d = rt = 30 \times 1.5 = 45$ mi
   ```
      125
    − 45
    ──────
    80 mi
   ```
6. (2) $40 \times \$6.50 + 6 \times \$9.75$
7.
   ```
      4534.4
    − 3789.6
    ──────────────
      744.8 miles
   ```
8. (3) $\$7.52 \div 3.2$
9.
   ```
             0.2833   to the nearest
       60)17.0000     thousandth = 0.283
   ```
10.
    ```
        0.268
      ×    45
      ────────
        1 340
       10 72
      ─────────
       12.060   to the nearest
               whole number = 12 hits
    ```

306   Answers and Solutions

11. 98.6°   104.4°
    + 5.8   − 3.9
    ─────   ─────
    104.4°   **100.5°**

12. **(2) D, E, B, A, C**
    A − 0.65 = 0.650 kg
    B − 1.05 = 1.050 kg
    C − 0.065 = 0.065 kg
    D − 1.65 = 1.650 kg
    E − 1.5   = 1.500 kg

Lesson 14 Exercise, pg. 81

1. **the weight of the fourth package**
2. **the length of one of the sides**
3. **the price per pound of the beef**
4. **the number of males and females**
5. **the number of hospitals**
6. **(5) Not enough information is given.** The amount they spend for food is missing.
7. **(5) Not enough information is given.** His overtime rate is missing.
8. **(2) 15.5**   20.0
              − 4.5
              ─────
              15.5
9. **(5) Not enough information is given.** The amount he spends for parking is missing.
10. **(4) 1440** $V = lwh$
                   $V = 16 \times 12 \times 7.5$
                   $V = $ **1440 cu ft**

Level 3 Review, pg. 83

1. **14 million**
2.   27.6
     − 25.9
     ─────
     1.7 million
3.   22.2
     − 14.0
     ─────
     8.2 million
4.   48.0
     − 28.8
     ─────
     19.2 million
5. **beef**
6. **Cheese.** 11.4 divides into 23.7 more than two times.
7.   74.3
     − 65.0
     ─────
     9.3 pounds
8. **1975**

9. Phil $d = rt$
         $d = 3.5 \times 1.75$
         $d = 6.125$ mi

   Sue $d = rt$
         $d = 4.5 \times 1.75$
         $d = 7.875$ mi

     6.125
   + 7.875
   ──────
   14.000 = **14 mi**

10. express →
    local →
    $d = rt$
    $d = 60 \times 2.25$
    $d = 135$ mi

    $d = rt$
    $d = 40 \times 2.25 = 90$
    $d = 90$ mi

      135
    − 90
    ─────
    45 mi

11. Sandy →
    Dick ←
    $d = rt = 38 \times 3.5 = 133$ mi
    $d = rt = 42 \times 2.5 = 105$ mi
    $d = rt = 54 \times 1 = 54$ mi

      105   133
    − 54   −51
    ─────   ─────
    51 mi   **82 mi**

12. Jack →
    Manny →
    $d = rt = 42 \times 2.5 = 105$ mi
    $d = rt = 15 \times 2.5 = 37.5$ mi

    105.0 mi
    − 37.5
    ──────
    67.5 mi

13. ← 420 mph   350 mph →
    $d = rt = 350 \times 1.5 = 525$ mi
    $d = rt = 420 \times 1.5 = 630$ mi

      525 mi
    + 630
    ──────
    1155 mi

14. Jose →
    Maria →
    $d = rt = 40 \times 5.5 = 220$ mi
    $d = rt = 30 \times 4.5 = 135$ mi

      220 mi
    − 135
    ──────
    85 mi

15. **(1) 2 × 6.5 + 2 × 4**
    $P = 2l + 2w$
    $P = 2 \times 6.5 + 2 \times 4$

16. **(4) $31.55**

$$\begin{array}{r} \$\ 4.75 \\ \times\quad 5 \\ \hline \$23.75 \end{array} \qquad \begin{array}{r} \$1.20 \\ \times\quad 6.5 \\ \hline 600 \\ 720 \\ \hline \$7.800 \end{array} \qquad \begin{array}{r} \$23.75 \\ +\ 7.80 \\ \hline \$31.55 \end{array}$$

17. **(5) Not enough information is given.** The price per pound of the beef is missing.

18. **(1)** $\dfrac{\$124.50 + \$190 + \$310}{3}$

19. **(3) 13.8**
    $P = a + b + c$
    $P = 4.2 + 4.4 + 5.2$
    $P = \mathbf{13.8}$

20. **(5) Not enough information is given.** Her regular hourly wage is missing.

Chapter 3 FRACTIONS

Level 1 Preview, pg. 87

1. $\dfrac{24 \div 8}{64 \div 8} = \dfrac{3}{8}$

2. $\dfrac{15}{4} = 4\overline{)15}\ \ 3\tfrac{3}{4}$

3. $\begin{aligned} 4\tfrac{7}{20} &= 4\tfrac{7}{20} \\ +9\tfrac{1}{2} &= 9\tfrac{10}{20} \\ \hline &\quad 13\tfrac{17}{20} \end{aligned}$

4. $\begin{aligned} 3\tfrac{1}{3} &= 3\tfrac{6}{18} \\ 1\tfrac{5}{9} &= 1\tfrac{10}{18} \\ +2\tfrac{5}{6} &= 2\tfrac{15}{18} \\ \hline &\quad 6\tfrac{31}{18} = 7\tfrac{13}{18} \end{aligned}$

5. $\begin{aligned} 9\tfrac{1}{2} &= 9\tfrac{4}{8} \\ -4\tfrac{3}{8} &= 4\tfrac{3}{8} \\ \hline &\quad 5\tfrac{1}{8} \end{aligned}$

6. $12\tfrac{1}{4} = 12\tfrac{4}{16} = 11\tfrac{4}{16} + \tfrac{16}{16} = 11\tfrac{20}{16}$
   $-10\tfrac{7}{16} = 10\tfrac{7}{16} = \qquad\qquad 10\tfrac{7}{16}$
   $\qquad\qquad\qquad\qquad\qquad\qquad\qquad 1\tfrac{13}{16}$

7. $\dfrac{\overset{2}{\cancel{4}}}{\underset{1}{\cancel{5}}} \times \dfrac{\overset{1}{\cancel{5}}}{\underset{3}{\cancel{6}}} = \dfrac{2}{3}$

8. $12 \times 4\tfrac{1}{2} \times 2\tfrac{2}{3} =$
   $\dfrac{\overset{6}{\cancel{12}}}{1} \times \dfrac{\overset{3}{\cancel{9}}}{\underset{1}{\cancel{2}}} \times \dfrac{8}{\underset{1}{\cancel{3}}} = \dfrac{144}{1} = \mathbf{144}$

9. $20 \div \dfrac{4}{5} =$
   $\dfrac{\overset{5}{\cancel{20}}}{1} \times \dfrac{5}{\underset{1}{\cancel{4}}} = \dfrac{25}{1} = \mathbf{25}$

10. $6\tfrac{1}{4} \div 1\tfrac{7}{8} =$
    $\dfrac{25}{4} \div \dfrac{15}{8} =$
    $\dfrac{\overset{5}{\cancel{25}}}{\underset{1}{\cancel{4}}} \times \dfrac{\overset{2}{\cancel{8}}}{\underset{3}{\cancel{15}}} = \dfrac{10}{3} = \mathbf{3\tfrac{1}{3}}$

11. $\dfrac{1}{3}$ **is larger.**
    $\dfrac{5}{16} = \dfrac{15}{48} \qquad \dfrac{1}{3} = \dfrac{16}{48}$

12. $\dfrac{45}{100} = \dfrac{9}{20}$

Lesson 1 Exercise pg. 90

1. $\dfrac{7}{10} \qquad \dfrac{9}{100} \qquad \dfrac{1}{12} \qquad \dfrac{14}{15}$

2. $\dfrac{18}{7} \qquad \dfrac{3}{3} \qquad \dfrac{25}{4}$

3. $5\tfrac{1}{2} \qquad 6\tfrac{9}{20} \qquad 8\tfrac{1}{2}$

4. a. $\dfrac{1}{6}$ b. $\dfrac{4}{9}$ c. $\dfrac{7}{8}$ d. $\dfrac{3}{10}$

5. $\dfrac{5}{8} \qquad \dfrac{15}{16} \qquad \dfrac{99}{100}$

6. $\dfrac{3}{2} \qquad \dfrac{13}{5} \qquad \dfrac{20}{10}$

7. a. $\dfrac{5 \div 5}{20 \div 5} = \dfrac{1}{4}$ \quad b. $\dfrac{32 \div 8}{56 \div 8} = \dfrac{4}{7}$
   c. $\dfrac{25 \div 5}{30 \div 5} = \dfrac{5}{6}$ \quad d. $\dfrac{90 \div 10}{200 \div 10} = \dfrac{9}{20}$

8. a. $\dfrac{40 \div 5}{55 \div 5} = \dfrac{8}{11}$ \quad b. $\dfrac{6 \div 6}{18 \div 6} = \dfrac{1}{3}$
   c. $\dfrac{45 \div 15}{75 \div 15} = \dfrac{3}{5}$ \quad d. $\dfrac{22 \div 2}{24 \div 2} = \dfrac{11}{12}$

9. a. $\dfrac{24 \div 8}{40 \div 8} = \dfrac{3}{5}$ \quad b. $\dfrac{19 \div 19}{38 \div 19} = \dfrac{1}{2}$
   c. $\dfrac{60 \div 12}{144 \div 12} = \dfrac{5}{12}$ \quad d. $\dfrac{50 \div 50}{1000 \div 50} = \dfrac{1}{20}$

308   Answers and Solutions

10. a. $\dfrac{25 \div 5}{45 \div 5} = \dfrac{5}{9}$ b. $\dfrac{48 \div 16}{64 \div 16} = \dfrac{3}{4}$
    c. $\dfrac{24 \div 24}{72 \div 24} = \dfrac{1}{3}$ d. $\dfrac{9 \div 9}{54 \div 9} = \dfrac{1}{6}$

Lesson 2 Exercise, pg. 92

1. a. $\dfrac{9 \times 4}{10 \times 4} = \dfrac{36}{40}$ b. $\dfrac{5 \times 3}{12 \times 3} = \dfrac{15}{36}$
   c. $\dfrac{2 \times 9}{5 \times 9} = \dfrac{18}{45}$ d. $\dfrac{2 \times 4}{3 \times 4} = \dfrac{8}{12}$

2. a. $\dfrac{4 \times 5}{9 \times 5} = \dfrac{20}{45}$ b. $\dfrac{3 \times 10}{20 \times 10} = \dfrac{30}{200}$
   c. $\dfrac{3 \times 7}{5 \times 7} = \dfrac{21}{35}$ d. $\dfrac{9 \times 2}{50 \times 2} = \dfrac{18}{100}$

3. a. $\dfrac{7 \times 3}{25 \times 3} = \dfrac{21}{75}$ b. $\dfrac{1 \times 8}{4 \times 8} = \dfrac{8}{32}$
   c. $\dfrac{2 \times 4}{9 \times 4} = \dfrac{8}{36}$ d. $\dfrac{5 \times 2}{8 \times 2} = \dfrac{10}{16}$

4. a. $\dfrac{6 \times 6}{7 \times 6} = \dfrac{36}{42}$ b. $\dfrac{7 \times 4}{12 \times 4} = \dfrac{28}{48}$
   c. $\dfrac{5 \times 9}{8 \times 9} = \dfrac{45}{72}$ d. $\dfrac{1 \times 18}{2 \times 18} = \dfrac{18}{36}$

5. a. $\dfrac{11}{2} = 5\dfrac{1}{2}$; $2\overline{)11}$, 10 , 1
   b. $\dfrac{17}{3} = 5\dfrac{2}{3}$; $3\overline{)17}$, 15 , 2
   c. $\dfrac{25}{8} = 3\dfrac{1}{8}$; $8\overline{)25}$, 24 , 1
   d. $\dfrac{36}{9} = 4$; $9\overline{)36}$

6. a. $\dfrac{52}{8} = 6\dfrac{4}{8} = 6\dfrac{1}{2}$; $8\overline{)52}$, 48 , 4
   b. $\dfrac{35}{10} = 3\dfrac{5}{10} = 3\dfrac{1}{2}$; $10\overline{)35}$, 30 , 5
   c. $\dfrac{28}{6} = 4\dfrac{4}{6} = 4\dfrac{2}{3}$; $6\overline{)28}$, 24 , 4
   d. $\dfrac{40}{12} = 3\dfrac{4}{12} = 3\dfrac{1}{3}$; $12\overline{)40}$, 36 , 4

7. a. $\dfrac{8}{5} = 1\dfrac{3}{5}$; $5\overline{)8}$, $\dfrac{5}{3}$
   b. $\dfrac{30}{6} = 5$; $6\overline{)30}$, $\dfrac{30}{0}$
   c. $\dfrac{18}{4} = 4\dfrac{2}{4} = 4\dfrac{1}{2}$; $4\overline{)18}$, $\dfrac{16}{2}$
   d. $\dfrac{26}{10} = 2\dfrac{6}{10} = 2\dfrac{3}{5}$; $10\overline{)26}$, $\dfrac{20}{6}$

8. a. $3\dfrac{7}{10} = \dfrac{37}{10}$ b. $2\dfrac{2}{3} = \dfrac{8}{3}$
   c. $7\dfrac{1}{2} = \dfrac{15}{2}$ d. $1\dfrac{5}{6} = \dfrac{11}{6}$

9. a. $10\dfrac{3}{8} = \dfrac{83}{8}$ b. $2\dfrac{3}{4} = \dfrac{11}{4}$
   c. $7\dfrac{2}{5} = \dfrac{37}{5}$ d. $12\dfrac{7}{20} = \dfrac{247}{20}$

10. a. $1\dfrac{14}{15} = \dfrac{29}{15}$ b. $16\dfrac{1}{2} = \dfrac{33}{2}$
    c. $4\dfrac{3}{5} = \dfrac{23}{5}$ d. $4\dfrac{5}{9} = \dfrac{41}{9}$

Lesson 3 Exercise, pg. 94

1. $\dfrac{5}{8}$
   $+ \dfrac{7}{8}$
   $\dfrac{12}{8} = 1\dfrac{4}{8} = 1\dfrac{1}{2}$

2. $3\dfrac{1}{6}$
   $+ 8\dfrac{5}{6}$
   $11\dfrac{6}{6} = 12$

3. $2\dfrac{3}{5}$
   $+ 1\dfrac{4}{5}$
   $3\dfrac{7}{5} = 4\dfrac{2}{5}$

4. $4\dfrac{11}{12}$
   $+ 4\dfrac{7}{12}$
   $8\dfrac{18}{12} = 9\dfrac{6}{12} = 9\dfrac{1}{2}$

5. $\dfrac{1}{4} = \dfrac{5}{20}$
   $+ \dfrac{3}{5} = \dfrac{12}{20}$
   $\dfrac{17}{20}$

6. $\dfrac{5}{12} = \dfrac{5}{12}$
   $+ \dfrac{3}{4} = \dfrac{9}{12}$
   $\dfrac{14}{12} = 1\dfrac{2}{12} = 1\dfrac{1}{6}$

7. $8\dfrac{2}{5} = 8\dfrac{6}{15}$
   $+ 1\dfrac{7}{15} = 1\dfrac{7}{15}$
   $9\dfrac{13}{15}$

8. $6\dfrac{1}{2} = 6\dfrac{4}{8}$
   $+ 5\dfrac{3}{8} = 5\dfrac{3}{8}$
   $11\dfrac{7}{8}$

9. $\dfrac{7}{8} = \dfrac{21}{24}$
   $+ \dfrac{2}{3} = \dfrac{16}{24}$
   $\dfrac{37}{24} = 1\dfrac{13}{24}$

10. $\dfrac{1}{6} = \dfrac{3}{18}$
    $+ \dfrac{5}{9} = \dfrac{10}{18}$
    $\dfrac{13}{18}$

11. $5\dfrac{3}{4} = 5\dfrac{27}{36}$
    $+\ 2\dfrac{2}{9} = 2\dfrac{8}{36}$
    $\phantom{+\ 2\dfrac{2}{9} =\ }7\dfrac{35}{36}$

12. $4\dfrac{3}{5} = 4\dfrac{18}{30}$
    $+\ 4\dfrac{1}{6} = 4\dfrac{5}{30}$
    $\phantom{+\ 4\dfrac{1}{6} =\ }8\dfrac{23}{30}$

13. $\dfrac{3}{10} = \dfrac{6}{20}$
    $\dfrac{1}{2} = \dfrac{10}{20}$
    $+\ \dfrac{3}{4} = \dfrac{15}{20}$
    $\phantom{+\ \dfrac{3}{4} =\ }\dfrac{31}{20} = 1\dfrac{11}{20}$

14. $\dfrac{3}{8} = \dfrac{9}{24}$
    $\dfrac{2}{3} = \dfrac{16}{24}$
    $+\ \dfrac{1}{12} = \dfrac{2}{24}$
    $\phantom{+\ \dfrac{1}{12} =\ }\dfrac{27}{24} = 1\dfrac{3}{24} = 1\dfrac{1}{8}$

15. $2\dfrac{5}{6} = 2\dfrac{20}{24}$
    $5\dfrac{3}{8} = 5\dfrac{9}{24}$
    $+\ 3\dfrac{1}{4} = 3\dfrac{6}{24}$
    $\phantom{+\ 3\dfrac{1}{4} =\ }10\dfrac{35}{24} = 11\dfrac{11}{24}$

16. $1\dfrac{5}{9} = 1\dfrac{10}{18}$
    $4\dfrac{1}{2} = 4\dfrac{9}{18}$
    $+\ 3\dfrac{2}{3} = 3\dfrac{12}{18}$
    $\phantom{+\ 3\dfrac{2}{3} =\ }8\dfrac{31}{18} = 9\dfrac{13}{18}$

Lesson 4 Exercise, pg. 96

1. $\dfrac{11}{12}$
   $-\ \dfrac{7}{12}$
   $\dfrac{4}{12} = \dfrac{1}{3}$

2. $\dfrac{7}{8}$
   $-\ \dfrac{3}{8}$
   $\dfrac{4}{8} = \dfrac{1}{2}$

3. $6\dfrac{9}{10}$
   $-\ 5\dfrac{7}{10}$
   $1\dfrac{2}{10} = 1\dfrac{1}{5}$

4. $9\dfrac{15}{16}$
   $-\ 3\dfrac{3}{16}$
   $6\dfrac{12}{16} = 6\dfrac{3}{4}$

5. $\dfrac{3}{4} = \dfrac{6}{8}$
   $-\ \dfrac{3}{8} = \dfrac{3}{8}$
   $\phantom{-\ \dfrac{3}{8} =\ }\dfrac{3}{8}$

6. $\dfrac{4}{5} = \dfrac{16}{20}$
   $-\ \dfrac{3}{4} = \dfrac{15}{20}$
   $\phantom{-\ \dfrac{3}{4} =\ }\dfrac{1}{20}$

7. $7\dfrac{1}{2} = 7\dfrac{5}{10}$
   $-\ 2\dfrac{1}{5} = 2\dfrac{2}{10}$
   $\phantom{-\ 2\dfrac{1}{5} =\ }5\dfrac{3}{10}$

8. $4\dfrac{7}{9} = 4\dfrac{14}{18}$
   $-\ 1\dfrac{1}{6} = 1\dfrac{3}{18}$
   $\phantom{-\ 1\dfrac{1}{6} =\ }3\dfrac{11}{18}$

9. $7 = 6\dfrac{8}{8}$
   $-\ 3\dfrac{5}{8} = 3\dfrac{5}{8}$
   $\phantom{-\ 3\dfrac{5}{8} =\ }3\dfrac{3}{8}$

10. $9 = 8\dfrac{12}{12}$
    $-\ 4\dfrac{7}{12} = 4\dfrac{7}{12}$
    $\phantom{-\ 4\dfrac{7}{12} =\ }4\dfrac{5}{12}$

11. $8 = 7\dfrac{10}{10}$
    $-\ 2\dfrac{7}{10} = 2\dfrac{7}{10}$
    $\phantom{-\ 2\dfrac{7}{10} =\ }5\dfrac{3}{10}$

12. $12 = 11\dfrac{16}{16}$
    $-\ 5\dfrac{13}{16} = 5\dfrac{13}{16}$
    $\phantom{-\ 5\dfrac{13}{16} =\ }6\dfrac{3}{16}$

13. $5\dfrac{1}{6} = 4\dfrac{1}{6} + \dfrac{6}{6} = 4\dfrac{7}{6}$
    $-\ 2\dfrac{5}{6} = 2\dfrac{5}{6} \phantom{+\dfrac{6}{6}} = 2\dfrac{5}{6}$
    $\phantom{-\ 2\dfrac{5}{6} = 2\dfrac{5}{6} +\dfrac{6}{6}\ \ }2\dfrac{2}{6} = 2\dfrac{1}{3}$

14. $9\dfrac{1}{3} = 8\dfrac{1}{3} + \dfrac{3}{3} = 8\dfrac{4}{3}$
    $-\ 3\dfrac{2}{3} = 3\dfrac{2}{3} \phantom{+\dfrac{3}{3}} = 3\dfrac{2}{3}$
    $\phantom{-\ 3\dfrac{2}{3} = 3\dfrac{2}{3} +\dfrac{3}{3}\ \ }5\dfrac{2}{3}$

15. $8\dfrac{5}{12} = 7\dfrac{5}{12} + \dfrac{12}{12} = 7\dfrac{17}{12}$
    $-\ 7\dfrac{7}{12} = \phantom{7\dfrac{5}{12} + \dfrac{12}{12} =\ }7\dfrac{7}{12}$
    $\phantom{-\ 7\dfrac{7}{12} = 7\dfrac{5}{12} + \dfrac{12}{12} =\ }\dfrac{10}{12} = \dfrac{5}{6}$

16. $4\frac{1}{5} = 3\frac{1}{5} + \frac{5}{5} = 3\frac{6}{5}$
    $-1\frac{4}{5} = = 1\frac{4}{5}$
    $ \mathbf{2\frac{2}{5}}$

17. $6\frac{1}{5} = 6\frac{4}{20} = 5\frac{4}{20} + \frac{20}{20} = 5\frac{24}{20}$
    $-3\frac{3}{4} = 3\frac{15}{20} = 3\frac{15}{20}$
    $ \mathbf{2\frac{9}{20}}$

18. $7\frac{1}{2} = 7\frac{4}{8} = 6\frac{4}{8} + \frac{8}{8} = 6\frac{12}{8}$
    $-1\frac{5}{8} = 1\frac{5}{8} = 1\frac{5}{8}$
    $ \mathbf{5\frac{7}{8}}$

19. $9\frac{1}{3} = 9\frac{4}{12} = 8\frac{4}{12} + \frac{12}{12} = 8\frac{16}{12}$
    $-5\frac{3}{4} = 5\frac{9}{12} = 5\frac{9}{12}$
    $ \mathbf{3\frac{7}{12}}$

20. $2\frac{1}{2} = 2\frac{3}{6} = 1\frac{3}{6} + \frac{6}{6} = 1\frac{9}{6}$
    $-1\frac{2}{3} = 1\frac{4}{6} = 1\frac{4}{6}$
    $ \mathbf{\frac{5}{6}}$

Lesson 5 Exercise, pg. 98

1. $\frac{2}{3} \times \frac{4}{5} = \mathbf{\frac{8}{15}}$

2. $\frac{5}{8} \times \frac{3}{4} = \mathbf{\frac{15}{32}}$

3. $\frac{3}{\cancel{4}} \times \frac{1}{\cancel{2}} \times \frac{\cancel{4}^{1}}{5} = \mathbf{\frac{3}{10}}$

4. $\frac{\cancel{9}^{3}}{\cancel{10}_{5}} \times \frac{\cancel{2}^{1}}{\cancel{3}_{1}} = \mathbf{\frac{3}{5}}$

5. $\frac{\cancel{3}^{1}}{\cancel{20}_{4}} \times \frac{\cancel{5}^{1}}{\cancel{12}_{4}} = \mathbf{\frac{1}{16}}$

6. $\frac{\cancel{2}^{1}}{3} \times \frac{\cancel{5}^{1}}{\cancel{6}_{3}} \times \frac{7}{\cancel{10}_{2}} = \mathbf{\frac{7}{18}}$

7. $6 \times \frac{3}{4} =$
   $\frac{\cancel{6}^{3}}{1} \times \frac{3}{\cancel{4}_{2}} = \frac{9}{2} = \mathbf{4\frac{1}{2}}$

8. $\frac{5}{6} \times 15 =$
   $\frac{5}{\cancel{6}_{2}} \times \frac{\cancel{15}^{5}}{1} = \frac{25}{2} = \mathbf{12\frac{1}{2}}$

9. $8 \times \frac{11}{12} =$
   $\frac{\cancel{8}^{2}}{1} \times \frac{11}{\cancel{12}_{3}} = \frac{22}{3} = \mathbf{7\frac{1}{3}}$

10. $\frac{3}{4} \times 3\frac{1}{5} =$
    $\frac{3}{\cancel{4}_{1}} \times \frac{\cancel{16}^{4}}{5} = \frac{12}{5} = \mathbf{2\frac{2}{5}}$

11. $2\frac{1}{4} \times 1\frac{2}{3} =$
    $\frac{\cancel{9}^{3}}{4} \times \frac{5}{\cancel{3}_{1}} = \frac{15}{4} = \mathbf{3\frac{3}{4}}$

12. $5\frac{1}{3} \times 1\frac{5}{16} =$
    $\frac{\cancel{16}^{1}}{\cancel{3}_{1}} \times \frac{\cancel{21}^{7}}{\cancel{16}_{1}} = \frac{7}{1} = \mathbf{7}$

13. $2\frac{1}{2} \times 1\frac{2}{5} \times 2\frac{2}{3} =$
    $\frac{\cancel{5}^{1}}{2} \times \frac{7}{\cancel{5}_{1}} \times \frac{\cancel{8}^{4}}{3} = \frac{28}{3} = \mathbf{9\frac{1}{3}}$

14. $\frac{5}{6} \times 3\frac{3}{5} \times 2\frac{2}{3} =$
    $\frac{\cancel{5}^{1}}{\cancel{6}_{1}} \times \frac{\cancel{18}^{3}}{\cancel{5}_{1}} \times \frac{8}{\cancel{3}_{1}} = \frac{8}{1} = \mathbf{8}$

15. $1\frac{1}{6} \times 1\frac{1}{3} \times 3\frac{3}{4} =$
    $\frac{7}{6} \times \frac{\cancel{4}^{1}}{\cancel{3}} \times \frac{\cancel{15}^{5}}{\cancel{4}_{1}} = \frac{35}{6} = \mathbf{5\frac{5}{6}}$

Lesson 6 Exercise, pg. 100

1. $\frac{2}{3} \div \frac{4}{9} =$
   $\frac{\cancel{2}^{1}}{\cancel{3}_{1}} \times \frac{\cancel{9}^{3}}{\cancel{4}_{2}} = \frac{3}{2} = \mathbf{1\frac{1}{2}}$

2. $\frac{9}{10} \div \frac{2}{5} =$
   $\frac{9}{\cancel{10}_{2}} \times \frac{\cancel{5}^{1}}{2} = \frac{9}{4} = \mathbf{2\frac{1}{4}}$

3. $\frac{3}{4} \div \frac{1}{8} =$
   $\frac{3}{\cancel{4}_{1}} \times \frac{\cancel{8}^{2}}{1} = \frac{6}{1} = \mathbf{6}$

4. $4 \div \frac{2}{3} =$
   $\frac{\cancel{4}^{2}}{1} \times \frac{3}{\cancel{2}_{1}} = \frac{6}{1} = \mathbf{6}$

5. $8 \div \frac{4}{5} =$
   $\frac{\cancel{8}^{2}}{1} \times \frac{5}{\cancel{4}_{1}} = \frac{10}{1} = \mathbf{10}$

6. $5 \div \frac{3}{4} =$
   $\frac{5}{1} \times \frac{4}{3} = \frac{20}{3} = \mathbf{6\frac{2}{3}}$

Answers and Solutions

7. $1\frac{1}{9} \div \frac{5}{6} =$

$\frac{\cancel{10}^2}{\cancel{9}_3} \times \frac{\cancel{6}^2}{\cancel{5}_1} = \frac{4}{3} = 1\frac{1}{3}$

8. $2\frac{5}{8} \div \frac{1}{4} =$

$\frac{21}{\cancel{8}_2} \times \frac{\cancel{4}^1}{1} = \frac{21}{2} = 10\frac{1}{2}$

9. $6\frac{1}{2} \div \frac{3}{8} =$

$\frac{13}{\cancel{2}_1} \times \frac{\cancel{8}^4}{3} = \frac{52}{3} = 17\frac{1}{3}$

10. $4\frac{1}{3} \div 5 =$

$\frac{13}{3} \div \frac{5}{1}$

$\frac{13}{3} \times \frac{1}{5} = \frac{13}{15}$

11. $10\frac{1}{2} \div 8 =$

$\frac{21}{2} \div \frac{8}{1}$

$\frac{21}{2} \times \frac{1}{8} = \frac{21}{16} = 1\frac{5}{16}$

12. $\frac{3}{8} \div 4 =$

$\frac{3}{8} \div \frac{4}{1}$

$\frac{3}{8} \times \frac{1}{4} = \frac{3}{32}$

13. $\frac{5}{9} \div 1\frac{1}{3} =$

$\frac{5}{9} \div \frac{4}{3} =$

$\frac{5}{\cancel{9}_3} \times \frac{\cancel{3}^1}{4} = \frac{5}{12}$

14. $4 \div 1\frac{3}{4} =$

$\frac{4}{1} \div \frac{7}{4} =$

$\frac{4}{1} \times \frac{4}{7} = \frac{16}{7} = 2\frac{2}{7}$

15. $\frac{3}{8} \div 2\frac{2}{3} =$

$\frac{3}{8} \div \frac{8}{3} =$

$\frac{3}{8} \times \frac{3}{8} = \frac{9}{64}$

16. $1\frac{1}{4} \div 2\frac{1}{2} =$

$\frac{5}{4} \div \frac{5}{2} =$

$\frac{\cancel{5}^1}{\cancel{4}_2} \times \frac{\cancel{2}^1}{\cancel{5}_1} = \frac{1}{2}$

17. $2\frac{3}{4} \div 1\frac{5}{8} =$

$\frac{11}{4} \div \frac{13}{8} =$

$\frac{11}{\cancel{4}_1} \times \frac{\cancel{8}^2}{13} = \frac{22}{13} = 1\frac{9}{13}$

18. $7\frac{1}{2} \div 3\frac{3}{4} =$

$\frac{15}{2} \div \frac{15}{4} =$

$\frac{\cancel{15}^1}{\cancel{2}_1} \times \frac{\cancel{4}^2}{\cancel{15}_1} = \frac{2}{1} = 2$

Lesson 7 Exercise, pg. 100

1. **a.** $\frac{8}{15} = \frac{16}{30}$ $\frac{1}{2} = \frac{15}{30}$ **b.** $\frac{2}{3} = \frac{6}{9}$ $\frac{5}{9} = \frac{5}{9}$

   $\frac{8}{15}$ is greater. $\frac{2}{3}$ is greater.

   c. $\frac{5}{8} = \frac{25}{40}$ $\frac{4}{5} = \frac{32}{40}$ **d.** $\frac{5}{12} = \frac{15}{36}$ $\frac{4}{9} = \frac{16}{36}$

   $\frac{4}{5}$ is greater. $\frac{4}{9}$ is greater.

2. $\frac{3}{5} = \frac{12}{20}$ $\frac{1}{2} = \frac{10}{20}$ $\frac{11}{20} = \frac{11}{20}$ $\frac{7}{10} = \frac{14}{20}$

   From least to greatest: $\frac{1}{2}, \frac{11}{20}, \frac{3}{5}, \frac{7}{10}$.

3. $\frac{5}{12} = \frac{20}{48}$ $\frac{1}{4} = \frac{12}{48}$ $\frac{5}{16} = \frac{15}{48}$ $\frac{1}{3} = \frac{16}{48}$

   From greatest to least: $\frac{5}{12}, \frac{1}{3}, \frac{5}{16}, \frac{1}{4}$.

4. $\frac{5}{8} = \frac{10}{16}$ $\frac{9}{16} = \frac{9}{16}$ $\frac{5}{8}$ inch is longer.

5. $\frac{3}{4} = \frac{12}{16}$ $\frac{13}{16} = \frac{13}{16}$ $\frac{13}{16}$ pound is heavier.

Lesson 8 Exercise, pg. 102

1. **a.** $0.8 = \frac{8}{10} = \frac{4}{5}$

   b. $0.04 = \frac{4}{100} = \frac{1}{25}$

   c. $0.35 = \frac{35}{100} = \frac{7}{20}$

   d. $0.005 = \frac{5}{1000} = \frac{1}{200}$

2. **a.** $0.065 = \frac{65}{1000} = \frac{13}{200}$

   b. $0.0075 = \frac{75}{10,000} = \frac{3}{400}$

   c. $0.002 = \frac{2}{1000} = \frac{1}{500}$

   d. $0.875 = \frac{875}{1000} = \frac{7}{8}$

3. **a.** $8.4 = 8\frac{4}{10} = 8\frac{2}{5}$

   b. $2.85 = 2\frac{85}{100} = 2\frac{17}{20}$

   c. $1.004 = 1\frac{4}{1000} = 1\frac{1}{250}$

   d. $6.3 = 6\frac{3}{10}$

4. **a.** $10.125 = 10\frac{125}{1000} = 10\frac{1}{8}$

   b. $3.009 = 3\frac{9}{1000}$

   c. $4.80 = 4\frac{80}{100} = 4\frac{4}{5}$

   d. $12.16 = 12\frac{16}{100} = 12\frac{4}{25}$

5. a. $\dfrac{9}{10} = 10\overline{)9.0}^{\mathbf{0.9}}$

   b. $\dfrac{4}{5} = 5\overline{)4.0}^{\mathbf{0.8}}$

   c. $\dfrac{9}{50} = 50\overline{)9.00}^{\mathbf{0.18}}$
   $\dfrac{5\ 0}{4\ 00}$
   $\dfrac{}{4\ 00}$

   d. $\dfrac{3}{8} = 8\overline{)3.00}^{0.37\frac{4}{8}} = 0.37\dfrac{1}{2}$ or $\mathbf{0.375}$

6. a. $\dfrac{1}{2} = 2\overline{)1.0}^{\mathbf{0.5}}$

   b. $\dfrac{2}{3} = 3\overline{)2.00}^{\mathbf{0.66}\frac{2}{3}}$

   c. $\dfrac{8}{25} = 25\overline{)8.00}^{\mathbf{0.32}}$
   $\dfrac{7\ 5}{50}$
   $\dfrac{}{50}$

   d. $\dfrac{1}{6} = 6\overline{)1.00}^{0.16\frac{4}{6}} = \mathbf{0.16}\dfrac{\mathbf{2}}{\mathbf{3}}$

7. a. $\dfrac{2}{9} = 9\overline{)2.00}^{\mathbf{0.22}\frac{2}{9}}$

   b. $\dfrac{3}{4} = 4\overline{)3.00}^{\mathbf{0.75}}$

   c. $\dfrac{5}{12} = 12\overline{)5.00}^{0.41\frac{8}{12}} = \mathbf{0.41}\dfrac{\mathbf{2}}{\mathbf{3}}$
   $\dfrac{4\ 8}{20}$
   $\dfrac{12}{8}$

   d. $\dfrac{7}{8} = 8\overline{)7.00}^{0.87\frac{4}{8}} = 0.87\dfrac{1}{2}$ or $\mathbf{0.875}$
   $\dfrac{6\ 4}{60}$
   $\dfrac{56}{4}$

Level 1 Review, pg. 103

1. $\dfrac{35 \div 7}{84 \div 7} = \dfrac{\mathbf{5}}{\mathbf{12}}$

2. $8\dfrac{4}{5} = \dfrac{\mathbf{44}}{\mathbf{5}}$

3. $8\dfrac{3}{4} = 8\dfrac{12}{16}$
   $+\ 7\dfrac{9}{16} = 7\dfrac{9}{16}$
   $15\dfrac{21}{16} = \mathbf{16}\dfrac{\mathbf{5}}{\mathbf{16}}$

4. $2\dfrac{3}{8} = 2\dfrac{15}{40}$
   $4\dfrac{1}{2} = 4\dfrac{20}{40}$
   $+\ 6\dfrac{3}{5} = 6\dfrac{24}{40}$
   $12\dfrac{59}{40} = \mathbf{13}\dfrac{\mathbf{19}}{\mathbf{40}}$

5. $7\dfrac{2}{3} = 7\dfrac{8}{12}$
   $-\ 3\dfrac{5}{12} = 3\dfrac{5}{12}$
   $4\dfrac{3}{12} = \mathbf{4}\dfrac{\mathbf{1}}{\mathbf{4}}$

6. $8\dfrac{1}{5} = 8\dfrac{4}{20} = 7\dfrac{4}{20} + \dfrac{20}{20} = 7\dfrac{24}{20}$
   $-\ 3\dfrac{3}{4} = 3\dfrac{15}{20} = 3\dfrac{15}{20}$
   $\mathbf{4}\dfrac{\mathbf{9}}{\mathbf{20}}$

7. $3\dfrac{3}{5} \times 10 =$

   $\dfrac{18}{\cancel{5}_{1}} \times \dfrac{\cancel{10}^{2}}{1} = \dfrac{36}{1} = \mathbf{36}$

8. $2\dfrac{1}{3} \times 4\dfrac{1}{8} \times \dfrac{1}{2} =$

   $\dfrac{7}{\cancel{3}_{1}} \times \dfrac{\cancel{33}^{11}}{8} \times \dfrac{1}{2} = \dfrac{77}{16} = \mathbf{4}\dfrac{\mathbf{13}}{\mathbf{16}}$

9. $6\dfrac{2}{3} \div 8 =$

   $\dfrac{20}{3} \div \dfrac{8}{1} =$

   $\dfrac{\cancel{20}^{5}}{3} \times \dfrac{1}{\cancel{8}_{2}} = \dfrac{\mathbf{5}}{\mathbf{6}}$

10. $9\dfrac{1}{3} \div 3\dfrac{1}{5} =$

    $\dfrac{28}{3} \div \dfrac{16}{5} =$

    $\dfrac{\cancel{28}^{7}}{3} \times \dfrac{5}{\cancel{16}_{4}} = \dfrac{35}{12} = \mathbf{2}\dfrac{\mathbf{11}}{\mathbf{12}}$

11. From least to greatest:
$$\frac{1}{2}, \frac{9}{16}, \frac{5}{8}$$
$$\frac{5}{8} = \frac{10}{16} \quad \frac{9}{16} \quad \frac{1}{2} = \frac{8}{16}$$

12. $\frac{11}{20} = $
```
      0.55
20)11.00
   10 0
    1 00
    1 00
       0
```

Lesson 9 Exercise, pg. 105

1. $15 - 9 = 6$ women
   a. $\frac{\text{men}}{\text{total}} = \frac{9}{15} = \frac{3}{5}$
   b. $\frac{\text{women}}{\text{total}} = \frac{6}{15} = \frac{2}{5}$

2. $45 + 36 = 81$ total
   a. $\frac{\text{union}}{\text{total}} = \frac{45}{81} = \frac{5}{9}$
   b. $\frac{\text{nonmembers}}{\text{total}} = \frac{36}{81} = \frac{4}{9}$

3. $33 - 15 = 18$ used cars
   a. $\frac{\text{new}}{\text{total}} = \frac{15}{33} = \frac{5}{11}$
   b. $\frac{\text{used}}{\text{total}} = \frac{18}{33} = \frac{6}{11}$

4. $12 + 2 + 6 = 20$ total
   a. $\frac{\text{cola}}{\text{total}} = \frac{12}{20} = \frac{3}{5}$
   b. $\frac{\text{lime}}{\text{total}} = \frac{2}{20} = \frac{1}{10}$
   c. $\frac{\text{orange}}{\text{total}} = \frac{6}{20} = \frac{3}{10}$

5. $45 + 35 + 20 = 100$ total
   a. $\frac{\text{for}}{\text{total}} = \frac{45}{100} = \frac{9}{20}$
   b. $\frac{\text{against}}{\text{total}} = \frac{35}{100} = \frac{7}{20}$
   c. $\frac{\text{undecided}}{\text{total}} = \frac{20}{100} = \frac{1}{5}$

Lesson 10 Exercise, pg. 107

1. a. $2\frac{2}{3}$ yd
      $3\overline{)8}$
   b. 2 yd 2 ft
      $3\overline{)8}$

2. a. $2\frac{15}{60} = 2\frac{1}{4}$ min
      $60\overline{)135}$
      $\underline{120}$
      15
   b. 2 min 15 sec
      $60\overline{)135}$
      $\underline{120}$
      15

3. a. $2\frac{2}{4} = 2\frac{1}{2}$ gal
      $4\overline{)10}$
   b. 2 gal 2 qt
      $4\overline{)10}$

4. a. $1\frac{4}{16} = 1\frac{1}{4}$ lb
      $16\overline{)20}$
   b. 1 lb 4 oz
      $16\overline{)20}$

5. a. $1\frac{8}{12} = 1\frac{2}{3}$ yr
      $12\overline{)20}$
      $\underline{12}$
      8
   b. 1 yr 8 mo
      $12\overline{)20}$
      $\underline{12}$
      8

6. a. $4\frac{1000}{2000} = 4\frac{1}{2}$ t
      $2000\overline{)9000}$
      $\underline{8000}$
      1000
   b. 4 t 1000 lb
      $2000\overline{)9000}$
      $\underline{8000}$
      1000

7. a. $3\frac{4}{12} = 3\frac{1}{3}$ ft
      $12\overline{)40}$
      $\underline{36}$
      4
   b. 3 ft 4 in
      $12\overline{)40}$
      $\underline{36}$
      4

8. a. $1\frac{40}{60} = 1\frac{2}{3}$ hr
      $60\overline{)100}$
      $\underline{60}$
      40
   b. 1 hr 40 min
      $60\overline{)100}$
      $\underline{60}$
      40

9. a. 10 mo $= \frac{10}{12} = \frac{5}{6}$ yr
   b. 45 min $= \frac{45}{60} = \frac{3}{4}$ hr
   c. 9 in. $= \frac{9}{12} = \frac{3}{4}$ ft

10. **a.** 8 oz = $\frac{8}{16} = \frac{1}{2}$ lb

 b. 20 in. = $\frac{20}{36} = \frac{5}{9}$ yd

 c. 2 qt = $\frac{2}{4} = \frac{1}{2}$ gal

11. **a.** 1800 lb = $\frac{1800}{2000} = \frac{9}{10}$ t

 b. 16 hr = $\frac{16}{24} = \frac{2}{3}$ da

 c. 1320 ft = $\frac{1320}{5280} = \frac{1}{4}$ mi

12. **a.** $4 \times 36 = $ **144** in.

 b. $2\frac{1}{2} \times 60 =$

 $\frac{5}{\cancel{2}} \times \frac{\cancel{60}^{30}}{1} = $ **150** sec

 c. $1\frac{1}{2} \times 2000 =$

 $\frac{3}{\cancel{2}} \times \frac{\cancel{2000}^{1000}}{1} = $ **3000** lb

Lesson 11 Exercise, pg. 108

1. $1\frac{1}{2}$ in.
2. $2\frac{1}{8}$ in.
3. $2\frac{11}{16}$ in.
4. $3\frac{1}{4}$ in.
5. $3\frac{7}{8}$ in.
6. $4\frac{3}{16}$ in.
7. $2\frac{1}{8} = 2\frac{1}{8} = 1\frac{9}{8}$
   $-1\frac{1}{2} = 1\frac{4}{8} = 1\frac{4}{8}$
   $\phantom{-1\frac{1}{2} = 1\frac{4}{8} = 1}\frac{5}{8}$ in.
8. $3\frac{7}{8} = 3\frac{7}{8}$
   $-3\frac{1}{4} = 3\frac{2}{8}$
   $\phantom{-3\frac{1}{4} = 3}\frac{5}{8}$ in.

Lesson 12 Exercise, pg. 109

1. **a.** $\left(\frac{1}{2}\right)^2 = \frac{1}{2} \times \frac{1}{2} = \frac{1}{4}$

 b. $\left(\frac{1}{9}\right)^2 = \frac{1}{9} \times \frac{1}{9} = \frac{1}{81}$

 c. $\left(\frac{2}{3}\right)^3 = \frac{2}{3} \times \frac{2}{3} \times \frac{2}{3} = \frac{8}{27}$

 d. $\left(\frac{1}{10}\right)^4 = \frac{1}{10} \times \frac{1}{10} \times \frac{1}{10} \times \frac{1}{10} = \frac{1}{10,000}$

2. **a.** $\left(\frac{3}{10}\right)^3 = \frac{3}{10} \times \frac{3}{10} \times \frac{3}{10} = \frac{27}{1000}$

 b. $\left(\frac{5}{6}\right)^2 = \frac{5}{6} \times \frac{5}{6} = \frac{25}{36}$

 c. $\left(\frac{7}{12}\right)^2 = \frac{7}{12} \times \frac{7}{12} = \frac{49}{144}$

 d. $\left(\frac{3}{4}\right)^3 = \frac{3}{4} \times \frac{3}{4} \times \frac{3}{4} = \frac{27}{64}$

3. **a.** $\sqrt{\frac{25}{36}} = \frac{5}{6}$ **b.** $\sqrt{\frac{4}{49}} = \frac{2}{7}$

 c. $\sqrt{\frac{1}{81}} = \frac{1}{9}$ **d.** $\sqrt{\frac{9}{100}} = \frac{3}{10}$

4. **a.** $\sqrt{\frac{64}{81}} = \frac{8}{9}$ **b.** $\sqrt{\frac{1}{144}} = \frac{1}{12}$

 c. $\sqrt{\frac{9}{16}} = \frac{3}{4}$ **d.** $\sqrt{\frac{36}{49}} = \frac{6}{7}$

Lesson 13 Exercise, pg. 111

1. $A = \frac{1}{2}bh$

 $A = \frac{1}{\cancel{2}} \times \frac{\cancel{8}^4}{1} \times \frac{10}{1}$

 $A = $ **40 sq in.**

2. $A = \frac{1}{2}bh$

 $A = \frac{1}{\cancel{2}} \times \frac{\cancel{30}^{15}}{1} \times \frac{18}{1}$

 $A = $ **270 sq ft**

3. $A = \frac{1}{2}bh$

 $A = \frac{1}{\cancel{2}} \times \frac{\cancel{14}^7}{1} \times \frac{8}{1}$

 $A = $ **56 sq yd**

Answers and Solutions

4. $A = \dfrac{1}{2}bh$

   $A = \dfrac{1}{\cancel{2}} \times \dfrac{\cancel{12}}{1} \times \dfrac{\cancel{22}^{11}}{1}$

   $A = \mathbf{132 \text{ sq ft}}$

5. $A = bh$
   $A = 12 \times 6$
   $A = \mathbf{72 \text{ sq in.}}$

6. $A = bh$
   $A = 10 \times 1\tfrac{1}{2}$

   $A = \dfrac{\cancel{10}^{5}}{1} \times \dfrac{3}{\cancel{2}}$

   $A = \mathbf{15 \text{ sq ft}}$

7. $P = 4s$
   $P = 4 \times 1\tfrac{1}{4}$

   $P = \dfrac{\cancel{4}}{1} \times \dfrac{5}{\cancel{4}}$

   $P = \mathbf{5 \text{ in.}}$

8. $A = s^2$
   $A = \left(1\tfrac{1}{4}\right)^2$
   $A = \dfrac{5}{4} \times \dfrac{5}{4}$
   $A = \dfrac{25}{16}$
   $A = \mathbf{1\tfrac{9}{16} \text{ sq in.}}$

9. $P = 2l + 2w$
   $P = 2\left(\tfrac{3}{4}\right) + 2\left(\tfrac{1}{2}\right)$
   $P = \dfrac{3}{2} + \dfrac{2}{2}$
   $P = \dfrac{5}{2}$
   $P = \mathbf{2\tfrac{1}{2} \text{ in.}}$

10. $A = lw$
    $A = \dfrac{3}{4} \times \dfrac{1}{2}$
    $A = \mathbf{\dfrac{3}{8} \text{ sq in.}}$

Lesson 14 Exercise, pg. 113

1. $15:36 = \mathbf{5:12}$
2. width : length $= 12:15 = \mathbf{4:5}$
3. town : country $= 16:24 = \mathbf{2:3}$
4. $4:16 = \mathbf{1:4}$
5. $20 - 12 = 8$ women
   men : women $= 12:8 = \mathbf{3:2}$
6. $\$450 + \$150 = \$600$ original price
   sale price : original price $= 450:600 = \mathbf{3:4}$
7. $\$220 + \$880 = \$1100$ total
   rent : total $= 220:1100 = \mathbf{1:5}$
8. $280 - 70 = 210$ lb, September weight.
   March weight : September weight $=$
   $280:210 = \mathbf{4:3}$
9. $\$5000 - \$1500 = \$3500$ still owed
   amount paid : amount owed $=$
   $1500:3500 = \mathbf{3:7}$
10. $40 + 16 = 56$ total
    right : total $= 40:56 = \mathbf{5:7}$

Lesson 15 Exercise, pg. 115

1. **a.** $\dfrac{c}{6} = \dfrac{7}{30}$ **b.** $\dfrac{9}{a} = \dfrac{3}{2}$

   $30c = 42$ $3a = 18$

   $c = 1\tfrac{12}{30} = \mathbf{1\tfrac{2}{5}}$ $a = \mathbf{6}$

   c. $\dfrac{5}{8} = \dfrac{m}{20}$ **d.** $\dfrac{4}{w} = \dfrac{18}{6}$

   $8m = 100$ $18w = 24$

   $m = 12\tfrac{4}{8} = \mathbf{12\tfrac{1}{2}}$ $w = 1\tfrac{6}{18} = \mathbf{1\tfrac{1}{3}}$

2. **a.** $\dfrac{9}{10} = \dfrac{27}{e}$ **b.** $\dfrac{x}{14} = \dfrac{4}{7}$

   $9e = 270$ $7x = 56$

   $e = \mathbf{30}$ $x = \mathbf{8}$

   c. $\dfrac{1}{s} = \dfrac{8}{5}$ **d.** $\dfrac{8}{15} = \dfrac{r}{2}$

   $8s = 5$ $15r = 16$

   $s = \mathbf{\dfrac{5}{8}}$ $r = \mathbf{1\tfrac{1}{15}}$

3. **a.** $3:n = 6:11$ **b.** $2:5 = t:60$

   $\dfrac{3}{n} = \dfrac{6}{11}$ $\dfrac{2}{5} = \dfrac{t}{60}$

   $6n = 33$ $5t = 120$

   $n = \mathbf{5\tfrac{1}{2}}$ $t = \mathbf{24}$

   c. $h:4 = 5:6$ **d.** $9:2 = 1:d$

   $\dfrac{h}{4} = \dfrac{5}{6}$ $\dfrac{9}{2} = \dfrac{1}{d}$

   $6h = 20$ $9d = 2$

   $h = \mathbf{3\tfrac{1}{3}}$ $d = \mathbf{\dfrac{2}{9}}$

4. **a.** $d:12 = 7:48$
$$\frac{d}{12} = \frac{7}{48}$$
$$48d = 84$$
$$d = 1\frac{3}{4}$$

b. $4:15 = 24:s$
$$\frac{4}{15} = \frac{24}{s}$$
$$4s = 360$$
$$s = 90$$

c. $6:v = 9:10$
$$\frac{6}{v} = \frac{9}{10}$$
$$9v = 60$$
$$v = 6\frac{2}{3}$$

d. $3:11 = y:5$
$$\frac{3}{11} = \frac{y}{5}$$
$$11y = 15$$
$$y = 1\frac{4}{11}$$

Lesson 16 Exercise, pg. 116

1. $\frac{15}{7500} = \frac{1}{500}$

2. $12 + 20 + 18 + 30 = 80$ total
   a. $\frac{20}{80} = \frac{1}{4}$
   b. $\frac{30}{80} = \frac{3}{8}$
   c. $12 + 18 = 30$
   $\frac{30}{80} = \frac{3}{8}$

3. $10 + 15 + 8 = 33$ total
   $\frac{15}{33} = \frac{5}{11}$

4. $5 + 7 + 3 = 15$ total
   a. $\frac{3}{15} = \frac{1}{5}$
   b. There are now 4 hardballs among the 12 balls remaining. $\frac{4}{12} = \frac{1}{3}$

5. $8 + 6 = 14$ total
   a. $\frac{8}{14} = \frac{4}{7}$
   b. The remaining cans include 7 tomato sauce and 5 green beans. $\frac{5}{12}$

6. **a.** $\frac{1}{540}$
   b. $\frac{3}{540} = \frac{1}{180}$

Level 2 Review, pg. 117

1. $\$240 + \$960 = \$1200$ total
   $\frac{\text{rent}}{\text{take-home}} = \frac{240}{1200} = \frac{1}{5}$

2. Total: $8 + 6 + 6 = 20$
   $\frac{\text{blue}}{\text{total}} = \frac{8}{20} = \frac{2}{5}$

3. $12\overline{)40} = 3\frac{4}{12} = 3\frac{1}{3}$ yr
   $\underline{36}$
   4

4. $\frac{12}{16} = \frac{3}{4}$ lb

5. $\frac{15}{16}$ in.

6. $2\frac{3}{8} = 2\frac{6}{16} = 1\frac{22}{16}$
   $-\frac{15}{16} = \frac{15}{16} = \frac{15}{16}$
   $1\frac{7}{16}$ in.

7. $\left(\frac{3}{8}\right)^2 = \frac{3}{8} \times \frac{3}{8} = \frac{9}{64}$

8. $\sqrt{\frac{49}{100}} = \frac{7}{10}$

9. $A = \frac{1}{2}bh$
   $A = \frac{1}{2} \times 12 \times 5$
   $A = \frac{1}{2} \times \frac{\overset{6}{\cancel{12}}}{1} \times \frac{5}{1} = 30$ sq in.

10. $A = lw$
    $A = 6\frac{5}{8} \times 4$
    $A = \frac{53}{\underset{2}{\cancel{8}}} \times \frac{\overset{1}{\cancel{4}}}{1} = \frac{53}{2} = 26\frac{1}{2}$ sq in.

11. $65 + 25 + 10 = 100$ total
    want : total $= 65 : 100 = $ **13 : 20**

12. Total: $4 + 6 = 10$
    women : total $= 4 : 10 = $ **2 : 5**

13. $8 : x = 5 : 12$
    $\frac{8}{x} = \frac{5}{12}$
    $5x = 96$
    $x = 19\frac{1}{5}$

14. Total: $5 + 7 + 3 = 15$.
    $\frac{\text{green}}{\text{total}} = \frac{5}{15} = \frac{1}{3}$

15. The bag now has 4 green + 5 blue + 3 black = 12 marbles.
    $\frac{3}{12} = \frac{1}{4}$

Answers and Solutions

Lesson 17 Exercise, pg. 120

1. $\dfrac{\text{length}}{\text{weight}} = \dfrac{6}{15} = \dfrac{16}{x}$
   $6x = 240$
   $x = \mathbf{40\ lb}$

2. $\dfrac{\text{girls}}{\text{boys}} = \dfrac{5}{4} = \dfrac{120}{x}$
   $5x = 480$
   $x = \mathbf{96\ boys}$

3. $\dfrac{\text{hours}}{\text{wages}} = \dfrac{8}{\$52.80} = \dfrac{20}{x}$
   $8x = \$1056$
   $x = \mathbf{\$132}$

4. $\dfrac{\text{acres}}{\text{bushels}} = \dfrac{12}{1440} = \dfrac{50}{x}$
   $12x = 72{,}000$
   $x = \mathbf{6{,}000\ bushels}$

5. $\dfrac{\text{width}}{\text{length}} = \dfrac{3}{5} = \dfrac{15}{x}$
   $3x = 75$
   $x = \mathbf{25\ in.}$

6. $\dfrac{\text{hours}}{\text{miles}} = \dfrac{1\frac{1}{2}}{72} = \dfrac{4}{x}$
   $1\frac{1}{2}x = 288$
   $x = 288 \div 1\frac{1}{2}$
   $x = \dfrac{288}{1} \div \dfrac{3}{2}$
   $x = \dfrac{\overset{96}{\cancel{288}}}{1} \times \dfrac{2}{\underset{1}{\cancel{3}}} = \mathbf{192\ mi}$

7. $\dfrac{\text{inches}}{\text{miles}} = \dfrac{\frac{1}{2}}{15} = \dfrac{2}{x}$
   $\dfrac{1}{2}x = 30$
   $x = 30 \div \dfrac{1}{2}$
   $x = \dfrac{30}{1} \times \dfrac{2}{1} = \mathbf{60\ mi}$

8. $\dfrac{\text{blue}}{\text{white}} = \dfrac{3}{2} = \dfrac{6}{x}$
   $3x = 12$
   $x = \mathbf{4\ gal}$

9. $\dfrac{\text{wins}}{\text{losses}} = \dfrac{3}{5} = \dfrac{15}{x}$
   $3x = 75$
   $x = \mathbf{25\ games}$

10. $\dfrac{\text{divorces}}{\text{pop.}} = \dfrac{5}{1000} = \dfrac{x}{15{,}000}$
    $1000x = 75{,}000$
    $x = \mathbf{75\ divorces}$

Lesson 18 Exercise, pg. 121

1. right   7
   wrong   +2
   total   9
   $\dfrac{\text{right}}{\text{total}} = \dfrac{7}{9} = \dfrac{x}{36}$
   $9x = 252$
   $x = \mathbf{28\ right}$

2. total   5
   men   −3
   women   2
   $\dfrac{\text{women}}{\text{total}} = \dfrac{2}{5} = \dfrac{x}{60}$
   $5x = 120$
   $x = \mathbf{24\ women}$

3. dom.   3
   imp.   +2
   total   5
   $\dfrac{\text{dom.}}{\text{total}} = \dfrac{3}{5} = \dfrac{42}{x}$
   $3x = 210$
   $x = \mathbf{70\ cars}$

4. total   500
   good   −497
   def.   3
   $\dfrac{\text{def.}}{\text{total}} = \dfrac{3}{500} = \dfrac{x}{15{,}000}$
   $500x = 45{,}000$
   $x = \mathbf{90\ defective}$

5. total   7
   car   −2
   other   5
   $\dfrac{\text{car}}{\text{other}} = \dfrac{2}{5} = \dfrac{96}{x}$
   $2x = 480$
   $x = \mathbf{\$240}$

6. total   8
   yes   −5
   no   3
   $\dfrac{\text{no}}{\text{yes}} = \dfrac{3}{5} = \dfrac{120}{x}$
   $3x = 600$
   $x = \mathbf{200\ people}$

7. men   9
   women   +4
   total   13
   $\dfrac{\text{women}}{\text{total}} = \dfrac{4}{13} = \dfrac{360}{x}$
   $4x = 4680$
   $x = \mathbf{1170\ people}$

8. total   10
   coffee   −7
   tea   3
   $\dfrac{\text{tea}}{\text{total}} = \dfrac{3}{10} = \dfrac{27}{x}$
   $3x = 270$
   $x = \mathbf{90\ people}$

Lesson 19 Exercise, pg. 122

1. **24 gallons is unnecessary.**

    $d = rt$

    $d = 48 \times 6\frac{1}{2}$

    $d = \dfrac{\overset{24}{\cancel{48}}}{1} \times \dfrac{13}{\underset{1}{\cancel{2}}}$

    $d = \mathbf{312\ miles}$

2. **48 with vaccinations is unnecessary.**

    boys 32
    girls + 28
    total 60

    $\dfrac{girls}{total} = \dfrac{28}{60} = \dfrac{7}{15}$

3. **10 inches wide is unnecessary.**

    $3\frac{1}{2} \times 12 =$

    $\dfrac{7}{\underset{1}{\cancel{2}}} \times \dfrac{\overset{6}{\cancel{12}}}{1} = \mathbf{42\ lb}$

4. **$3.80 and $1.89 a pound are unnecessary.**

    $4\frac{1}{4}$
    $+ 5\frac{3}{4}$
    $\overline{9\frac{4}{4}} = \mathbf{10\ lb}$

5. **8 that he guessed is unnecessary.**

    right 54
    wrong + 6
    total 60

    $\dfrac{right}{total} = \dfrac{54}{60} = \dfrac{9}{10}$

6. **$1.12 is unnecessary.**

    $16\overline{)248}\ \ 15\frac{8}{16} = \mathbf{15\frac{1}{2}\ mi}$

    $\dfrac{16}{88}$
    $\dfrac{80}{8}$

7. **7½ feet high is unnecessary.**

    $A = lw$
    $A = 60 \times 20$
    $A = \mathbf{1200\ sq\ ft}$

8. **3 gallons of thinner is unnecessary.**

    $\dfrac{gray}{green} = \dfrac{2}{5} = \dfrac{x}{10}$

    $5x = 20$

    $x = \mathbf{4\ gallons\ of\ gray\ paint}$

Lesson 20 Exercise, pg. 124

1. $\dfrac{3}{\underset{1}{\cancel{8}}} \times \dfrac{\overset{1200}{\cancel{9600}}}{1} = \mathbf{3600\ voters}$

2. $209 \div 9\frac{1}{2} =$

    $\dfrac{209}{1} \div \dfrac{19}{2} =$

    $\dfrac{\overset{11}{\cancel{209}}}{1} \times \dfrac{2}{\underset{1}{\cancel{19}}} = \mathbf{22\ mi}$

3. $\dfrac{1}{\underset{1}{\cancel{4}}} \times \dfrac{\overset{18}{\cancel{72}}}{1} = \$18$

    $\begin{array}{r}\$72 \\ -\ 18 \\ \hline \$54\end{array}$

4. $24 \div \dfrac{3}{4} =$

    $\dfrac{24}{1} \div \dfrac{3}{4} =$

    $\dfrac{\overset{8}{\cancel{24}}}{1} \times \dfrac{4}{\underset{1}{\cancel{3}}} = \mathbf{32\ lots}$

5. $10\frac{1}{2} \div 3 =$

    $\dfrac{21}{2} \div \dfrac{3}{1} =$

    $\dfrac{\overset{7}{\cancel{21}}}{2} \times \dfrac{1}{\underset{1}{\cancel{3}}} = \dfrac{7}{2} = \mathbf{3\frac{1}{2}\ bu}$

6. $\dfrac{.30}{.45} = \dfrac{2}{3}$

7. $\dfrac{.48}{.64} = \dfrac{3}{4}$

8. $\dfrac{.25}{1.50} = \dfrac{1}{6}$

9. $\dfrac{.45}{.90} = \dfrac{1}{2}$

10. **(4) 40**

    $10 \div \dfrac{1}{4} =$

    $\dfrac{10}{1} \times \dfrac{4}{1} = \mathbf{40}$

11. **(1) $12(5 \times 2\frac{1}{2})$**

    Length in feet $= (5 \times 2\frac{1}{2})$.

    Multiply by 12 for length in inches.

Answers and Solutions   319

12. (3) $\dfrac{500}{90}$

$\dfrac{\text{words}}{\text{minutes}} = \dfrac{90}{1} = \dfrac{500}{m}$

$90m = 500$

$m = \dfrac{500}{90}$

13. (4) $\dfrac{10 \times 275}{75}$

report $= 10 \times 275$ words

$\dfrac{\text{words}}{\text{minutes}} = \dfrac{75}{1} = \dfrac{10 \times 275}{m}$

$75m = 10 \times 275$

$m = \dfrac{10 \times 275}{75}$

14. (3) $10 \times 6 \times 2\tfrac{1}{2}$

$V = lwh$

$V = 10 \times 6 \times 2\tfrac{1}{2}$

15. (4) C, E, D, B, A

A — $1\tfrac{1}{2} = 1\tfrac{8}{16}$

B — $1\tfrac{5}{8} = 1\tfrac{10}{16}$

C — $2\tfrac{3}{8} = 2\tfrac{6}{16}$

D — $2\tfrac{3}{16} = 2\tfrac{3}{16}$

E — $2\tfrac{1}{4} = 2\tfrac{4}{16}$

Level 3 Review, pg. 126

1. $\dfrac{\text{won}}{\text{lost}} = \dfrac{8}{5} = \dfrac{x}{15}$

   $5x = 120$

   $x = 24$

2. $\dfrac{\text{cash}}{\text{total}} = \dfrac{4}{7} = \dfrac{x}{280}$

   $7x = 1120$

   $x = 160$

3. $\dfrac{\text{defective}}{\text{total}} = \dfrac{3}{100} = \dfrac{x}{20{,}000}$

   $100x = 60{,}000$

   $x = 600$

4. mgmt   3 $\dfrac{\text{mgmt}}{\text{total}} = \dfrac{3}{17} = \dfrac{x}{340}$

   labor + 14

   total   17 $17x = 1020$

   $x = 60$

5. girls   5 $\dfrac{\text{boys}}{\text{total}} = \dfrac{4}{9} = \dfrac{84}{x}$

   boys   + 4

   total   9 $4x = 756$

   $x = 189$

6. total   9 $\dfrac{\text{rail}}{\text{truck}} = \dfrac{2}{7} = \dfrac{x}{140}$

   rail   − 2

   truck   7 $7x = 280$

   $x = 40$

7. $c = nr$

   $c = 10 \times \$12.50$

   $c = \$125.00$

8. $d = rt$

   $d = 3\tfrac{1}{2} \times 36$

   $d = \dfrac{7}{2} \times \dfrac{36}{1}$

   $d = 126$ miles

9. $\dfrac{\text{food}}{\text{total}} = \dfrac{\$360}{\$960} = \dfrac{3}{8}$

10. $\dfrac{1}{5} \times \dfrac{\$2400}{1} = \$480$

11. (2) $\dfrac{4 \times 10}{3}$

12. (3) $25 \times 60 \times 8$

13. (3) $\dfrac{310 + 165}{450}$

14. (5) $\tfrac{1}{2} \times 6 \times 4$

    $A = \tfrac{1}{2}bh$

    $A = \tfrac{1}{2} \times 6 \times 4$

15. (4) $20 \times 8 \times 5$

Chapter 4 PERCENTS

Level 1 Preview, pg. 129

1. $2\% = .02 = 0.02$

2. $0.035 = 0.035 = 3.5\%$

3. $84\% = \dfrac{84}{100} = \dfrac{21}{25}$

4. $\dfrac{7}{12} \times \dfrac{100}{1} = \dfrac{175}{3} = 58\tfrac{1}{3}\%$

5. $3.5\% = 0.035$

   0.035

   $\underline{\times700}$

   $24.500 = 24.5$

6. $37\frac{1}{2}\% = \frac{3}{8}$

$\frac{3}{\cancel{8}_1} \times \frac{\cancel{144}^{18}}{1} = 54$

7. $\frac{32}{80} = \frac{2}{5}$ $\frac{2}{\cancel{5}_1} \times \frac{\cancel{100}^{20}}{1} = 40\%$

8. $\frac{120}{48} = \frac{5}{2}$ $\frac{5}{\cancel{2}_1} \times \frac{\cancel{100}^{50}}{1} = 250\%$

9. $75\% = \frac{3}{4}$

$72 \div \frac{3}{4} =$

$\frac{\cancel{72}^{24}}{1} \times \frac{4}{\cancel{3}_1} = 96$

10. $12\frac{1}{2}\% = \frac{1}{8}$

$20 \div \frac{1}{8} =$

$\frac{20}{1} \times \frac{8}{1} = 160$

Lesson 1 Exercise, pg. 132

1. a. 75% = .75 = **0.75**
   b. 4% = .04 = **0.04**
   c. 62.5% = 62.5 = **0.625**
   d. 7% = .07 = **0.07**
2. a. 60% = .60 = **0.6**
   b. 150% = 1.50 = **1.5**
   c. 1% = .01 = **0.01**
   d. 300% = 3.00 = **3**
3. a. $8\frac{3}{4}\%$ = $.08\frac{3}{4}$ = **$0.08\frac{3}{4}$**
   b. 12.6% = 12.6 = **0.126**
   c. 0.6% = 00.6 = **0.006**
   d. 15% = .15 = **0.15**
4. a. 0.1% = 00.1 = **0.001**
   b. 0.25% = 00.25 = **0.0025**
   c. 2.5% = 02.5 = **0.025**
   d. 200% = 2.00 = **2**
5. a. 0.46 = .46 = **46%**
   b. 0.08 = .08 = **8%**
   c. 0.045 = .045 = **4.5%**
   d. $0.08\frac{1}{3}$ = $.08\frac{1}{3}$ = **$8\frac{1}{3}\%$**

6. a. 0.9 = 0.90 = **90%**
   b. 0.25 = 0.25 = **25%**
   c. 0.005 = 0.005 = **0.5%**
   d. 0.05 = 0.05 = **5%**
7. a. 0.0825 = 0.0825 = **8.25%**
   b. 0.4 = 0.40 = **40%**
   c. 0.675 = 0.675 = **67.5%**
   d. 1.2 = 1.20 = **120%**
8. a. 4.75 = 4.75 = **475%**
   b. 0.625 = 0.625 = **62.5%**
   c. 8 = 8.00 = **800%**
   d. $0.66\frac{2}{3}$ = $0.66\frac{2}{3}$ = **$66\frac{2}{3}\%$**

Lesson 2 Exercise, pg. 134

1. a. $15\% = \frac{15}{100} = \frac{3}{20}$
   b. $85\% = \frac{85}{100} = \frac{17}{20}$
   c. $96\% = \frac{96}{100} = \frac{24}{25}$
   d. $60\% = \frac{60}{100} = \frac{3}{5}$
2. a. $275\% = \frac{275}{100} = \frac{11}{4} = 2\frac{3}{4}$
   b. $8\% = \frac{8}{100} = \frac{2}{25}$
   c. $450\% = \frac{450}{100} = \frac{9}{2} = 4\frac{1}{2}$
   d. $42\% = \frac{42}{100} = \frac{21}{50}$
3. a. 1.5% = 01.5 = 0.015
   $\frac{15}{1000} = \frac{3}{200}$
   b. 4.8% = 04.8 = 0.048
   $\frac{48}{1000} = \frac{6}{125}$
   c. 12.5% = 12.5 = 0.125
   $\frac{125}{1000} = \frac{1}{8}$
   d. 6.25% = 06.25 = 0.0625
   $\frac{625}{10,000} = \frac{1}{16}$

4. **a.** $6\tfrac{1}{4}\% = \dfrac{6\tfrac{1}{4}}{100}$

   $6\tfrac{1}{4} \div 100 =$

   $\dfrac{25}{4} \div \dfrac{100}{1} =$

   $\dfrac{\cancel{25}^{1}}{4} \times \dfrac{1}{\cancel{100}_{4}} = \dfrac{1}{16}$

   b. $18\tfrac{3}{4}\% = \dfrac{18\tfrac{3}{4}}{100}$

   $18\tfrac{3}{4} \div 100 =$

   $\dfrac{75}{4} \div \dfrac{100}{1} =$

   $\dfrac{\cancel{75}^{3}}{4} \times \dfrac{1}{\cancel{100}_{4}} = \dfrac{3}{16}$

   c. $13\tfrac{1}{3}\% = \dfrac{13\tfrac{1}{3}}{100}$

   $13\tfrac{1}{3} \div 100 =$

   $\dfrac{40}{3} \div \dfrac{100}{1} =$

   $\dfrac{\cancel{40}^{2}}{3} \times \dfrac{1}{\cancel{100}_{5}} = \dfrac{2}{15}$

   d. $56\tfrac{1}{4}\% = \dfrac{56\tfrac{1}{4}}{100}$

   $56\tfrac{1}{4} \div 100 =$

   $\dfrac{225}{4} \div \dfrac{100}{1} =$

   $\dfrac{\cancel{225}^{9}}{4} \times \dfrac{1}{\cancel{100}_{4}} = \dfrac{9}{16}$

5. **a.** $\dfrac{3}{\cancel{5}_{1}} \times \dfrac{\cancel{100}^{20}}{1} = \mathbf{60\%}$

   b. $\dfrac{12}{\cancel{25}_{1}} \times \dfrac{\cancel{100}^{4}}{1} = \mathbf{48\%}$

   c. $\dfrac{1}{\cancel{2}_{1}} \times \dfrac{\cancel{100}^{50}}{1} = \mathbf{50\%}$

   d. $\dfrac{2}{3} \times \dfrac{100}{1} = \dfrac{200}{3} = \mathbf{66\tfrac{2}{3}\%}$

6. **a.** $\dfrac{23}{\cancel{100}_{1}} \times \dfrac{\cancel{100}^{1}}{1} = \mathbf{23\%}$

   b. $\dfrac{3}{\cancel{10}_{1}} \times \dfrac{\cancel{100}^{10}}{1} = \mathbf{30\%}$

   c. $\dfrac{5}{\cancel{8}_{2}} \times \dfrac{\cancel{100}^{25}}{1} = \dfrac{125}{2} = \mathbf{62\tfrac{1}{2}\%}$

   d. $\dfrac{3}{\cancel{16}_{4}} \times \dfrac{\cancel{100}^{25}}{1} = \dfrac{75}{4} = \mathbf{18\tfrac{3}{4}\%}$

7. **a.** $\dfrac{2}{7} \times \dfrac{100}{1} = \dfrac{200}{7} = \mathbf{28\tfrac{4}{7}\%}$

   b. $\dfrac{5}{\cancel{6}_{3}} \times \dfrac{\cancel{100}^{50}}{1} = \dfrac{250}{3} = \mathbf{83\tfrac{1}{3}\%}$

   c. $\dfrac{19}{\cancel{20}_{1}} \times \dfrac{\cancel{100}^{5}}{1} = \mathbf{95\%}$

   d. $\dfrac{4}{\cancel{5}_{1}} \times \dfrac{\cancel{100}^{20}}{1} = \mathbf{80\%}$

Lesson 3 Exercise, pg. 136

1. **a.** $6\% = 0.06$

   $\begin{array}{r} 150 \\ \times\ 0.06 \\ \hline 9.00 = \mathbf{9} \end{array}$

   b. $40\% = 0.4$

   $\begin{array}{r} 160 \\ \times\ 0.4 \\ \hline 64.0 = \mathbf{64} \end{array}$

   c. $125\% = 1.25$

   $\begin{array}{r} 1.25 \\ \times\ 36 \\ \hline 45.00 = \mathbf{45} \end{array}$

   d. $1.9\% = 0.019$

   $\begin{array}{r} 0.019 \\ \times\ 200 \\ \hline 3.800 = \mathbf{3.8} \end{array}$

2. **a.** $5.4\% = 0.054$

   $\begin{array}{r} 0.054 \\ \times\ 80 \\ \hline 4.320 = \mathbf{4.32} \end{array}$

   b. $0.8\% = 0.008$

   $\begin{array}{r} 0.008 \\ \times\ 50 \\ \hline 0.400 = \mathbf{0.4} \end{array}$

   c. $6.25\% = 0.0625$

   $\begin{array}{r} 0.0625 \\ \times\ 300 \\ \hline 18.7500 = \mathbf{18.75} \end{array}$

   d. $10.4\% = 0.104$

   $\begin{array}{r} 0.104 \\ \times\ 500 \\ \hline 52.000 = \mathbf{52} \end{array}$

3. **a.** $66\tfrac{2}{3}\% = \dfrac{2}{3}$

   $\dfrac{2}{\cancel{3}_{1}} \times \dfrac{\cancel{240}^{80}}{1} = \mathbf{160}$

   b. $12\tfrac{1}{2}\% = \dfrac{1}{8}$

   $\dfrac{1}{\cancel{8}_{1}} \times \dfrac{\cancel{400}^{50}}{1} = \mathbf{50}$

   c. $62\tfrac{1}{2}\% = \dfrac{5}{8}$

   $\dfrac{5}{\cancel{8}_{1}} \times \dfrac{\cancel{48}^{6}}{1} = \mathbf{30}$

   d. $83\tfrac{1}{3}\% = \dfrac{5}{6}$

   $\dfrac{5}{\cancel{6}_{1}} \times \dfrac{\cancel{120}^{20}}{1} = \mathbf{100}$

4. **a.** $33\frac{1}{3}\% = \frac{1}{3}$ **b.** $16\frac{2}{3}\% = \frac{1}{6}$

$\frac{1}{\cancel{3}} \times \frac{\cancel{18}^{6}}{1} = 6 \qquad \frac{1}{\cancel{6}} \times \frac{\cancel{96}^{16}}{1} = 16$

c. $37\frac{1}{2}\% = \frac{3}{8}$ **d.** $87\frac{1}{2}\% = \frac{7}{8}$

$\frac{3}{\cancel{8}} \times \frac{\cancel{24}^{3}}{1} = 9 \qquad \frac{7}{\cancel{8}} \times \frac{\cancel{1200}^{150}}{1} = 1050$

5. **a.** $75\% = \frac{3}{4}$ **b.** $8.5\% = 0.085$

$\frac{3}{\cancel{4}} \times \frac{\cancel{84}^{21}}{1} = 63 \qquad \begin{array}{r} 0.085 \\ \times\ 400 \\ \hline 34.000 \end{array} = 34$

c. $90\% = 0.9$ **d.** $50\% = \frac{1}{2}$

$\begin{array}{r} 130 \\ \times\ 0.9 \\ \hline 117.0 \end{array} = 117 \qquad \frac{1}{\cancel{2}} \times \frac{\cancel{28}^{14}}{1} = 14$

6. **a.** $250\% = 2.5$ **b.** $60\% = \frac{3}{5}$

$\begin{array}{r} 36 \\ \times\ 2.5 \\ \hline 18\ 0 \\ 72\ \ \\ \hline 90.0 \end{array} = 90 \qquad \frac{3}{\cancel{5}} \times \frac{\cancel{200}^{40}}{1} = 120$

c. $25\% = \frac{1}{4}$ **d.** $35\% = 0.35$

$\frac{1}{\cancel{4}} \times \frac{\cancel{116}^{29}}{1} = 29 \qquad \begin{array}{r} 260 \\ \times\ 0.35 \\ \hline 13\ 00 \\ 78\ 0\ \ \\ \hline 91.00 \end{array} = 91$

Lesson 4 Exercise, pg. 138

1. $\frac{28}{70} = \frac{2}{5}$

$\frac{2}{\cancel{5}} \times \frac{\cancel{100}^{20}}{1} = 40\%$

2. $\frac{104}{160} = \frac{13}{20}$

$\frac{13}{\cancel{20}} \times \frac{\cancel{100}^{5}}{1} = 65\%$

3. $\frac{16}{20} = \frac{4}{5}$

$\frac{4}{\cancel{5}} \times \frac{\cancel{100}^{20}}{1} = 80\%$

4. $\frac{45}{135} = \frac{1}{3}$

$\frac{1}{3} \times \frac{100}{1} = \frac{100}{3} = 33\frac{1}{3}\%$

5. $\frac{24}{32} = \frac{3}{4}$

$\frac{3}{\cancel{4}} \times \frac{\cancel{100}^{25}}{1} = 75\%$

6. $\frac{30}{48} = \frac{5}{8}$

$\frac{5}{\cancel{8}} \times \frac{\cancel{100}^{25}}{1} = \frac{125}{2} = 62\frac{1}{2}\%$

7. $\frac{225}{375} = \frac{3}{5}$

$\frac{3}{\cancel{5}} \times \frac{\cancel{100}^{20}}{1} = 60\%$

8. $\frac{27}{72} = \frac{3}{8}$

$\frac{3}{\cancel{8}} \times \frac{\cancel{100}^{25}}{1} = \frac{75}{2} = 37\frac{1}{2}\%$

9. $\frac{36}{24} = \frac{3}{2}$

$\frac{3}{\cancel{2}} \times \frac{\cancel{100}^{50}}{1} = 150\%$

10. $\frac{15}{120} = \frac{1}{8}$

$\frac{1}{\cancel{8}} \times \frac{\cancel{100}^{25}}{1} = \frac{25}{2} = 12\frac{1}{2}\%$

11. $\frac{36}{40} = \frac{9}{10}$

$\frac{9}{\cancel{10}} \times \frac{\cancel{100}^{10}}{1} = 90\%$

12. $\frac{150}{75} = \frac{2}{1}$

$\frac{2}{1} \times \frac{100}{1} = 200\%$

Lesson 5 Exercise, pg. 139

1. 60% = 0.6

   $0.6 \overline{)24.0}$ = **40**

   $\underline{24}$

   0 0

2. 25% = $\frac{1}{4}$

   $18 \div \frac{1}{4} =$

   $\frac{18}{1} \times \frac{4}{1} =$ **72**

3. $12\frac{1}{2}\% = \frac{1}{8}$

   $15 \div \frac{1}{8} =$

   $\frac{15}{1} \times \frac{8}{1} =$ **120**

4. 4% = 0.04

   $0.04 \overline{)9.60}$ = **240**

5. $66\frac{2}{3}\% = \frac{2}{3}$

   $52 \div \frac{2}{3} =$

   $\frac{\overset{26}{\cancel{52}}}{1} \times \frac{3}{\cancel{2}} =$ **78**

6. $87\frac{1}{2}\% = \frac{7}{8}$

   $112 \div \frac{7}{8} =$

   $\frac{\overset{16}{\cancel{112}}}{1} \times \frac{8}{\cancel{7}} =$ **128**

7. 2.5% = 0.025

   $0.025 \overline{)8.000}$ = **320**

   $\underline{7\ 5}$

   50

   $\underline{50}$

   00

8. 35% = 0.35

   $0.35 \overline{)140.00}$ = **400**

   $\underline{140}$

   0 00

9. 90% = 0.9

   $0.9 \overline{)23.4}$ = **26**

10. $62\frac{1}{2}\% = \frac{5}{8}$

    $45 \div \frac{5}{8} =$

    $\frac{\overset{9}{\cancel{45}}}{1} \times \frac{8}{\cancel{5}} =$ **72**

11. 8.5% = 0.085

    $0.085 \overline{)17.000}$ = **200**

    $\underline{17\ 0}$

    000

12. $16\frac{2}{3}\% = \frac{1}{6}$

    $36 \div \frac{1}{6} =$

    $\frac{36}{1} \times \frac{6}{1} =$ **216**

4. $\frac{9}{\cancel{16}_4} \times \frac{\overset{25}{\cancel{100}}}{1} = \frac{225}{4} = 56\frac{1}{4}\%$

5. **Type 2**

   $\frac{24}{72} = \frac{1}{3}$

   $\frac{1}{3} \times \frac{100}{1} = \frac{100}{3} = 33\frac{1}{3}\%$

6. **Type 1**

   4.5% = 0.045

   2400

   $\underline{\times\ 0.045}$

   108.000 = **108**

7. **Type 1**

   $83\frac{1}{3}\% = \frac{5}{6}$

   $\frac{5}{\cancel{6}_1} \times \frac{\overset{55}{\cancel{330}}}{1} =$ **275**

8. **Type 3**

   65% = 0.65

   $0.65 \overline{)52.00}$ = **80**

   $\underline{52\ 0}$

   00

9. **Type 2**

   $\frac{105}{60} = \frac{7}{4}$

   $\frac{7}{\cancel{4}_1} \times \frac{\overset{25}{\cancel{100}}}{1} =$ **175%**

10. **Type 3**

    $16\frac{2}{3}\% = \frac{1}{6}$

    $25 \div \frac{1}{6} =$

    $\frac{25}{1} \times \frac{6}{1} =$ **150**

11. **Type 1**

    8% = 0.08

    250

    $\underline{\times\ 0.08}$

    20.00 = **20**

12. **Type 2**

    $\frac{28}{42} = \frac{2}{3}$

    $\frac{2}{3} = 66\frac{2}{3}\%$

13. **Type 2**

    $\frac{200}{800} = \frac{1}{4}$

    $\frac{1}{4} =$ **25%**

Level 1 Review, pg. 139

1. 9.6% = 09.6 = **0.096**

2. 0.0145 = .0145 = **1.45%**

3. $8\frac{1}{3}\% = \frac{8\frac{1}{3}}{100}$

   $8\frac{1}{3} \div 100 =$

   $\frac{25}{3} \div \frac{100}{1} =$

   $\frac{\overset{1}{\cancel{25}}}{3} \times \frac{1}{\cancel{100}_4} = \frac{1}{12}$

14. **Type 3**
 $20\% = \frac{1}{5}$
 $14 \div \frac{1}{5} =$
 $\frac{14}{1} \times \frac{5}{1} = \textbf{70}$

15. **Type 1**
 $33\frac{1}{3}\% = \frac{1}{3}$
 $\frac{1}{\cancel{3}} \times \frac{\cancel{45}^{15}}{1} = \textbf{15}$

16. **Type 3**
 $12\frac{1}{2}\% = \frac{1}{8}$
 $60 \div \frac{1}{8} =$
 $\frac{60}{1} \times \frac{8}{1} = \textbf{480}$

17. **Type 1**
 $1.5\% = 0.015$
 $\begin{array}{r} 0.015 \\ \times\ 700 \\ \hline 10.500 \end{array} = \textbf{10.5}$

18. **Type 2**
 $\frac{180}{90} = \frac{2}{1}$
 $\frac{2}{1} \times \frac{100}{1} = \textbf{200\%}$

19. **Type 1**
 $60\% = 0.6$
 $\begin{array}{r} 110 \\ \times\ 0.6 \\ \hline 66.0 \end{array} = \textbf{66}$

20. **Type 2**
 $\frac{12}{400} = \frac{3}{100}$
 $\frac{3}{\cancel{100}} \times \frac{\cancel{100}^1}{1} = \textbf{3\%}$

Lesson 6 Exercise, pg. 141

1. $80\% = 0.8$
 $\begin{array}{r} 60 \\ \times\ 0.8 \\ \hline 48.0 \end{array} = \textbf{48 right}$

2. $21\% = 0.21$
 $\begin{array}{r} \$320 \\ \times\ 0.21 \\ \hline \$67.20 \end{array}$

3. $6\% = 0.06$
 $\begin{array}{r} \$78,000 \\ \times\ 0.06 \\ \hline \$4680.00 \end{array}$

4. $65\% = 0.65$
 $\begin{array}{r} 320 \\ \times\ 0.65 \\ \hline 208.00 \end{array} = \textbf{208 women}$

5. $5\% = 0.05$
 $\begin{array}{r} \$36,400 \\ \times\ 0.05 \\ \hline \$1820.00 \end{array}$

6. $1.5\% = 0.015$
 $\begin{array}{r} \$430 \\ \times\ 0.015 \\ \hline \$6.450 \end{array} = \textbf{\$6.45}$

7. $4.5\% = 0.045$
 $\begin{array}{r} \$119 \\ \times\ 0.045 \\ \hline \$5.355 \end{array}$ to the nearest cent = **\$5.36**

8. $16\frac{2}{3}\% = \frac{1}{6}$
 $\frac{1}{\cancel{6}} \times \frac{\cancel{420}^{70}}{1} = \textbf{70 voters}$

Lesson 7 Exercise, pg. 143

1. $\frac{\text{right}}{\text{total}} = \frac{34}{40} = \frac{17}{20}$
 $\frac{17}{\cancel{20}} \times \frac{\cancel{100}^5}{1} = \textbf{85\%}$

2. $\frac{\text{late}}{\text{total}} = \frac{6}{48} = \frac{1}{8}$
 $\frac{1}{\cancel{8}} \times \frac{\cancel{100}^{25}}{1} = \frac{25}{2} = \textbf{12}\frac{1}{2}\textbf{\%}$

3. $\frac{\text{tax}}{\text{total}} = \frac{1.80}{22.50} = \frac{6}{75} = \frac{2}{25}$
 $\frac{2}{\cancel{25}} \times \frac{\cancel{100}^4}{1} = \textbf{8\%}$

4. $\frac{\text{cash}}{\text{total}} = \frac{280}{320} = \frac{7}{8}$
 $\frac{7}{\cancel{8}} \times \frac{\cancel{100}^{25}}{1} = \frac{175}{2} = \textbf{87}\frac{1}{2}\textbf{\%}$

5. $\frac{1992\ \text{value}}{1990\ \text{value}} = \frac{2560}{6400} = \frac{64}{160} = \frac{2}{5}$
 $\frac{2}{\cancel{5}} \times \frac{\cancel{100}^{20}}{1} = \textbf{40\%}$

6. $\frac{\text{rent}}{\text{total}} = \frac{252}{1200} = \frac{63}{300} = \frac{21}{100}$
 $\frac{21}{\cancel{100}} \times \frac{\cancel{100}^1}{1} = \textbf{21\%}$

7. $\frac{\text{sale price}}{\text{purchase price}} = \frac{84,000}{48,000} = \frac{7}{4}$
 $\frac{7}{\cancel{4}} \times \frac{\cancel{100}^{25}}{1} = \textbf{175\%}$

8. $\frac{\text{weight at 30}}{\text{weight at 20}} = \frac{200}{160} = \frac{5}{4}$
 $\frac{5}{\cancel{4}} \times \frac{\cancel{100}^{25}}{1} = \textbf{125\%}$

Lesson 8 Exercise, pg. 144

1. 80% = 0.8
$$0.8 \overline{)16.0} = 20 \text{ problems}$$

2. 9% = 0.09
$$0.09 \overline{)\$630.00} = \$70.00$$

3. 75% = 0.75
$$0.75 \overline{)420.00} = 560 \text{ miles}$$
   $$\begin{array}{r} 375 \\ \hline 45\,0 \\ 45\,0 \\ \hline 00 \end{array}$$

4. 15% = 0.15
$$0.15 \overline{)24.00} = 160 \text{ employees}$$
   $$\begin{array}{r} 15 \\ \hline 9\,0 \\ 9\,0 \\ \hline 00 \end{array}$$

5. 60% = 0.6
$$0.6 \overline{)\$1440.0} = \$2400$$

6. 22% = 0.22
$$0.22 \overline{)\$528\,0.00} = \$24,000$$
   $$\begin{array}{r} 44 \\ \hline 88 \\ 88 \\ \hline 0\,0\,00 \end{array}$$

7. $66\frac{2}{3}\% = \frac{2}{3}$
$$180 \div \frac{2}{3} =$$
$$\frac{\overset{90}{\cancel{180}}}{1} \times \frac{3}{\cancel{2}} = \textbf{270 lb}$$

8. $37\frac{1}{2}\% = \frac{3}{8}$
$$450 \div \frac{3}{8} =$$
$$\frac{\overset{150}{\cancel{450}}}{1} \times \frac{8}{\cancel{3}} = \textbf{1200 people}$$

Lesson 9 Exercise, pg. 146

These problems are solved by substituting values into the equation for finding interest, $i = prt$. Notice how these solutions are set up.

1. $i = \dfrac{\overset{6}{\cancel{\$600}}}{1} \times \dfrac{4}{\underset{1}{\cancel{100}}} \times 1 = \textbf{\$24}$

2. $i = \dfrac{\overset{15}{\cancel{\$1500}}}{1} \times \dfrac{10}{\underset{1}{\cancel{100}}} \times 1 = \textbf{\$150}$

3. $i = \dfrac{\overset{4}{\cancel{\$400}}}{1} \times \dfrac{8.5}{\underset{1}{\cancel{100}}} \times 1 = \textbf{\$34}$
   or $i = \$400 \times 0.085 \times 1 = \textbf{\$34}$

4. $i = \dfrac{\overset{16}{\cancel{\$1600}}}{1} \times \dfrac{10.5}{\underset{1}{\cancel{100}}} \times 1 = \textbf{\$168}$
   or $i = \$1600 \times 0.105 \times 1 = \textbf{\$168}$

5. $i = \dfrac{\overset{40}{\cancel{\$4000}}}{1} \times \dfrac{12}{\underset{1}{\cancel{100}}} \times 1 = \textbf{\$480}$

6. $i = \dfrac{\overset{12}{\cancel{\$1200}}}{1} \times \dfrac{5\frac{1}{4}}{\underset{1}{\cancel{100}}} \times 1 = \textbf{\$63}$
   or $i = \$1200 \times 0.0525 \times 1 = \textbf{\$63}$

7. $i = \dfrac{\overset{\overset{5}{\cancel{20}}}{\cancel{2000}}}{\underset{1}{\cancel{100}}} \times \dfrac{35}{\underset{1}{\cancel{4}}} \times 1 = \textbf{\$175}$
   or $i = \$2000 \times 0.0875 \times 1 = \textbf{\$175}$

8. 6 mo = $\dfrac{6}{12} = \dfrac{1}{2}$ yr
   $i = \dfrac{\overset{5}{\cancel{\$500}}}{1} \times \dfrac{\overset{4}{\cancel{8}}}{\underset{1}{\cancel{100}}} \times \dfrac{1}{\cancel{2}} = \textbf{\$20}$

9. 8 mo = $\dfrac{8}{12} = \dfrac{2}{3}$ yr
   $i = \dfrac{\overset{10}{\cancel{1000}}}{1} \times \dfrac{\overset{3}{\cancel{9}}}{\underset{1}{\cancel{100}}} \times \dfrac{2}{\cancel{3}} = \textbf{\$60}$

10. 1 yr 3 mo = $1\dfrac{3}{12} = 1\dfrac{1}{4}$ yr = 1.25 yr
    $i = \dfrac{\overset{1}{\cancel{400}}}{1} \times \dfrac{5\frac{1}{2}}{\underset{1}{\cancel{100}}} \times \dfrac{5}{4} = \textbf{\$27.50}$
    or $i = \$400 \times 0.055 \times 1.25 = \textbf{\$27.50}$

11. 9 mo = $\dfrac{9}{12} = \dfrac{3}{4}$ yr
    $i = \dfrac{\overset{8}{\cancel{800}}}{1} \times \dfrac{4.5}{\underset{1}{\cancel{100}}} \times \dfrac{3}{\cancel{4}} = \textbf{\$27}$

12. $i = \dfrac{\overset{25}{\cancel{2500}}}{1} \times \dfrac{15}{\underset{1}{\cancel{100}}} \times \dfrac{3}{1} = \textbf{\$1125}$

13. 1 yr 6 mo = $1\dfrac{6}{12} = 1\dfrac{1}{2}$ yr
    $i = \dfrac{\overset{\overset{15}{\cancel{30}}}{\cancel{3000}}}{1} \times \dfrac{9.5}{\underset{1}{\cancel{100}}} \times \dfrac{3}{\cancel{2}} = \textbf{\$427.50}$

14. 2 yr 4 mo = $2\frac{4}{12} = 2\frac{1}{3}$ yr

$i = \frac{\cancel{1600}^{16}}{1} \times \frac{\cancel{18}^{6}}{\cancel{100}_{1}} \times \frac{7}{\cancel{3}_{1}} = \$672$

15. 10 mo = $\frac{10}{12} = \frac{5}{6}$ yr

$i = \frac{\cancel{1800}^{18\,\,3}}{1} \times \frac{3\frac{1}{2}}{\cancel{100}_{1}} \times \frac{5}{\cancel{6}_{1}} = \$52.50$

Level 2 Review, pg. 147

1. **Type 1**
   15% = 0.15 $55
   × 0.15
   $8.25

2. **Type 2**
   $\frac{\text{by car}}{\text{total}} = \frac{20}{24} = \frac{5}{6}$

   $\frac{5}{\cancel{6}_{3}} \times \frac{\cancel{100}^{50}}{1} = \frac{250}{3} = 83\frac{1}{3}\%$

3. **Type 3**
   75% = $\frac{3}{4}$

   $\$1350 \div \frac{3}{4} =$

   $\frac{\cancel{\$1350}^{450}}{1} \times \frac{4}{\cancel{3}_{1}} = \$1800$

4. **Type 3**
   1.5% = 0.015

   $0.015\overline{)36.000}$ → 2 400 parts
   30
   6 0
   6 0
   000

5. **Type 2**
   $\frac{\text{down payment}}{\text{total price}} = \frac{700}{3500} = \frac{1}{5}$

   $\frac{1}{\cancel{5}_{1}} \times \frac{\cancel{100}^{20}}{1} = 20\%$

6. **Type 2**
   $\frac{\text{agreed}}{\text{total}} = \frac{65}{500} = \frac{13}{100}$

   $\frac{\cancel{13}}{\cancel{100}_{1}} \times \frac{\cancel{100}^{1}}{1} = 13\%$

7. **Type 1**
   2% = 0.02   1250
                  × 0.02
                  25.00 = **25 drivers**

8. **Type 1**
   6% = 0.06 $6850
                × 0.06
                $411.00

9. **Type 1**
   18% = 0.18 $6.50
                × 0.18
                $1.17 00 = **$1.17**

10. **Type 2**
    $\frac{\text{tax}}{\text{total}} = \frac{0.75}{12.50} = \frac{3}{50}$

    $\frac{3}{\cancel{50}_{1}} \times \frac{\cancel{100}^{2}}{1} = 6\%$

11. **Type 3**
    4% = 0.04 $1 24
                  $0.04\overline{)\$4.96}$

12. **Type 2**
    $\frac{\text{food}}{\text{total}} = \frac{115}{345} = \frac{1}{3}$

    $\frac{1}{3} = 33\frac{1}{3}\%$

13. **Type 3**
    60% = 0.6 $ 7 8,00 0
                  $0.6\overline{)\$46,8\,00.0}$

14. $i = \frac{\cancel{2400}^{24}}{1} \times \frac{14.5}{\cancel{100}_{1}} \times 1 = \$348$

    or $i = \$2400 \times 0.145 \times 1 = \$348$

15. 1 yr 8 mo = $1\frac{8}{12} = 1\frac{2}{3}$ yr

    $i = \frac{360}{1} \times \frac{7}{100} \times 1\frac{2}{3}$

    $i = \frac{\cancel{360}}{1} \times \frac{7}{\cancel{100}} \times \frac{\cancel{5}}{\cancel{3}} = \$42$

16. 6 mo = $\frac{6}{12} = \frac{1}{2}$ yr

    $i = \frac{\cancel{800}^{8\,\,4}}{1} \times \frac{3}{\cancel{100}_{1}} \times \frac{1}{\cancel{2}_{1}} = \$12$

Lesson 10 Exercise, pg. 149

1. 40% = 0.4 $2000 $2000
                × 0.4   + 800
                $800.0 $2800

2. $35\% = 0.35$

$$\begin{array}{r} 3800 \\ \times\ 0.35 \\ \hline 1330.00 \end{array} \quad \begin{array}{r} 3800 \\ -\ 1330 \\ \hline \textbf{2470 people} \end{array}$$

3. $150\% = 1.5$

$$\begin{array}{r} \$4.50 \\ \times\ 1.5 \\ \hline \$6.75\ 0 \end{array} \quad \begin{array}{r} \$4.50 \\ +\ 6.75 \\ \hline \textbf{\$11.25} \end{array}$$

4. $18\% = 0.18$

$$\begin{array}{r} 0.18 \\ \times\ 60{,}000 \\ \hline \$10{,}800.00 \end{array} \quad \begin{array}{r} \$60{,}000 \\ +\ 10{,}800 \\ \hline \textbf{\$70{,}800} \end{array}$$

5. $\begin{array}{r}\$49 \\ +\ 12 \\ \hline \$61\end{array} \quad \begin{array}{r}\$61 \\ \times\ 0.07 \\ \hline \$4.27\end{array} \quad \begin{array}{r}\$61.00 \\ +\ 4.27 \\ \hline \textbf{\$65.27}\end{array}$

6. $40\% = 0.4$ $\begin{array}{r}\$28 \\ \times\ 0.4 \\ \hline \$11.2\end{array} \quad \begin{array}{r}\$28.00 \\ +\ 11.20 \\ \hline \$39.20\end{array}$ to the nearest dollar = **$39.00**

7. $65\% = 0.65$ $\begin{array}{r}0.65 \\ \times\ \$50\ 00 \\ \hline \$3250.00\end{array} \quad \begin{array}{r}\$5000 \\ -\ 3250 \\ \hline \textbf{\$1750}\end{array}$

8. $\begin{array}{r}\$390 \\ \times\ 12 \\ \hline 780 \\ 390 \\ \hline \$4680\end{array} \quad 10\% = 0.1 \quad \begin{array}{r}\$4680 \\ \times\ 0.1 \\ \hline \$468.0\end{array} = \textbf{\$468}$

9. $15\% = 0.15$ $\begin{array}{r}\$4680 \\ \times\ 0.15 \\ \hline 234\ 00 \\ 468\ 0 \\ \hline \$702.00\end{array}$

10. $8\% = 0.08$ $\begin{array}{r}\$390 \\ \times\ 0.08 \\ \hline \$31.20\end{array} \quad \begin{array}{r}\$390.00 \\ +\ 31.20 \\ \hline \$421.20\end{array}$

Lesson 11 Exercise, pg. 151

1. $\begin{array}{r}\$1.32 \\ -\ 1.21 \\ \hline \$0.11\end{array} \quad \dfrac{\text{change}}{\text{original}} = \dfrac{11}{132} = \dfrac{1}{12}$

$$\dfrac{1}{\cancel{12}_3} \times \dfrac{\cancel{100}^{25}}{1} = \dfrac{25}{3} = 8\dfrac{1}{3}\%$$

2. $\begin{array}{r}\$360 \\ -\ 306 \\ \hline \$\ 54\end{array} \quad \dfrac{\text{change}}{\text{original}} = \dfrac{54}{360} = \dfrac{6}{40} = \dfrac{3}{20}$

$$\dfrac{3}{\cancel{20}_1} \times \dfrac{\cancel{100}^{5}}{1} = \textbf{15}\%$$

3. $\begin{array}{r}15{,}000 \\ -\ 12{,}000 \\ \hline 3{,}000\end{array} \quad \dfrac{\text{change}}{\text{original}} = \dfrac{3{,}000}{12{,}000} = \dfrac{1}{4}$

$$\dfrac{1}{4} = \textbf{25}\%$$

4. $\begin{array}{r}\$24 \\ -\ 16 \\ \hline \$\ 8\end{array} \quad \dfrac{\text{change}}{\text{original}} = \dfrac{8}{16} = \dfrac{1}{2}$

$$\dfrac{1}{2} = \textbf{50}\%$$

5. $\begin{array}{r}84 \\ +\ 126 \\ \hline 210\end{array} \quad \dfrac{\text{women}}{\text{total}} = \dfrac{84}{210} = \dfrac{12}{30} = \dfrac{2}{5}$

$$\dfrac{2}{5} = \textbf{40}\%$$

6. $\begin{array}{r}\$270 \\ +\ 630 \\ \hline \$900\end{array} \quad \dfrac{\text{food}}{\text{total}} = \dfrac{270}{900} = \dfrac{3}{10}$

$$\dfrac{3}{10} = \textbf{30}\%$$

7. $\begin{array}{r}\$4800 \\ -\ 1800 \\ \hline \$3000\end{array} \quad \dfrac{\text{change}}{\text{original}} = \dfrac{3000}{4800} = \dfrac{5}{8}$

$$\dfrac{5}{\cancel{8}_2} \times \dfrac{\cancel{100}^{25}}{1} = \dfrac{125}{2} = \textbf{62}\dfrac{1}{2}\%$$

8. $\begin{array}{r}792 \\ -\ 720 \\ \hline 72\end{array} \quad \dfrac{\text{change}}{\text{original}} = \dfrac{72}{720} = \dfrac{1}{10}$

$$\dfrac{1}{10} = \textbf{10}\%$$

9. $\dfrac{900}{6000} = \dfrac{3}{20} \quad \dfrac{3}{\cancel{20}_1} \times \dfrac{\cancel{100}^{5}}{1} = \textbf{15}\%$

10. $\begin{array}{r}\$\ 85 \\ 12\overline{)1020}\end{array} \quad \begin{array}{r}\$115 \\ +\ 85 \\ \hline \$200\end{array} \quad \dfrac{200}{1600} = \dfrac{1}{8}$

$$\dfrac{1}{\cancel{8}_2} \times \dfrac{\cancel{100}^{25}}{1} = \dfrac{25}{2} = \textbf{12}\dfrac{1}{2}\%$$

Lesson 12 Exercise, pg. 153

1. $60\% = 0.6$ $\quad \begin{array}{r}\$\ 45\ 0 \\ 0.6\overline{)\$270.0}\end{array}$

$$\begin{array}{r}\$450 \\ -\ 270 \\ \hline \textbf{\$180}\ \textbf{needed}\end{array}$$

2. $75\% = \dfrac{3}{4} \quad 900 \div \dfrac{3}{4} =$

$$\dfrac{\cancel{900}^{300}}{1} \times \dfrac{4}{\cancel{3}_1} = 1200$$

$$\begin{array}{r}1200 \\ -\ 900 \\ \hline \textbf{300 signatures}\end{array}$$

3. $40\% = 0.4 \quad \begin{array}{r}28\ 0 \\ 0.4\overline{)112.0}\end{array}$

$$\begin{array}{r}280 \\ -\ 112 \\ \hline \textbf{168}\end{array}$$

4. $85\% = 0.85$

$$0.85 \overline{)102.00}$$
$$\underline{85}$$
$$17\,0$$
$$\underline{17\,0}$$
$$00$$

Quotient: 1 20

$120
− 102
$ 18 saved

5. $70\% = 0.7$

$$0.7\overline{)420.0}$$

Quotient: 60 0

600
− 420
180 failed

6. $90\% = 0.9$

$$0.9\overline{)216.0}$$

Quotient: 24 0

240
− 216
24 pounds lost

7. $65\% = 0.65$

$$0.65\overline{)338.00}$$

Quotient: 5 20

520
− 338
182 miles to go

8. $21\% = 0.21$

$$0.21\overline{)\$346\,5.00}$$

Quotient: $ 16,5 00

$16,500
− 3,465
$13,035 net

Lesson 13 Exercise, pg. 155

1. $\dfrac{p}{1200} = \dfrac{15}{100}$

$100p = 18,000$
$p = 180$

1200
− 180
1020 people

2. $1216
+ 64
$1280

$\dfrac{\$64}{\$1280} = \dfrac{r}{100}$

$1280r = \$6400$
$r = 5$
The rate is 5%.

3. $\dfrac{380}{w} = \dfrac{80}{100}$

$80w = 38,000$
$w = 475$

475
− 380
95 **patients**

4. $\dfrac{\$104}{w} = \dfrac{65}{100}$

$65w = \$10,400$
$w = \$160$

$160
− 104
$ 56 **saved**

5. $49
+ 9
$58

$\dfrac{p}{\$58} = \dfrac{6.5}{100}$

$100p = \$377$
$p = \$3.77$

$58.00
+ 3.77
$61.77

6. 13,500
− 12,000
1,500

$\dfrac{1500}{12,000} = \dfrac{r}{100}$

$12,000r = 150,000$
$r = 12.5$
The rate is 12.5%.

Answers 7 to 12 show, first, how to solve using proportion, then how to solve using the percent formula.

7. a. $\dfrac{w}{150} = \dfrac{120}{100}$

$100w = 18,000$
$w = 180$

150
+180
330 **drivers**

b. part = rate × whole
part = 120% × 150

$120\% = 1.2$

150
× 1.2
300
150
180.0

150
+180
330 **drivers**

8. a. $\dfrac{\$437.50}{w} = \dfrac{70}{100}$

$70w = \$43,750$
$w = \$625$

$625.00
− 437.50
$187.50

b. whole = part ÷ rate
whole = $437.50 ÷ 70%

$$0.70\overline{)\$437.50}$$

Quotient: $ 6 25.

420
17 5
14 0
3 50
3 50
0

$625.00
− 437.50
$187.50

9. **a.** $\dfrac{\$1200}{\$8000} = \dfrac{r}{100}$

$\$8000r = \$120{,}000$

$r = 15$

The rate is 15%.

b. rate $= \dfrac{\text{part}}{\text{whole}}$

$\dfrac{\$1200}{\$8000} = \dfrac{3}{20} \quad \dfrac{3}{\cancel{20}} \times \dfrac{\cancel{100}^{5}}{1} = 15\%$

10. **a.** $\$1224 \div 12 = \$102$

$\begin{array}{r}\$102\\ +118\\ \hline 220\end{array}$

$\dfrac{\$220}{\$1600} = \dfrac{r}{100}$

$\$1600r = \$22{,}000$

$r = 13.75$

The rate is 13.75% or $13\dfrac{3}{4}\%$.

b. $\$1224 \div 12 = \$102$

$\begin{array}{r}\$102\\ +118\\ \hline \$220\end{array}$

rate $= \dfrac{\text{part}}{\text{whole}}$

$\dfrac{\$220}{\$1600} = \dfrac{11}{80} \quad \dfrac{11}{\cancel{80}_4} \times \dfrac{\cancel{100}^{5}}{1} = \dfrac{55}{4} = 13\dfrac{3}{4}\%$

Lesson 14 Exercise, pg. 156

1. **B.** 15% of the yearly rent of $6000.
   - A. $35 \times 30 = \$1050$
   - B. $15\% = 0.15 \quad 0.15 \times \$6000 = \$900$
   - C. $\$25 \times 30 = \$750$
     $\$750 + \$250 = \$1000$
   - D. $\$285 \times 4 = \$1140$

2. **D.** $250 a week for half a year.
   - A. $50\% = \dfrac{1}{2} \quad \dfrac{1}{2} \times \$10{,}000 = \$5{,}000$
   - B. $\$800 \times 6 = \$4800$
   - C. $\$1500 \times 3 = \$4500$
   - D. $\dfrac{1}{2}$ yr $= \dfrac{52}{2} = 26$ weeks
     $\$250 \times 26 = \$6500$

3. **B.** 1300 more people per year for three years.
   - A. $15\% = 0.15 \quad 0.15 \times 25{,}000 = 3750$
   - B. $1300 \times 3 = 3900$
   - C. $5\% = 0.05 \quad 0.05 \times 25{,}000 = 1250$
     $1250 + 1000 + 1000 = 3250$
   - D. 3 yr $= 3 \times 12 = 36$ mo.
     $36 \times 80 = 2880$

4. **C.** $150 for each 100 pounds.
   - A. $\begin{array}{r}500\\ -100\\ \hline 400\text{ lb}\end{array}\quad 10\overline{)400}\;\;40$
     $\begin{array}{r}\$12.50\\ \times\;40\\ \hline \$500.00\end{array}\quad \begin{array}{r}\$300\\ +500\\ \hline \$800\end{array}$
   - B. $50\overline{)500}\;\;10$
     $\begin{array}{r}\$65\\ \times 10\\ \hline \$650\end{array}\quad \begin{array}{r}\$650\\ +100\\ \hline \$750\end{array}$
     $5\% = 0.05$
     $\begin{array}{r}\$750\\ \times .05\\ \hline \$37.50\end{array}\quad \begin{array}{r}\$750.00\\ +37.50\\ \hline \$787.50\end{array}$
   - C. $100\overline{)500}\;\;5$
     $\begin{array}{r}\$150\\ \times 5\\ \hline \$750\end{array}$
   - D. $\begin{array}{r}\$1.25\\ \times 500\\ \hline \$625.00\end{array}\quad \begin{array}{r}\$625\\ +150\\ \hline \$775\end{array}$

Level 3 Review, pg. 157

Each problem is solved, first, using the percent formula and then using proportion.

1. **a.** $\begin{array}{r}56\\ -21\\ \hline 35\end{array}\quad \dfrac{\text{right}}{\text{total}} = \dfrac{35}{56} = \dfrac{5}{8}$

   $\dfrac{5}{\cancel{8}_2} \times \dfrac{\cancel{100}^{25}}{1} = \dfrac{125}{2} = 62\dfrac{1}{2}\%$

   b. $\begin{array}{r}56\\ -21\\ \hline 35\end{array}\quad \dfrac{35}{56} = \dfrac{r}{100}$

   $56r = 3500$
   $r = 62.5$

   The rate is 62.5% or $62\dfrac{1}{2}\%$.

2. **a.** $\begin{array}{r}\$85{,}000\\ -25{,}000\\ \hline \$60{,}000\end{array}\quad \dfrac{\text{increase}}{\text{original}} = \dfrac{60{,}000}{25{,}000} = \dfrac{12}{5}$

   $\dfrac{12}{\cancel{5}_1} \times \dfrac{\cancel{100}^{20}}{1} = 240\%.$

   b. $\begin{array}{r}\$85{,}000\\ -25{,}000\\ \hline \$60{,}000\end{array}\quad \dfrac{\$60{,}000}{\$25{,}000} = \dfrac{r}{100}$

   $25{,}000r = \$6{,}000{,}000$
   $r = 240$

   The rate is 240%.

3. **a.** $80\% = 0.8 \quad 0.8\overline{)\$192.0}\;\;\$240$

   $\begin{array}{r}\$240\\ -192\\ \hline \$\;48\text{ saved}\end{array}$

   b. $\dfrac{\$192}{w} = \dfrac{80}{100}$

   $80w = \$19{,}200$
   $w = \$240$

   $\begin{array}{r}\$240\\ -192\\ \hline \$\;48\text{ saved}\end{array}$

4. **a.** $75\% = \frac{3}{4}$ $54 \div \frac{3}{4} =$

$$\frac{\overset{18}{\cancel{54}}}{1} \times \frac{4}{\cancel{3}} = 72$$

$$\begin{array}{r} 72 \\ -\ 54 \\ \hline 18 \text{ miles} \end{array}$$

b. $\frac{54}{w} = \frac{75}{100}$

$75w = 5400$

$w = 72$

$$\begin{array}{r} 72 \\ -\ 54 \\ \hline 18 \text{ miles} \end{array}$$

5. $\begin{array}{r} \$24.00 \\ -\ 20.40 \\ \hline \$\ 3.60 \end{array}$ $\frac{\text{decrease}}{\text{increase}} = \frac{3.60}{24.00} = \frac{3}{20}$

$$\frac{3}{\cancel{20}} \times \frac{\overset{5}{\cancel{100}}}{1} = \mathbf{15\%}$$

b. $\begin{array}{r} \$24.00 \\ -\ 20.40 \\ \hline \$\ 3.60 \end{array}$ $\frac{\$3.60}{24} = \frac{r}{100}$

$\$24r = \$360$

$r = 15$

The rate is **15%**.

6. **a.** $\begin{array}{r} \$390 \\ +\ 260 \\ \hline \$650 \end{array}$ $\frac{\text{saved}}{\text{total}} = \frac{390}{650} = \frac{3}{5}$

$$\frac{3}{\cancel{5}} \times \frac{\overset{20}{\cancel{100}}}{1} = \mathbf{60\%}$$

b. $\begin{array}{r} \$260 \\ +\ 390 \\ \hline \$650 \end{array}$ $\frac{\$390}{\$650} = \frac{r}{100}$

$\$650r = \$39{,}000$

$r = 60$

The rate is **60%**.

7. **a.** $85\% = 0.85$

$$0.85\overline{)136.00} \quad 1\ 60 \text{ lb}$$
$$\begin{array}{r} 85 \\ \hline 51\ 0 \\ 51\ 0 \\ \hline 00 \end{array}$$

$\begin{array}{r} 160 \\ -\ 136 \\ \hline \mathbf{24 \text{ lb lost}} \end{array}$

b. $\frac{136}{w} = \frac{85}{100}$

$85w = 13{,}600$

$w = 160$

$\begin{array}{r} 160 \\ -\ 136 \\ \hline \mathbf{24 \text{ lb lost}} \end{array}$

8. **a.** $75\% = 0.75$

$$0.75\overline{)675.00} \quad 9\ 00$$
$$\begin{array}{r} 675 \\ \hline 0\ 00 \end{array}$$

$\begin{array}{r} 900 \\ -\ 675 \\ \hline \mathbf{225 \text{ empty seats}} \end{array}$

b. $\frac{675}{w} = \frac{75}{100}$

$75w = 67{,}500$

$w = 900$

$\begin{array}{r} 900 \\ -\ 675 \\ \hline \mathbf{225 \text{ empty seats}} \end{array}$

9. **a.** $15\% = 0.15$ $\begin{array}{r} 0.15 \\ \times\ \$400 \\ \hline \$60.00 \end{array}$ $\begin{array}{r} \$400 \\ -\ 60 \\ \hline \mathbf{\$340} \end{array}$

b. $\frac{p}{\$400} = \frac{15}{100}$

$100p = \$6000$

$p = \$60$ $\begin{array}{r} \$400 \\ -\ 60 \\ \hline \mathbf{\$340} \end{array}$

10. **a.** $25\% = 0.25$ $\begin{array}{r} 0.25 \\ \times\ \$700 \\ \hline \$175.00 \end{array}$ $\begin{array}{r} \$700 \\ -\ 175 \\ \hline \mathbf{\$525} \end{array}$

b. $\frac{p}{\$700} = \frac{25}{100}$

$100p = \$17{,}500$

$p = \$175$ $\begin{array}{r} \$700 \\ -\ 175 \\ \hline \mathbf{\$525} \end{array}$

11. **a.** $50\% = 0.5$ $\begin{array}{r} \$700 \\ \times\ 0.5 \\ \hline \$350.0 \end{array}$ $\begin{array}{r} \$700 \\ -\ 350 \\ \hline \$350 \end{array}$ $\begin{array}{r} \$525 \\ -\ 350 \\ \hline \mathbf{\$175} \end{array}$

b. $\frac{p}{\$700} = \frac{50}{100}$

$100p = \$35{,}000$

$p = \$350$ $\begin{array}{r} \$700 \\ -\ 350 \\ \hline \$350 \end{array}$ $\begin{array}{r} \$525 \\ -\ 350 \\ \hline \mathbf{\$175} \end{array}$

12. **a.** $\begin{array}{r} \$700 \\ +\ 400 \\ \hline \$1100 \end{array}$ $\begin{array}{r} \$1100 \\ \times\ .8 \\ \hline \$880.0 \end{array}$ $\begin{array}{r} \$1100 \\ -\ 880 \\ \hline \$\ 220 \end{array}$ $\begin{array}{r} \$220 \\ \times\ .05 \\ \hline \$11.00 \end{array}$ $\begin{array}{r} \$220 \\ +\ 11 \\ \hline \mathbf{\$231} \end{array}$

b. $\begin{array}{r} \$700 \\ +\ 400 \\ \hline \$1100 \end{array}$ $\frac{p}{\$1100} = \frac{80}{100}$

$100p = \$88{,}000$

$p = \$880$

$\begin{array}{r} \$1100 \\ -\ 880 \\ \hline \$\ 220 \end{array}$ $\frac{p}{\$220} = \frac{5}{100}$

$100p = \$1100$

$p = \$11$ tax

$\begin{array}{r} \$220 \\ +\ 11 \\ \hline \mathbf{\$231} \end{array}$

Answers and Solutions

13. **a.** 6% = 0.06 $100,000 $20,000
                      × 0.06   + 6,000
                      $6000.00   **$26,000**

   b. $\frac{p}{\$100,000} = \frac{6}{100}$

   $100p = \$600,000$
   $p = \$6000$ $20,000
                + 6,000
                $26,000

14. **C.** $\frac{1}{4}$ **down and $50 a month for ten months.**

   A. $33\frac{1}{3}\% = \frac{1}{3}$ $\frac{1}{3} \times \frac{\cancel{\$600}^{200}}{1} = \$200$

   $40 × 12 = $480. $480
                      + 200
                      $680

   B. $250 × 3 = $750

   C. $\frac{1}{4} \times \frac{\cancel{\$600}^{150}}{1} = \$150$

   $50 × 10 = $500. $500
                      + 150
                      $650

   D. 1 yr 6 mo = 18 mo
   $45 × 18 = $810

15. **D. $550 more each quarter.**
   A. $40 × 52 = $2080
   B. $2000
   C. 10% = 0.1
      0.1 × $1200 = $120 a month
      $120 × 12 = $1440
   D. $550 × 4 = $2200

Chapter 5 GRAPHS

Lesson 1 Exercise, pg. 160

1. a. **5.5 million**   b. **4.5 million**
   c. **14 million**   d. **4.5 million**
2. **Iowa**
3.   4.5 million
    + 3.5 million
     8.0 million
4. (2) **slightly more than the production in Iowa**
         5.5 million
         4.5
       + 4.5
       14.5 million
5. $\frac{3.5}{14} = $ 0.25 = **25%**
   14)3.50
       2 8
       —
        70
        70
        —
         0

6. $175
   × 4.5
   87.5
   700
   $787.5 million

7. **1 billion bushels of corn**
8. a. **4 bil. bu.**   b. **6.5 bil. bu.**
   c. **9 bil. bu.**   d. **8 bil. bu.**
9. a. **0.5 bil. bu.** b. **2.5 bil. bu.**
   c. **1 bil. bu**   d. **2 bil. bu.**
10. **1985**
11. **1980**
12. $\frac{1}{2} \div 4 = \frac{1}{2} \times \frac{1}{4} = \frac{1}{8}$
13.   6.5 bil. bu.
    − 4.0
    2.5 bil. bu.
14. $\frac{4}{8} = \frac{1}{2} = $ **50%**
15. **1980**

Lesson 2 Exercise, pg. 163

1. a. **26%**   b. **13%**   c. **27%**   d. **3%**
2. **education**
3. $\frac{4}{100} = \frac{1}{25}$
4.   4%
      3
    + 13
    $20\% = \frac{1}{5}$
5. (2) **a little over** $\frac{1}{2}$
    26%
   + 27
   53% is a little over 50%
6. 4% = 0.04 $960 billion
                × 0.04
                $38.40 = **$38.4 billion**
7. **1970**
8. a. **71.7%** b. **62.0%** c. **3.2%** d. **8.1%**
9. **divorced**
10.   8.1%
    − 3.2
    4.9%
11. (2) $\frac{1}{4}$
12. 62.0% = 0.62   0.62
                   × 300,000
                   186,000.00 = **186,000**
13. (1) **The percent of adults who were married or widowed decreased, and the percent who were single or divorced increased.**

Lesson 3 Exercise, pg. 165

1. **(2) $83**
2. **(4) $178**
3. **(3) 1984**
4. **(4) $80**

    1985 is about $215
    1980 is about − 135
    The difference is $ 80

5. **(1) $\frac{1}{2}$**

    1978 is about $110.
    1985 is about $215.
    $\frac{110}{215}$ is close to $\frac{100}{200} = \frac{1}{2}$.

6. **(2) $2\frac{1}{2}$ times**

    1976 is about $80.
    1984 is about $200.

    $2\frac{40}{80} = 2\frac{1}{2}$

    $80 \overline{)200}$
       $\underline{160}$
        40

7. **(3) 1980 to 1982**
8. **(2) $1400**

    The 1980 price is about $180.
    $180 × 8 = $1440, which is close to $1400.

Lesson 4 Exercise, pg. 167

1. a. **145,000**   b. **267,000**   c. **47,000**
   d. **160,000**
2.   94
     76
   309
   $\underline{+ 160}$
   639 thousand
3. **1978**
4. **1984**
5. **1982**
6. **(4) $\frac{1}{2}$**

    413 is about 400.
    816 is about 800.
    $\frac{400}{800} = \frac{1}{2}$

7. **1984**
8. **(2) 50%**

    309 is about 300.
    639 is about 600.
    $\frac{300}{600} = \frac{1}{2} = 50\%$

9. **the South**

Lesson 5 Exercise, pg. 168

1. a. **9%**   b. **14%**   c. **8%**   d. **5%**
2. **1978**
3. **1982**
4.   14%
   $\underline{- 5}$
     9%
5. $\frac{6\%}{12\%} = \frac{1}{2}$
6.   14%
   $\underline{- 8}$
     6%
7. 8% = 0.08   200,000
                 $\underline{\times\ 0.08}$
                 16,000.00 = **16,000 people**
8. **(3) 1979 to 1982**
    For these years the line rises from left to right.
9. **(4) 1988, 1989, 1990**
10. **billions of dollars**
11. **(1) $128 billion**
12. **(4) $119 billion**
13. **(2) $135 billion**
14. **(2) $25 billion**
    $100 billion
    $\underline{- 75}$
    $ 25 billion
15. **(4) $30 billion**
    $166 billion
    $\underline{- 136}$
    $ 30 billion
16. **(1) 1979**
17. **(3) 1982−1983**

Graphs Review, pg. 170

1. **(1) $22\frac{1}{2}$ gallons**

    $4\frac{1}{2} \times 5 =$

    $\frac{9}{2} \times \frac{5}{1} = \frac{45}{2} = 22\frac{1}{2}$

2. **(3) 2 times**

         2 times
    $20\overline{)40}$

3. **(4) 10 gallons**

    $7\frac{1}{2} \times 5 = \ \ \ 37\frac{1}{2}$

    $5\frac{1}{2} \times 5 = \underline{-27\frac{1}{2}}$
                   10 gal.

Answers and Solutions

4. **(2) 13.9%**
   26.3%
   −12.4
   ―――
   13.9%

5. **(3) a little over $\frac{1}{4}$**
   14.7%
   +12.4
   ―――
   27.1% is a little more than 25% = $\frac{1}{4}$

6. **(4) 50%**
   26.3%
   +24.3
   ―――
   50.6% is about 50%

7. **(1) $372**
   12.4% = 0.124   0.124
                   × 3,000
                   ―――――
                   $372.000

8. **(2) $\frac{1}{4}$**
   $\frac{200}{800} = \frac{1}{4}$

9. **(1) 1979–1980**

10. **(4) 1976–1978**

11. **(4) married couple**

12. $\frac{1}{4}$
    In 1990, 1-person, non-family households were about 25% or $\frac{1}{4}$ of the total.

13. **(1) 1978**

14. **(2) a little over 500,000**
    The difference is about half a million or 500,000.

15. **(3) The number of female students increased steadily while the number of male students gradually decreased.**

Review Test, pg. 175

1. **(3) $136.32**
   6.5% = 0.065 $128 $128.00
                × 0.065   +   8.32
                ―――――   ―――――
                   640        $136.32
                  7 68
                ―――――
                $8.320 = $8.32

2. **(2) $66\frac{2}{3}$%** $\frac{120}{180} = \frac{2}{3} = 66\frac{2}{3}$%

3. **(3) 729** $V = s^3$
                 $V = 9^3$
                 $V = 9 \times 9 \times 9$
                 $V = 729$ cu in.

4. **(2) 17 : 13**   150
                    − 85
                   ―――
                     65
   strike : not strike = 85 : 65 = 17 : 13

5. **(3) 0.307** $\quad$ 0.3066 to the nearest
   75)23.0000 thousandth = 0.307
      22 5
      ―――
         50
          0
      ―――
        500
        450
      ―――
        500
        450

6. **(5) $\frac{8}{27}$** $\left(\frac{2}{3}\right)^3 = \frac{2}{3} \times \frac{2}{3} \times \frac{2}{3} = \frac{8}{27}$

7. **(3) 18,576 ft**   20,320   18,576
                       18,008   3)55,728
                     + 17,400
                     ―――――
                       55,728

8. **(1) $3\frac{1}{5}$** $\frac{\text{flour}}{\text{sugar}}$ $\frac{5}{2} = \frac{8}{x}$
                            $5x = 16$
                            $x = 3\frac{1}{5}$

9. **(5) $66,100** $62,400 $ 66,100
                     76,600   3)$198,300
                   + 59,300
                   ―――――
                   $198,300

10. **(4) 211** $14^2 − 8^0 + 16^1 =$
                  $14 \times 14 − 1 + 16 =$
                  $196 − 1 + 16 = 211$

11. **(4) $\frac{1}{50}$** $\frac{\text{Smiths}}{\text{total}} = \frac{10}{500} = \frac{1}{50}$

12. **(2) $\frac{3}{50}$**   8 $\frac{\text{3 families}}{\text{total}} = \frac{30}{500} = \frac{3}{50}$
                            12
                          + 10
                          ―――
                            30

13. **(4) $9 \times 10 \times 10 \times 10 \times 10$**

14. **(4) 342**
    Area of top rectangle: $A = lw$
    $A = 10 \times 9 = 90$ ft^2
    Area of bottom rectangle: $A = lw$
    $A = 21 \times 12 = 252$ ft^2
    Sum: $90 + 252 = 342$ ft^2

15. **(4) 4.71 m** $C = \pi d$
                     $C = 3.14 \times 1.5$
                     $C = 4.71$ m

16. **(3) $182** $553
                  − 371
                  ―――
                   $182

17. **(5) $327**   New York $500
                               392
                             + 153
                             ―――
                              $1045
                   Illinois $371   1045
                             278   − 718
                           + 69   ―――
                           ――― $ 327
                            $718

18. **(1) 2.5 cm**   4.1 cm
                    − 1.6
                    ―――
                     2.5 cm

19. **(3) 13**
$\begin{array}{r} 38{,}734 \\ -38{,}526 \\ \hline 208 \end{array}$ $\begin{array}{r} 13 \\ 16\overline{)208} \\ \underline{16} \\ 48 \\ \underline{48} \end{array}$

20. **(2) 21**
$\begin{array}{r} 39{,}028 \\ -38{,}734 \\ \hline 294 \end{array}$ $\begin{array}{r} 21 \\ 14\overline{)294} \\ \underline{28} \\ 14 \\ \underline{14} \end{array}$

21. **(4) $34.50**
$\begin{array}{r} 16 \\ +14 \\ \hline 30 \end{array}$ $\begin{array}{r} \$1.15 \\ \times\ 30 \\ \hline \$34.50 \end{array}$

22. **(4) 35 × 8.50 + 6 × 12.75**

23. **(2) $20\frac{1}{4}$** $A = s^2$
$A = \left(4\frac{1}{2}\right)^2$
$A = 4\frac{1}{2} \times 4\frac{1}{2}$
$A = \frac{9}{2} \times \frac{9}{2} = \frac{81}{4} = 20\frac{1}{4}$

24. **(5) 520** $V = lwh$
$V = 6.5 \times 4 \times 20$
$V = 520 \text{ cm}^3$

25. **(2) B, A, D, E, C**
$A = 0.850 \text{ m}$
$B = 0.095 \text{ m}$
$C = 1.200 \text{ m}$
$D = 0.900 \text{ m}$
$E = 1.070 \text{ m}$

26. **(1) California**

27. **(4) over $\frac{1}{2}$**
$\begin{array}{r} 29.4 \\ 27.0 \\ 20.2 \\ +13.0 \\ \hline 89.6 \end{array}$ $\frac{89.6}{170.7}$ is more than $\frac{1}{2}$ because the numerator is more than half the denominator.

28. **(4) $\frac{5}{12}$** $15 + 12 + 9 = 36$ total
$\frac{15}{36} = \frac{5}{12}$

29. **(3) $\frac{3}{11}$** $13 + 11 + 9 = 33$ remaining
$\frac{9}{33} = \frac{3}{11}$

30. **(5) 45%** $\begin{array}{r} \$29 \\ -20 \\ \hline \$\ 9 \text{ markup} \end{array}$
$\frac{\text{markup}}{\text{original}} = \frac{9}{20} = \frac{9}{\cancel{20}} \times \frac{\cancel{100}^5}{1} = 45\%$
or $\frac{9}{20} = \frac{r}{100}$ $20r = 900$ $r = 45$
The markup is 45%.

31. **(3) 55%** $\begin{array}{r} 165 \\ +135 \\ \hline 300 \end{array}$ $\frac{\text{women}}{\text{total}} = \frac{165}{300} = \frac{55}{100} = 55\%$
or $\frac{165}{300} = \frac{r}{100}$
$300r = 16{,}500$
$r = 55$
55% are women.

32. **(1) 882** $\begin{array}{r} \text{children}\ \ 5 \\ \text{adults}\ +2 \\ \hline \text{total}\ \ \ \ 7 \end{array}$
$\frac{\text{children}}{\text{total}} = \frac{5}{7} = \frac{630}{x}$
$5x = 4410$
$x = 882$ tickets sold

33. **(2) $\frac{6 \times 380}{50}$**

34. **(3) 30% of the amount Jeff paid for the car.**
(1) $24 \times \$50 = \$1200$
(2) $9 \times \$60 = \$540$ $\$540 + \$500 = \$1040$
(3) $30\% = 0.3$ $0.3 \times \$4500 = \$1350$
(4) $44 \times \$30 = \$1320$
(5) $8 \times \$80 = \$640$ $\$640 + \$400 = \$1040$

35. **(3) 32** $304 \div 9\frac{1}{2} =$
$\frac{304}{1} \div \frac{19}{2} =$
$\frac{\cancel{304}^{16}}{1} \times \frac{2}{\cancel{19}_1} = 32$

36. **(2) $\frac{1}{4}$**   26.5% is close to 25% $= \frac{25}{100} = \frac{1}{4}$

37. **(4) 23.2%** $\begin{array}{r} 11.9\% \\ +26.5 \\ \hline 38.4\% \end{array}$ $\begin{array}{r} 61.6\% \\ -38.4 \\ \hline 23.2\% \end{array}$

38. **(5) Not enough information is given.**
You do not know what percent of the population is 18 to 40.

39. **(3) 79,500** $26.5\% = 0.265$ $\begin{array}{r} 0.265 \\ \times 300{,}000 \\ \hline 79{,}500.00 \end{array}$

40. **(1) 600 + 60 × 80**

41. **(4) 20** $\begin{array}{r} 20 \\ 150\overline{)3000} \end{array}$

42. **(2) 2000** $5\% = 0.05$ $\begin{array}{r} 120{,}000 \\ \times\ \ \ 0.05 \\ \hline 6000.00 \end{array}$
$\frac{1}{\cancel{3}_1} \times \frac{\cancel{6000}^{2000}}{1} = 2000$

43. **(5) Not enough information is given.**
You do not know the relationship between the new jobs and the change in population.

44. **(3) 126,400**
    In 2 years Elk Electronics will have
    $2 \times 2000 = 4000$ jobs.
    2% = 0.02   120,000
    $\times\ \ 0.02$
    $\overline{2400.00}$ new jobs at Paulson's.

    4,000 jobs at Elk
    2,400 jobs at Paulson's
    +120,000 existing jobs
    126,400 total

45. **(2) 1300 more**
    2000 new jobs at Elk Electronics
    − 700 lost jobs
    1300 more

46. **(5)** $\dfrac{20 \times 6}{3}$

47. **(4) 40**   80% = 0.8 $0.8\overline{)160.0}$ = 200
    old weight   200
    current weight − 160
    40 lb lost

48. **(2) won 35, lost 14**
    won   5 $\dfrac{\text{won}}{\text{played}} = \dfrac{5}{7} = \dfrac{x}{49}$
    lost   +2
    played   7 $7x = 245$
                        $x = 35$ won
                        $49 - 35 = 14$ lost

49. **(3) $0.98**   first oz   = $0.29
                    next 3 oz = 3 × $.23 =   0.69
                    total $0.98

50. **(2) $1.52**   first oz   = $0.29
                    second oz   =   0.23
                    certified fee = 1.00
                    total $1.52

Chapter 6 ALGEBRA

Level 1 Preview, pg. 184

1. **point** C
2. $(12) + (-9) + (-15) =$
   $12 - 9 - 15 =$
   $12 - 24 = \mathbf{-12}$
3. $(-14) - (-18) =$
   $14 + 18 = \mathbf{+4}$
4. $(8)(-5)(-3) = \mathbf{+120}$
5. $\dfrac{-108}{-12} = \mathbf{+9}$
6. $a + c = -15 + (-30) = -15 - 30 = \mathbf{-45}$
7. $xy = (10)(-0.4) = \mathbf{-4}$
8. $(-5y) + (-3y) + (+y) =$
   $-5y - 3y + y = -8y + y = \mathbf{-7y}$
9. $(11w) - (-4w) = 11w + 4w = \mathbf{15w}$
10. $(3ab)(-4a) = \mathbf{-12a^2b}$
11. $\dfrac{-4x^2y^2}{-8xy} = \dfrac{xy}{2}$
12. $3m - mn = 3(-5) - (-5)(-2) =$
    $-15 - (+10) = -15 - 10 = \mathbf{-25}$

Lesson 1 Exercise, pg. 186

1. **I**   2. **C**   3. **F**   4. **A**
5. **G**   6. **B**   7. **E**   8. **H**

Lesson 2 Exercise, pg. 188

1. $+6 + (-10) = \mathbf{-4}$ 2. $+15 + (-4) = \mathbf{+11}$
3. $-8 + (-12) = \mathbf{-20}$
4. $+7 + (-6) = \mathbf{+1}$ 5. $-13 + 13 = \mathbf{0}$
6. $-14 + 18 = \mathbf{+4}$
7. $(+9) + (-15) = \mathbf{-6}$
8. $(-8) + (-11) = \mathbf{-19}$
9. $(-12) + (+12) = \mathbf{0}$
10. $(-24) + (+7) = \mathbf{-17}$
11. $(-3) + (+14) = \mathbf{+11}$
12. $(-19) + (-19) = \mathbf{-38}$
13. $+3 + (-9) + 7 = +10 + (-9) = \mathbf{+1}$
14. $-4 + (-6) + (-3) = \mathbf{-13}$
15. $+2 + 8 + (-10) = +10 + (-10) = \mathbf{0}$
16. $(+3) + (+11) + (+5) = +3 + 11 + 5 = \mathbf{+19}$
17. $(-9) + (-1) + (+4) = -10 + 4 = \mathbf{-6}$
18. $+7 + (-4) + (-3) + 2 = +9 + (-7) = \mathbf{+2}$
19. $-8 + (-1) + (-6) + 10 = -15 + 10 = \mathbf{-5}$
20. $(-8) + (+12) + (+16) + (-11) =$
    $-19 + 28 = \mathbf{+9}$
21. $(+7) + (-12) + (+3) + (+2) =$
    $+12 + (-12) = \mathbf{0}$

Lesson 3 Exercise, pg. 190

1. $(+8) - (+9) = +8 + (-9) = \mathbf{-1}$
2. $(+7) - (-6) = +7 + 6 = \mathbf{+13}$
3. $(-9) - (+3) = -9 + (-3) = \mathbf{-12}$
4. $(-3) - (-14) = -3 + 14 = \mathbf{11}$
5. $(-10) - (+15) = -10 + (-15) = \mathbf{-25}$
6. $(+6) - (-11) = +6 + 11 = \mathbf{+17}$
7. $(+20) - (-4) = +20 + 4 = \mathbf{24}$
8. $(+2) - (+18) = +2 + (-18) = \mathbf{-16}$ ✓
9. $(-16) - (-5) = -16 + 5 = \mathbf{-11}$
10. $(-11) - (+15) = -11 + (-15) = \mathbf{-26}$
11. $(+7) - (-8) = +7 + 8 = \mathbf{+15}$
12. $(-9) - (+9) = -9 + (-9) = \mathbf{-18}$
13. $(+8) - (+7) + (6) = +8 + (-7) + 6 =$
    $+14 + (-7) = \mathbf{+7}$
14. $(-4) + (-5) - (+3) =$
    $-4 + (-5) + (-3) = \mathbf{-12}$

15. $(-10) - (-7) - (-1) = -10 + 7 + 1 = -10 + 8 = \mathbf{-2}$

16. $(+11) - (-14) + (-3) = +11 + 14 + (-3) = +25 + (-3) = \mathbf{+22}$

Lesson 4 Exercise, pg. 192

1. $(+6)(-10) = \mathbf{-60}$
2. $(+12)(-1) = \mathbf{-12}$
3. $(-5)(-13) = \mathbf{+65}$
4. $(+8)\left(+\frac{1}{2}\right) = \mathbf{+4}$
5. $\left(-\frac{1}{3}\right)(-15) = \mathbf{+5}$
6. $(-9)(+8) = \mathbf{-72}$
7. $\left(-\frac{3}{4}\right)(+24) = \mathbf{-18}$
8. $(+20)(-3) = \mathbf{-60}$
9. $(-15)(-2) = \mathbf{+30}$
10. $(+4)(-10)(+8) = \mathbf{-320}$
11. $(+5)(-9)\left(-\frac{2}{3}\right) = \mathbf{+30}$
12. $\left(-\frac{1}{2}\right)(-12)(-5) = \mathbf{-30}$
13. $(-8)(+12)(-1) = \mathbf{+96}$
14. $(-4)(+10)\left(+\frac{3}{4}\right)(+2) = \mathbf{-60}$
15. $(+5)(-1)(+9)(-6) = \mathbf{+270}$

Lesson 5 Exercise, pg. 193

1. $\frac{-24}{-8} = \mathbf{+3}$
2. $\frac{-30}{+10} = \mathbf{-3}$
3. $\frac{+5}{-10} = \mathbf{-\frac{1}{2}}$
4. $\frac{+8}{+2} = \mathbf{4}$
5. $\frac{-12}{+12} = \mathbf{-1}$
6. $\frac{-15}{-20} = \mathbf{\frac{3}{4}}$
7. $\frac{+100}{-20} = \mathbf{-5}$
8. $\frac{-20}{-40} = \mathbf{\frac{1}{2}}$
9. $\frac{63}{-9} = \mathbf{-7}$
10. $\frac{-48}{-12} = \mathbf{4}$
11. $\frac{-80}{+100} = \mathbf{-\frac{4}{5}}$
12. $\frac{+18}{-6} = \mathbf{-3}$
13. $\frac{150}{-25} = \mathbf{-6}$
14. $\frac{-35}{49} = \mathbf{-\frac{5}{7}}$
15. $\frac{-20}{-24} = \mathbf{\frac{5}{6}}$
16. $\frac{-96}{+12} = \mathbf{-8}$

Lesson 6 Exercise, pg. 193

1. $a + b = (-6) + (-4) = \mathbf{-10}$
2. $m + n + p = (3) + (9) + (-7) = 12 + (-7) = \mathbf{5}$
3. $x + y = (15) + (-15) = \mathbf{0}$
4. $r + s + t = (-1) + (-8) + (7) = -9 + 7 = \mathbf{-2}$
5. $a - b = (-6) - (-7) = -6 + 7 = \mathbf{1}$
6. $m - n = (14) - (-4) = 14 + 4 = \mathbf{18}$
7. $w - y = (-6) - (+7) = -6 + (-7) = \mathbf{-13}$
8. $c - d = (15) - (-9) = 15 + 9 = \mathbf{24}$
9. $ab = (-7)(-6) = \mathbf{+42}$
10. $pqr = (5)(-3)(-10) = \mathbf{150}$
11. $mn = (-9)(+9) = \mathbf{-81}$
12. $xyz = (-1)\left(-\frac{1}{2}\right)\left(-\frac{1}{4}\right) = \mathbf{-\frac{1}{8}}$
13. $\frac{a}{b} = \frac{-14}{-2} = \mathbf{+7}$
14. $\frac{x}{y} = \frac{-12}{15} = \mathbf{-\frac{4}{5}}$
15. $\frac{m}{n} = \frac{23}{-1} = \mathbf{-23}$
16. $\frac{c}{d} = \frac{+18}{-24} = \mathbf{-\frac{3}{4}}$

Lesson 7 Exercise, pg. 195

1. $5m + 3m = \mathbf{8m}$
2. $2c + (-6c) = \mathbf{-4c}$
3. $8x + (-x) = \mathbf{7x}$
4. $-4p + 5p = \mathbf{p}$
5. $18xy + (-19xy) = \mathbf{-xy}$
6. $-6st + (-5st) = \mathbf{-11st}$
7. $9w + (-3w) + 4w = \mathbf{10w}$
8. $-8y + (-3y) + (-y) = \mathbf{-12y}$
9. $-d + 5d + (-4d) = \mathbf{0}$
10. $(-8e) + (-6e) + (+11e) = -14e + 11e = \mathbf{-3e}$
11. $(7n) + (-12n) + (-4n) = +7n + (-16n) = \mathbf{-9n}$
12. $(-a) + (5a) + (-9a) + (+3a) = -10a + 8a = \mathbf{-2a}$
13. $(-7w) + (-2w) + (-3w) + (11w) = -12w + 11w = \mathbf{-w}$

Lesson 8 Exercise, pg. 196

1. $(+5p) - (+3p) = 5p + (-3p) = \mathbf{2p}$
2. $(+5p) - (-3p) = 5p + 3p = \mathbf{8p}$
3. $(-5p) - (-3p) = -5p + 3p = \mathbf{-2p}$
4. $(-6a) - (-3a) = -6a + 3a = \mathbf{-3a}$
5. $(+4c) - (-4c) = +4c + 4c = \mathbf{8c}$
6. $(-7mn) - (+3mn) = -7mn + (-3mn) = \mathbf{-10mn}$
7. $(-9f) - (-f) + (-2f) = -9f + f + (-2f) = -11f + f = \mathbf{-10f}$
8. $(10y) - (3y) - (-7y) = 10y + (-3y) + 7y = 17y + (-3y) = \mathbf{14y}$

Answers and Solutions 337

9. $(-8cd) + (-2cd) - (3cd) = -8cd + (-2cd) + (-3cd) = $ **$-13cd$**

10. $(-12t) - (-3t) + (9t) = -12t + 3t + 9t = -12t + 12t = $ **0**

11. $(11u) - (12u) + (-u) = 11u + (-12u) + (-u)$
    $11u + (-13u) = $ **$-2u$**

12. $(-9xy) + (-xy) + (-3xy) = $ **$-13xy$**

Lesson 9 Exercise, pg. 197

1. $c^2 \cdot c^3 = $ **c^5**
2. $m^4 \cdot m = $ **m^5**
3. $x^3 \cdot x^5 = $ **x^8**
4. $a \cdot a = $ **a^2**
5. $a \cdot b = $ **ab**
6. $(-a^2b^2)(a^3b^4) = $ **$-a^5b^6$**
7. $(2x)(-3x) = $ **$-6x^2$**
8. $(-4y^2)(-2y^5) = $ **$8y^7$**
9. $(-5p)(2p^3) = $ **$-10p^4$**
10. $(5yz)(-6yz) = $ **$-30y^2z^2$**
11. $(-mn^2)(-4m^2n^3) = $ **$4m^3n^5$**
12. $(-2rs)(-9r^3s) = $ **$18r^4s^2$**

Lesson 10 Exercise, pg. 198

1. $\dfrac{m^6}{m^2} = $ **m^4**
2. $\dfrac{st}{s} = $ **t**
3. $\dfrac{c^5}{c^2} = $ **c^3**
4. $\dfrac{a^5}{a^4} = $ **a**
5. $\dfrac{x^3y^2}{xy} = $ **x^2y**
6. $\dfrac{x^6}{x^6} = $ **1**
7. $\dfrac{-12x^2}{3x} = $ **$-4x$**
8. $\dfrac{20m^3n}{-4m} = $ **$-5m^2n$**
9. $\dfrac{-36a^5}{-9a^4} = $ **$4a$**
10. $\dfrac{-4m^2n^2}{16mn} = $ **$-\dfrac{mn}{4}$**
11. $\dfrac{-18c^3d^4}{-12c^3d} = $ **$\dfrac{3d^3}{2}$**
12. $\dfrac{+24x}{+30x} = $ **$\dfrac{4}{5}$**

Lesson 11 Exercises, pg. 199

1. $ab - c = (-4)(-3) - (-5) = +12 + 5 = $ **$+17$**
2. $m(m - n) = -2(-2 - 7) = -2(-2 + (-7)) = -2(-9) = $ **$+18$**
3. $x^2y = (-3)^2(-4) = (+9)(-4) = $ **-36**
4. $s(s + t) - t = -5(-5 + 1) - 1 = -5(-4) - 1 = +20 - 1 = $ **$+19$**
5. $e + ef = 6 + (6)(-4) = 6 + (-24) = $ **-18**
6. $p + pq = -6 + (-6)(4) = -6 + (-24) = $ **-30**
7. $ab^2 = (-2)(-5)^2 = (-2)(+25) = $ **-50**
8. $x(x - y) = -8(-8 - 2) = -8(-8 + (-2)) = -8(-10) = $ **$+80$**

9. $g(g - h) = 6(6 - (-7)) = 6(6 + 7) = 6(13) = $ **78**

10. $(j - k)^2 = (-3 - 1)^2 = (-3 + (-1))^2 = (-4)^2 = $ **$+16$**

11. $\dfrac{a + b}{2} = \dfrac{-6 + (-2)}{2} = \dfrac{-8}{2} = $ **-4**

12. $\dfrac{m - n}{n} = \dfrac{8 - (-4)}{-4} = \dfrac{8 + 4}{-4} = \dfrac{+12}{-4} = $ **-3**

Level 1 Review, pg. 200

1. **point B**
2. $(-13) + (8) + (-7) = -20 + 8 = $ **-12**
3. $(-9) - (-11) - (+4) = -9 + 11 + (-4) = -13 + 11 = $ **-2**
4. $(-6)\left(-\dfrac{2}{3}\right)(-5) = $
   $\left(-\dfrac{\cancel{6}^2}{1}\right)\left(-\dfrac{2}{\cancel{3}_1}\right)\left(-\dfrac{5}{1}\right) = $ **-20**
5. $\dfrac{+24}{-36} = $ **$-\dfrac{2}{3}$**
6. $p + q = (-20) + (-14) = $ **-34**
7. $cd = \left(\dfrac{\cancel{24}^{12}}{1}\right)\left(-\dfrac{1}{\cancel{2}_1}\right) = $ **-12**
8. $\dfrac{a}{e} = \dfrac{144}{-12} = $ **-12**
9. $(4m) + (-2m) + (-7m) = 4m + (-9m) = $ **$-5m$**
10. $(-n) - (-6n) + (-4n) = -n + 6n + (-4n) = -5n + 6n = $ **n**
11. $(-9x^2y)(-2xy) = $ **$18x^3y^2$**
12. $\dfrac{12a^2c}{-3a} = $ **$-4ac$**
13. $ab - 4b = (-7)(-3) - 4(-3) = +21 - (-12) = +21 + 12 = $ **$+33$**
14. $c^2d = (-5)^2(10) = (+25)(10) = $ **250**
15. $x(x + y) = -4(-4 + 9) = -4(+5) = $ **-20**

Lesson 12 Exercise, pg. 204

1. $m + 11 = 30$
   $-11 \quad -11$
   $\overline{m } = 19$

2. $\dfrac{8w}{8} = \dfrac{56}{8}$
   $w = 7$

3. $c - 12 = 5$
   $+ 12 \quad +12$
   $\overline{c } = 17$

4. $16 = f - 4$
   $+4 + 4$
   $\overline{20 = f}$

5. $\dfrac{4}{1} \cdot \dfrac{c}{4} = 5 \cdot 4$
   $c = 20$

6. $\dfrac{6n}{6} = \dfrac{9}{6}$
   $n = 1\dfrac{3}{6} = 1\dfrac{1}{2}$

7. $9 \cdot 2 = \dfrac{y}{9} \cdot \dfrac{9}{1}$
   $18 = y$

8. $\dfrac{12}{18} = \dfrac{18p}{18}$
   $\dfrac{2}{3} = p$

338   Answers and Solutions

9. $14 = a + 3$
   $\underline{-3 -3}$
   $11 = a$

10. $g - 9 = 41$
    $\underline{+9 +9}$
    $g = 50$

11. $e + 6 = -8$
    $\underline{-6 -6}$
    $e = -14$

12. $\dfrac{12}{1} \cdot \dfrac{n}{12} = 1 \cdot 12$
    $n = 12$

13. $15 = i - 8$
    $\underline{+8 +8}$
    $23 = i$

14. $\dfrac{200}{25} = \dfrac{25r}{25}$
    $8 = r$

15. $\dfrac{4}{3} \cdot \dfrac{3}{4} s = \dfrac{24}{1} \cdot \dfrac{4}{3}$
    $s = 32$

16. $2 \cdot 10 = \dfrac{1}{2} w \cdot \dfrac{2}{1}$
    $20 = w$

17. $21 = d + 16$
    $\underline{-16 -16}$
    $5 = d$

18. $5 \cdot 10 = \dfrac{z}{5} \cdot \dfrac{5}{1}$
    $50 = z$

19. $\dfrac{24f}{24} = \dfrac{12}{24}$
    $f = \dfrac{1}{2}$

20. $p + 14 = 4$
    $\underline{-14 -14}$
    $p = -10$

21. $\dfrac{8}{3} \cdot \dfrac{3}{8} x = \dfrac{15}{1} \cdot \dfrac{8}{3}$
    $x = 40$

7. $2n - 11 = 3$
   $\underline{+11 +11}$
   $\dfrac{2n}{2} = \dfrac{14}{2}$
   $n = 7$

8. $\dfrac{3}{4} a + 5 = 17$
   $\underline{-5 -5}$
   $\dfrac{4}{3} \cdot \dfrac{3}{4} a = \dfrac{12}{1} \cdot \dfrac{4}{3}$
   $a = 16$

9. $\dfrac{s}{10} - 6 = 3$
   $\underline{+6 +6}$
   $\dfrac{10}{1} \cdot \dfrac{s}{10} = 9 \cdot 10$
   $s = 90$

10. $5y + 7 = -3$
    $\underline{-7 -7}$
    $\dfrac{5y}{5} = \dfrac{-10}{5}$
    $y = -2$

11. $2 = \dfrac{w}{9} + 11$
    $\underline{-11 -11}$
    $9 \cdot -9 = \dfrac{w}{9} \cdot \dfrac{9}{1}$
    $-81 = w$

12. $\dfrac{4}{5} f - 7 = 1$
    $\underline{+7 +7}$
    $\dfrac{5}{4} \cdot \dfrac{4}{5} f = \dfrac{8}{1} \cdot \dfrac{5}{4}$
    $f = 10$

13. $8z + 3 = 9$
    $\underline{-3 -3}$
    $\dfrac{8z}{8} = \dfrac{6}{8}$
    $z = \dfrac{3}{4}$

14. $2 = 9p - 4$
    $\underline{+4 +4}$
    $\dfrac{6}{9} = \dfrac{9p}{9}$
    $\dfrac{2}{3} = p$

15. $-4 = 7t + 3$
    $\underline{-3 -3}$
    $\dfrac{-7}{7} = \dfrac{7t}{7}$
    $-1 = t$

16. $10b - 8 = 12$
    $\underline{+8 +8}$
    $\dfrac{10b}{10} = \dfrac{20}{10}$
    $b = 2$

Lesson 13 Exercise, pg. 205

1. $6m + 5 = 47$
   $\underline{-5 = -5}$
   $\dfrac{6m}{6} = \dfrac{42}{6}$
   $m = 7$

2. $3x - 2 = 28$
   $\underline{+2 = +2}$
   $\dfrac{3x}{3} = \dfrac{30}{3}$
   $x = 10$

3. $\dfrac{c}{4} + 1 = 8$
   $\underline{-1 = -1}$
   $\dfrac{4}{1} \cdot \dfrac{c}{4} = 7 \cdot 4$
   $c = 28$

4. $17 = 7a + 3$
   $\underline{-3 -3}$
   $\dfrac{14}{7} = \dfrac{7a}{7}$
   $2 = a$

5. $50 = 9d - 4$
   $\underline{+4 +4}$
   $\dfrac{54}{9} = \dfrac{9d}{9}$
   $6 = d$

6. $8 = \dfrac{x}{7} - 2$
   $\underline{+2 +2}$
   $7 \cdot 10 = \dfrac{x}{7} \cdot \dfrac{7}{1}$
   $70 = x$

Answers and Solutions

17. $19 = 4s + 1$
$ -1 -1$
$ \dfrac{18}{4} = \dfrac{4s}{4}$
$ 4\dfrac{2}{4} = s$
$ 4\dfrac{1}{2} = s$

18. $\dfrac{n}{6} - 5 = 1$
$\phantom{\dfrac{n}{6}} +5 +5$
$\dfrac{6}{1} \cdot \dfrac{n}{6} = 6 \cdot 6$
$\phantom{\dfrac{6}{1} \cdot} n = \mathbf{36}$

Lesson 14 Exercise, pg. 206

1. $9a - 2a = 21$
$\dfrac{7a}{7} = \dfrac{21}{7}$
$a = \mathbf{3}$

2. $5m = 18 + 2m$
$-2m -2m$
$\dfrac{3m}{3} = \dfrac{18}{3}$
$m = \mathbf{6}$

3. $8r = 15 + 3r$
$-3r -3r$
$\dfrac{5r}{5} = \dfrac{15}{5}$
$r = \mathbf{3}$

4. $12 - 5x = 7x$
$ +5x +5x$
$\dfrac{12}{12} = \dfrac{12x}{12}$
$\mathbf{1} = x$

5. $16 = 13c + 7c$
$16 = 20c$
$\dfrac{16}{20} = \dfrac{20c}{20}$
$\dfrac{4}{5} = c$

6. $3p + 7 = 10p$
$-3p -3p$
$\dfrac{7}{7} = \dfrac{7p}{7}$
$\mathbf{1} = p$

7. $6y - y = 10$
$\dfrac{5y}{5} = \dfrac{10}{5}$
$y = \mathbf{2}$

8. $3t = 9 + 2t$
$-2t -2t$
$t = \mathbf{9}$

9. $4 - 2n = 6n$
$+2n +2n$
$\dfrac{4}{8} = \dfrac{8n}{8}$
$\dfrac{1}{2} = n$

10. $5x + 4 = 3x + 20$
$-3x -3x$
$2x + 4 = 20$
$ -4 -4$
$\dfrac{2x}{2} = \dfrac{16}{2}$
$x = \mathbf{8}$

11. $8w - 5 = 7w + 13$
$-7w -7w$
$w - 5 = 13$
$+5 +5$
$w = \mathbf{18}$

12. $3p + 12 = 8p - 23$
$-3p -3p$
$12 = 5p - 23$
$+23 +23$
$\dfrac{35}{5} = \dfrac{5p}{5}$
$\mathbf{7} = p$

13. $7c - 3c = c + 27$
$4c = c + 27$
$-c -c$
$\dfrac{3c}{3} = \dfrac{27}{3}$
$c = \mathbf{9}$

14. $9m - 12 = m + 20$
$-m -m$
$8m - 12 = 20$
$+12 +12$
$\dfrac{8m}{8} = \dfrac{32}{8}$
$m = \mathbf{4}$

15. $2d - 8 = 7d + 12$
$-2d -2d$
$-8 = 5d + 12$
$-12 -12$
$\dfrac{-20}{5} = \dfrac{5d}{5}$
$\mathbf{-4} = d$

Lesson 15 Exercise, pg. 207

1. $4(m - 3) = 20$
$4m - 12 = 20$
$+12 +12$
$\dfrac{4m}{4} = \dfrac{32}{4}$
$m = \mathbf{8}$

2. $5(a + 2) = 15$
$5a + 10 = 15$
$-10 -10$
$\dfrac{5a}{5} = \dfrac{5}{5}$
$a = \mathbf{1}$

3. $9 = 2(x - 3)$
$9 = 2x - 6$
$+6 +6$
$\dfrac{15}{2} = \dfrac{2x}{2}$
$7\dfrac{1}{2} = x$

4. $3(c + 4) = 2c + 17$
   $3c + 12 = 2c + 17$
   $\underline{-2c \qquad\quad -2c}$
   $c + 12 = 17$
   $\underline{\;-12 \qquad -12}$
   $c = 5$

5. $8n - 7 = 6(n - 1)$
   $8n - 7 = 6n - 6$
   $\underline{-6n \qquad\quad -6n}$
   $2n - 7 = -6$
   $\underline{\;+7 \qquad\quad +7}$
   $\dfrac{2n}{2} = \dfrac{1}{2}$
   $n = \dfrac{1}{2}$

6. $9(p + 2) = p + 20$
   $9p + 18 = p + 20$
   $\underline{-p \qquad\quad -p}$
   $8p + 18 = 20$
   $\underline{\;-18 \qquad -18}$
   $\dfrac{8p}{8} = \dfrac{2}{8}$
   $p = \dfrac{1}{4}$

7. $4(a - 5) = 3(a + 2)$
   $4a - 20 = 3a + 6$
   $\underline{-3a \qquad\quad -3a}$
   $a - 20 = 6$
   $\underline{\;+20 \qquad +20}$
   $a = 26$

8. $6(d - 1) = 3(d + 2)$
   $6d - 6 = 3d + 6$
   $\underline{-3d \qquad\quad -3d}$
   $3d - 6 = 6$
   $\underline{\;+6 \qquad\quad +6}$
   $\dfrac{3d}{3} = \dfrac{12}{3}$
   $d = 4$

9. $5(y + 2) = 3(y - 8)$
   $5y + 10 = 3y - 24$
   $\underline{-3y \qquad\quad -3y}$
   $2y + 10 = -24$
   $\underline{\;-10 \qquad\quad -10}$
   $\dfrac{2y}{2} = \dfrac{-34}{2}$
   $y = -17$

Lesson 16 Exercise, pg. 207

1. $a + 6 > 9$
   $\underline{\;-6 \qquad -6}$
   $a > 3$

2. $c - 12 \leq 3$
   $\underline{\;+12 \qquad +12}$
   $c \leq 15$

3. $\dfrac{2}{1} \cdot \dfrac{n}{2} < 7 \cdot 2$
   $n < 14$

4. $\dfrac{16r}{16} \geq \dfrac{20}{16}$
   $r \geq \dfrac{5}{4}$ or $1\dfrac{1}{4}$

5. $6m - 2 < 22$
   $\underline{\;+2 \qquad +2}$
   $\dfrac{6m}{6} < \dfrac{24}{6}$
   $m < 4$

6. $\dfrac{5}{3} \cdot \dfrac{3}{5} x \leq \dfrac{18}{1} \cdot \dfrac{5}{3}$
   $x \leq 30$

7. $3p - 4 > p + 6$
   $\underline{-p \qquad\quad -p}$
   $2p - 4 > 6$
   $\underline{\;+4 \qquad +4}$
   $\dfrac{2p}{2} > \dfrac{10}{2}$
   $p > 5$

8. $9w + 2 \geq w + 10$
   $\underline{-w \qquad\quad -w}$
   $8w + 2 \geq 10$
   $\underline{\;-2 \qquad\quad -2}$
   $\dfrac{8w}{8} \geq \dfrac{8}{8}$
   $w \geq 1$

9. $3(m - 2) < 9$
   $3m - 6 < 9$
   $\underline{\;+6 \qquad +6}$
   $\dfrac{3m}{3} < \dfrac{15}{3}$
   $m < 5$

10. $\dfrac{1}{2}y - 3 \geq 1$
    $\underline{\;+3 \qquad +3}$
    $\dfrac{2}{1} \cdot \dfrac{1}{2}y \geq 4 \cdot 2$
    $y \geq 8$

11. $4 > 2(n - 9)$
    $4 > 2n - 18$
    $\underline{+18 \qquad\quad +18}$
    $\dfrac{22}{2} > \dfrac{2n}{2}$
    $11 > n$ or $n < 11$

12. $8t - 5 \leq 2t + 1$
    $\underline{-2t \qquad\quad -2t}$
    $6t - 5 \leq 1$
    $\underline{\;+5 \qquad\quad +5}$
    $\dfrac{6t}{6} \leq \dfrac{6}{6}$
    $t \leq 1$

Lesson 17 Exercise, pg. 209

1.
$$\begin{array}{r} x + 5 \\ x + 2 \\ \hline 2x + 10 \\ x^2 + 5x \\ \hline x^2 + 7x + 10 \end{array}$$

2.
$$\begin{array}{r} x + 3 \\ x + 1 \\ \hline x + 3 \\ x^2 + 3x \\ \hline x^2 + 4x + 3 \end{array}$$

3.
$$\begin{array}{r} x + 2 \\ x + 6 \\ \hline 6x + 12 \\ x^2 + 2x \\ \hline x^2 + 8x + 12 \end{array}$$

4.
$$\begin{array}{r} x - 4 \\ x - 3 \\ \hline -3x + 12 \\ x^2 - 4x \\ \hline x^2 - 7x + 12 \end{array}$$

5.
$$\begin{array}{r} x - 1 \\ x - 8 \\ \hline -8x + 8 \\ x^2 - x \\ \hline x^2 - 9x + 8 \end{array}$$

6.
$$\begin{array}{r} x - 4 \\ x - 5 \\ \hline -5x + 20 \\ x^2 - 4x \\ \hline x^2 - 9x + 20 \end{array}$$

7.
$$\begin{array}{r} x + 5 \\ x - 2 \\ \hline -2x - 10 \\ x^2 + 5x \\ \hline x^2 + 3x - 10 \end{array}$$

8.
$$\begin{array}{r} x - 6 \\ x + 7 \\ \hline +7x - 42 \\ x^2 - 6x \\ \hline x^2 + x - 42 \end{array}$$

9.
$$\begin{array}{r} x + 12 \\ x - 10 \\ \hline -10x - 120 \\ x^2 + 12x \\ \hline x^2 + 2x - 120 \end{array}$$

10.
$$\begin{array}{r} x + 8 \\ x - 8 \\ \hline -8x - 64 \\ x^2 + 8x \\ \hline x^2 \quad\;\; - 64 \end{array}$$

11.
$$\begin{array}{r} x + 10 \\ x - 10 \\ \hline -10x - 100 \\ x^2 + 10x \\ \hline x^2 \quad\;\; - 100 \end{array}$$

12.
$$\begin{array}{r} x - 3 \\ x + 3 \\ \hline +3x - 9 \\ x^2 - 3x \\ \hline x^2 \quad\;\; - 9 \end{array}$$

13.
$$\begin{array}{r} x - 7 \\ x + 3 \\ \hline +3x - 21 \\ x^2 - 7x \\ \hline x^2 - 4x - 21 \end{array}$$

14.
$$\begin{array}{r} x + 6 \\ x - 9 \\ \hline -9x - 54 \\ x^2 + 6x \\ \hline x^2 - 3x - 54 \end{array}$$

15.
$$\begin{array}{r} x - 11 \\ x + 6 \\ \hline +6x - 66 \\ x^2 - 11x \\ \hline x^2 - 5x - 66 \end{array}$$

Lesson 18 Exercise, pg. 210

1. $3x + 9 = 3(x + 3)$
2. $8w - 12 = 4(2w - 3)$
3. $10c - 5d = 5(2c - d)$
4. $4x - 16 = 4(x - 4)$
5. $7a + 21 = 7(a + 3)$
6. $5y - 20z = 5(y - 4z)$
7. $36f - 12 = 12(3f - 1)$
8. $9m + 21 = 3(3m + 7)$
9. $6w - 9 = 3(2w - 3)$
10. $c^2 + 8c = c(c + 8)$
11. $m^2 + 6m = m(m + 6)$
12. $x^2 - 5x = x(x - 5)$
13. $p^2 + 10p = p(p + 10)$
14. $a^2 - a = a(a - 1)$
15. $n^2 - 2n = n(n - 2)$
16. $e^2 + 7e = e(e + 7)$
17. $t^2 - 3t = t(t - 3)$
18. $y^2 - y = y(y - 1)$

Lesson 19 Exercise, pg. 213

1. $x^2 + 6x + 8 = (x + 4)(x + 2)$
2. $x^2 + 8x + 7 = (x + 7)(x + 1)$
3. $x^2 + 13x + 40 = (x + 8)(x + 5)$
4. $x^2 - 13x + 36 = (x - 4)(x - 9)$
5. $x^2 - 12x + 36 = (x - 6)(x - 6)$
6. $x^2 - 10x + 9 = (x - 1)(x - 9)$
7. $x^2 + 5x - 14 = (x + 7)(x - 2)$
8. $x^2 + 2x - 8 = (x - 2)(x + 4)$
9. $x^2 + x - 56 = (x + 8)(x - 7)$
10. $x^2 - 36 = (x + 6)(x - 6)$
11. $x^2 - 4 = (x + 2)(x - 2)$
12. $x^2 - 81 = (x + 9)(x - 9)$
13. $x^2 - 3x - 10 = (x - 5)(x + 2)$
14. $x^2 - 4x - 96 = (x + 8)(x - 12)$
15. $x^2 - x - 12 = (x - 4)(x + 3)$

Lesson 20 Exercise, pg. 214

1. $x^2 + 7x + 10 = 0$
$(x + 2)(x + 5) = 0$
$$\begin{array}{ll} x + 2 = 0 & x + 5 = 0 \\ \underline{-2 = -2} & \underline{-5 \;\; -5} \\ x\;\;\;\; = -2 \text{ and } x\;\;\;\; = -5 \end{array}$$

2. $x^2 + 10x + 9 = 0$
$(x + 1)(x + 9) = 0$
$$\begin{array}{ll} x + 1 = 0 & x + 9 = 0 \\ \underline{-1 = -1} & \underline{-9 \;\; -9} \\ x\;\;\;\; = -1 \text{ and } x\;\;\;\; = -9 \end{array}$$

3. $x^2 + 15x + 56 = 0$
$(x + 8)(x + 7) = 0$
$$\begin{array}{ll} x + 8 = 0 & x + 7 = 0 \\ \underline{-8 = -8} & \underline{-7 \;\; -7} \\ x\;\;\;\; = -8 \text{ and } x\;\;\;\; = -7 \end{array}$$

4. $x^2 - 7x + 12 = 0$
$(x - 3)(x - 4) = 0$
$$\begin{array}{ll} x - 3 = 0 & x - 4 = 0 \\ \underline{+3 = +3} & \underline{+4 \;\; +4} \\ x\;\;\;\; = 3 \text{ and } x\;\;\;\; = +4 \end{array}$$

5. $x^2 - 17x + 60 = 0$
$(x - 5)(x - 12) = 0$
$$\begin{array}{ll} x - 5 = 0 & x - 12 = 0 \\ \underline{+5 = +5} & \underline{+12 \;\; +12} \\ x\;\;\;\; = 5 \text{ and } x\;\;\;\; = 12 \end{array}$$

6. $x^2 - 12x + 27 = 0$
   $(x - 9)(x - 3) = 0$
   $x - 9 = 0 \qquad x - 3 = 0$
   $\underline{+9 = +9} \qquad \underline{+3 \quad +3}$
   $x \quad = 9$ and $x \quad = 3$

7. $x^2 + 2x - 24 = 0$
   $(x + 6)(x - 4) = 0$
   $x + 6 = 0 \qquad x - 4 = 0$
   $\underline{-6 = -6} \qquad \underline{+4 \quad +4}$
   $x \quad = -6$ and $x \quad = 4$

8. $x^2 + x - 72 = 0$
   $(x - 8)(x + 9) = 0$
   $x - 8 = 0 \qquad x + 9 = 0$
   $\underline{+8 = +8} \qquad \underline{-9 \quad -9}$
   $x \quad = 8$ and $x \quad = -9$

9. $x^2 + 7x - 30 = 0$
   $(x + 10)(x - 3) = 0$
   $x + 10 = 0 \qquad x - 3 = 0$
   $\underline{-10 = -10} \qquad \underline{+3 \quad +3}$
   $x \quad = -10$ and $x \quad = +3$

10. $x^2 - 4x - 21 = 0$
    $(x - 7)(x + 3) = 0$
    $x - 7 = 0 \qquad x + 3 = 0$
    $\underline{+7 = +7} \qquad \underline{-3 \quad -3}$
    $x \quad = 7$ and $x \quad = -3$

11. $x^2 - 5x - 6 = 0$
    $(x + 1)(x - 6) = 0$
    $x + 1 = 0 \qquad x - 6 = 0$
    $\underline{-1 = -1} \qquad \underline{+6 \quad +6}$
    $x \quad = -1$ and $x \quad = 6$

12. $x^2 - 10x - 24 = 0$
    $(x - 12)(x + 2) = 0$
    $x - 12 = 0 \qquad x + 2 = 0$
    $\underline{+12 = +12} \qquad \underline{-2 \quad -2}$
    $x \quad = 12$ and $x \quad = -2$

13. $x^2 - 49 = 0$
    $(x + 7)(x - 7) = 0$
    $x + 7 = 0 \qquad x - 7 = 0$
    $\underline{-7 = -7} \qquad \underline{+7 \quad +7}$
    $x \quad = -7$ and $x \quad = 7$

14. $x^2 - 1 = 0$
    $(x + 1)(x - 1) = 0$
    $x + 1 = 0 \qquad x - 1 = 0$
    $\underline{-1 = -1} \qquad \underline{+1 \quad +1}$
    $x \quad = -1$ and $x \quad = 1$

15. $x^2 - 144 = 0$
    $(x + 12)(x - 12) = 0$
    $x + 12 = 0 \qquad x - 12 = 0$
    $\underline{-12 = -12} \qquad \underline{+12 \quad +12}$
    $x \quad = -12$ and $x \quad = 12$

Lesson 21 Exercise, pg. 216

1. $\sqrt{8} = \sqrt{4} \cdot \sqrt{2} = \mathbf{2\sqrt{2}}$
2. $\sqrt{75} = \sqrt{25} \cdot \sqrt{3} = \mathbf{5\sqrt{3}}$
3. $\sqrt{18} = \sqrt{9} \cdot \sqrt{2} = \mathbf{3\sqrt{2}}$
4. $\sqrt{24} = \sqrt{4} \cdot \sqrt{6} = \mathbf{2\sqrt{6}}$
5. $\sqrt{72} = \sqrt{36} \cdot \sqrt{2} = \mathbf{6\sqrt{2}}$
6. $\sqrt{54} = \sqrt{9} \cdot \sqrt{6} = \mathbf{3\sqrt{6}}$
7. $\sqrt{500} = \sqrt{100} \cdot \sqrt{5} = \mathbf{10\sqrt{5}}$
8. $\sqrt{45} = \sqrt{9} \cdot \sqrt{5} = \mathbf{3\sqrt{5}}$
9. $\sqrt{96} = \sqrt{16} \cdot \sqrt{6} = \mathbf{4\sqrt{6}}$
10. $\sqrt{128} = \sqrt{64} \cdot \sqrt{2} = \mathbf{8\sqrt{2}}$

Level 2 Review, pg. 216

1. $24 = c - 6$
   $\underline{+6 \quad +6}$
   $30 = c$

2. $\dfrac{8m}{8} = \dfrac{12}{8}$
   $m = \dfrac{3}{2}$ or $1\dfrac{1}{2}$

3. $10n - 2 = 3$
   $\underline{\quad +2 \quad +2}$
   $\dfrac{10n}{10} = \dfrac{5}{10}$
   $n = \dfrac{1}{2}$

4. $15 = 9x - 3$
   $\underline{+3 \qquad +3}$
   $\dfrac{18}{9} = \dfrac{9x}{9}$
   $2 = x$

5. $9y - 2y = 35$
   $\dfrac{7y}{7} = \dfrac{35}{7}$
   $y = 5$

6. $4a + 2 + 3a = a + 20$
   $7a + 2 \quad = a + 20$
   $\underline{-a \qquad\qquad -a}$
   $6a + 2 \quad = \quad 20$
   $\underline{\;\; -2 \qquad\qquad -2}$
   $\dfrac{6a}{6} = \dfrac{18}{6}$
   $a \quad = \quad 3$

7. $4(p - 7) = p + 5$
   $4p - 28 = p + 5$
   $\underline{-p \qquad\quad -p}$
   $3p - 28 = \quad 5$
   $\underline{\;\; +28 \quad\; +28}$
   $\dfrac{3p}{3} = \dfrac{33}{3}$
   $p = 11$

8. $7x - 3 = 5(x + 5)$
   $7x - 3 = 5x + 25$
   $\underline{-5x \qquad -5x}$
   $2x - 3 = \quad 25$
   $\underline{\;\; +3 \qquad\; +3}$
   $\dfrac{2x}{2} = \dfrac{28}{2}$
   $x \quad = \quad 14$

9. $6w - 4 \leq 17$
$+4 +4$
$\dfrac{6w}{6} \leq \dfrac{21}{6}$
$w \leq 3\dfrac{3}{6}$
$w \leq 3\dfrac{1}{2}$

10. $6n - 5 > 4n + 21$
$-4n -4n$
$2n - 5 > 21$
$+5 +5$
$\dfrac{2n}{2} > \dfrac{26}{2}$
$n > 13$

11. $(a + 9)(a + 1) = a^2 + 10a + 9$
$a + 9$
$a + 1$
$+\, \overline{a + 9}$
$a^2 + 9a$
$\overline{a^2 + 10a + 9}$

12. $(y - 8)(y + 7) = y^2 - y - 56$
$y - 8$
$y + 7$
$+\, \overline{7y - 56}$
$y^2 - 8y$
$\overline{y^2 - y - 56}$

13. $m^2 + 12m = m(m + 12)$

14. $c^2 - 6c = c(c - 6)$

15. $x^2 + 11x + 30 = (x + 5)(x + 6)$

16. $x^2 + 6x - 16 = (x + 8)(x - 2)$

17. $x^2 - 100 = 0$
$(x + 10)(x - 10) = 0$
$x + 10 = 0 \quad x - 10 = 0$
$-10 -10 \quad +10 +10$
$\overline{x = -10} \text{ and } \overline{x = +10}$

18. $x^2 - 6x - 27 = 0$
$(x - 9)(x + 3) = 0$
$x - 9 = 0 \quad x + 3 = 0$
$+9 +9 \quad -3 -3$
$\overline{x = 9} \text{ and } \overline{x = -3}$

19. $\sqrt{12} = \sqrt{4} \cdot \sqrt{3} = 2\sqrt{3}$

20. $\sqrt{28} = \sqrt{4} \cdot \sqrt{7} = 2\sqrt{7}$

Lesson 22 Exercise, pg. 219

1. $x + 2$
2. $10 - x$
3. $x + 13$ or $13 + x$
4. $x + 12$
5. $8x$
6. $\dfrac{x}{4}$
7. $x - 19$
8. $40 + x$ or $x + 40$
9. $6x$
10. $\dfrac{3}{8}x$
11. $\dfrac{16}{x}$
12. $x - 1$
13. $14 - x$
14. $x + 100$ or $100 + x$
15. $x - 3$

Lesson 23 Exercise, pg. 220

1. $x + 5x$
2. $\dfrac{1}{3}x - \dfrac{1}{6}x$
3. $6(x + 8)$
4. $10x + 1$
5. $\dfrac{1}{2}x - 9$
6. $20 - 2x$
7. $\dfrac{(x + 7)}{3}$ or $\dfrac{x + 7}{3}$
8. $4(x + 2x)$
9. $\dfrac{3}{4}(x - 11)$
10. $\dfrac{3}{4}x - 11$

Lesson 24 Exercise, pg. 221

1. $\dfrac{1}{4}p$
2. a. $x - 5$
   b. $x + 3$
3. $4r$
4. $t - 45$
5. $\dfrac{l}{3}$
6. $2x$
7. $0.22g$
8. $1\dfrac{1}{2}w$ or $\dfrac{3}{2}w$
9. $\dfrac{s}{35}$
10. $x - 32$

Lesson 25 Exercise, pg. 222

1. $x + 15 = 21$
$-15 -15$
$\overline{x = 6}$

2. $\dfrac{6x}{6} = \dfrac{72}{6}$
$x = 12$

344 Answers and Solutions

3. $\dfrac{25}{1} \cdot \dfrac{x}{25} = 6 \cdot 25$
   $x = \mathbf{150}$

4. $30 = x - 27$
   $\underline{+27 \quad\quad +27}$
   $57 = x$

5. $4x + 9 = 33$
   $\underline{-9 \quad -9}$
   $\dfrac{4x}{4} = \dfrac{24}{4}$
   $x = \mathbf{6}$

6. $7x - 6 = 78$
   $\underline{+6 \quad +6}$
   $\dfrac{7x}{7} = \dfrac{84}{7}$
   $x = \mathbf{12}$

7. $\dfrac{x}{5} + 2 = 8$
   $\underline{-2 \quad -2}$
   $5 \cdot \dfrac{x}{5} = 6 \cdot 5$
   $x = \mathbf{30}$

8. $77 = 8x - 3$
   $\underline{+3 \quad\quad +3}$
   $\dfrac{80}{8} = \dfrac{8x}{8}$
   $10 = x$

9. $\dfrac{1}{2}x - 5 = 35$
   $\underline{+5 \quad +5}$
   $2 \cdot \dfrac{1}{2}x = 40 \cdot 2$
   $x = \mathbf{80}$

10. $7x - x = 78$
    $\dfrac{6x}{6} = \dfrac{78}{6}$
    $x = \mathbf{13}$

11. $9x - 3 = 2x + 25$
    $\underline{-2x \quad\quad -2x}$
    $7x - 3 = 25$
    $\underline{+3 \quad\quad +3}$
    $\dfrac{7x}{7} = \dfrac{28}{7}$
    $x = \mathbf{4}$

12. $10x - 5 = 2x + 19$
    $\underline{-2x \quad\quad -2x}$
    $8x - 5 = 19$
    $\underline{+5 \quad\quad +5}$
    $\dfrac{8x}{8} = \dfrac{24}{8}$
    $x = \mathbf{3}$

13. $12x = 3x + 45$
    $\underline{-3x \quad -3x}$
    $\dfrac{9x}{9} = \dfrac{45}{9}$
    $x = \mathbf{5}$

14. $5(x - 2) = 30$
    $5x - 10 = 30$
    $\underline{+10 \quad +10}$
    $\dfrac{5x}{5} = \dfrac{40}{5}$
    $x = \mathbf{8}$

15. $7(x - 1) = 4(x + 2)$
    $7x - 7 = 4x + 8$
    $\underline{-4x \quad\quad -4x}$
    $3x - 7 = 8$
    $\underline{+7 \quad\quad +7}$
    $\dfrac{3x}{3} = \dfrac{15}{3}$
    $x = \mathbf{5}$

Lesson 26 Exercise, pg. 224

1. $d = rt$
   $\dfrac{210}{42} = \dfrac{42t}{42}$
   $\mathbf{5\ hr} = t$

2. $d = rt$
   $\dfrac{1194}{3} = \dfrac{r \cdot 3}{3}$
   $\mathbf{398\ mph} = r$

3. $d = rt$
   $\dfrac{27}{6} = \dfrac{r \cdot 6}{6}$
   $\mathbf{4\dfrac{1}{2}\ mph} = r$

4. $d = rt$
   $\dfrac{135}{54} = \dfrac{54t}{54}$
   $\mathbf{2\dfrac{1}{2}\ hr} = t$

5. $d = rt$
   $\dfrac{496}{8} = \dfrac{r \cdot 8}{8}$
   $\mathbf{62\ mph} = r$

6. $c = nr$
   $\dfrac{4.17}{3} = \dfrac{3r}{3}$
   $\mathbf{\$1.39} = r$

7. $c = nr$
   $\dfrac{528}{22} = \dfrac{n \cdot 22}{22}$
   $\mathbf{24\ sweaters} = n$

8. $c = nr$
   $\dfrac{420}{12} = \dfrac{12r}{12}$
   $\mathbf{\$35} = r$

9. $c = nr$
   $\dfrac{5625}{4.5} = \dfrac{4.5r}{4.5}$
   $\mathbf{\$1250} = r$
   $4.5\overline{)5625.0}$

Answers and Solutions

10. $c = nr$
$$\frac{3380}{6.50} = \frac{n \cdot 6.50}{6.50}$$
$$\mathbf{5\ 20} = n$$
$6.50\overline{)3380.00}$

11. $i = prt$
$$\frac{51}{850} = \frac{850 \cdot r \cdot 1}{850}$$
$$r = .06 = \mathbf{6\%}$$
$850\overline{)51.00}$

12. $8\% = 0.08$
$$i = prt$$
$$14 = 350 \cdot 0.08 \cdot t$$
$$\frac{14}{28} = \frac{28t}{28}$$
$$\tfrac{1}{2}\ \text{yr} = t = \mathbf{6\ months}$$

13. $12\% = 0.12$
$$i = prt$$
$$\frac{4800}{0.12} = \frac{p \cdot 0.12 \cdot 1}{0.12}$$
$$\mathbf{\$40{,}0\ 00} = p$$
$0.12\overline{)480\ 0.00}$

14. $6\ \text{mo} = \tfrac{1}{2}\ \text{yr}$
$$i = prt$$
$$27 = 600 \cdot r \cdot \tfrac{1}{2}$$
$$\frac{27}{300} = \frac{300r}{300}$$
$$\frac{9}{100} = \mathbf{9\%}$$

15. $9\% = 0.09$
$$i = prt$$
$$36 = 600 \cdot .09t$$
$$\frac{36}{54} = \frac{54t}{54}$$
$$\tfrac{2}{3} = t$$
$$\tfrac{2}{\cancel{3}_1} \times \tfrac{\cancel{12}^4}{1} = \mathbf{8\ months}$$

Lesson 27 Exercise, pg. 226

1. small no. $= x$
   large no. $= 3x$
   $3x - 10 = x + 18$
   $\ -x \ -x$
   $2x - 10 = \ 18$
   $\ +10\ +10$
   $\frac{2x}{2} = \frac{28}{2}$
   $x = \mathbf{14}$
   $3x = 3(14) = \mathbf{42}$

2. small no. $= x$
   large no. $= x + 7$
   $3x - 4 = 2(x + 7)$
   $3x - 4 = 2x + 14$
   $-2x -2x$
   $x - 4 = 14$
   $+4 +4$
   $x = \mathbf{18}$
   $x + 7 = 18 + 7 = \mathbf{25}$

3. daughter's age $= x$
   Louise's age $= x + 26$
   $x + 26 = 4x + 2$
   $-x -x$
   $26 = 3x + 2$
   $-2 - 2$
   $\frac{24}{3} = \frac{3x}{3}$
   $\mathbf{8} = x$
   $x + 26 = 8 + 26 = \mathbf{34}$

4. Grandfather's age $= x$
   Douglas's age $= x - 46$
   $3(x - 46) = x - 8$
   $3x - 138 = x - 8$
   $-x -x$
   $2x - 138 = - 8$
   $+138 +138$
   $\frac{2x}{2} = \frac{130}{2}$
   $x = \mathbf{65}$
   $x - 46 = 65 - 46 = \mathbf{19}$

5. deductions $= x$
   take-home $= 5x$
   $x + 5x = \$324$
   $\frac{6x}{6} = \frac{\$324}{6}$
   $x = \mathbf{\$54}$
   $5x = 5(\$54) = \mathbf{\$270}$

6. no. of men $= x$
   no. of women $= x + 15$
   $x + x + 15 = 47$
   $2x + 15 = 47$
   $-15 -15$
   $\frac{2x}{2} = \frac{32}{2}$
   $x = \mathbf{16\ men}$
   $x + 15 = 16 + 15 = \mathbf{31\ women}$

7. Assistant's wages $= x$
   Joe's wages $= x + 3$
   $10x + 10(x + 3) = 350$
   $10x + 10x + 30 = 350$
   $20x + 30 = 350$
   $-30 -30$
   $\frac{20x}{20} = \frac{320}{20}$
   $x = \mathbf{\$16}$
   $x + 3 = \$16 + 3 = \mathbf{\$19}$

8. Jim's hours $= x$
   Carmen's hours $= x + 5$
   George's hours $= 2(x + 5)$
   $x + x + 5 + 2(x + 5) = 95$
   $2x + 5 + 2x + 10 = 95$
   $4x + 15 = 95$
   $ -15 -15$
   $\dfrac{4x}{4} = \dfrac{80}{4}$
   $x = \textbf{20 hrs}$
   $x + 5 = 20 + 5 = \textbf{25 hrs}$
   $2(x + 5) = 2(20 + 5) = \textbf{50 hrs}$

9.

	age now	age in 5 yrs
Andy	x	$x + 5$
Chris	$x + 3$	$x + 3 + 5 = x + 8$

   $3(x + 5) = 2(x + 8) + 6$
   $3x + 15 = 2x + 16 + 6$
   $3x + 15 = 2x + 22$
   $-2x -2x$
   $x + 15 = 22$
   $ -15 -15$
   $x = 7$
   $x + 3 = 7 + 3 = \textbf{10 yrs}$
   Andy is 7 years old and Chris is 10 years old.

10. food $= 3x$
    car $= 2x$
    rent $= 4x$
    $3x + 2x + 4x = 648$
    $\dfrac{9x}{9} = \dfrac{648}{9}$
    $x = \textbf{\$72}$
    $4x = 4(\$72) = \textbf{\$288}$

Level 3 Review, pg. 226

1. $x/15$ or $\dfrac{x}{15}$
2. $20x$
3. $4(x - 7)$
4. $\dfrac{2x - 3}{5}$
5. $p + 1$
6. $x - 12$
7. $t - 8$
8. $0.3m$
9. $6x + 5 = 17$
   $ -5 -5$
   $\dfrac{6x}{6} = \dfrac{12}{6}$
   $x = 2$
10. $43 = 9x - 2$
    $+2 +2$
    $\dfrac{45}{9} = \dfrac{9x}{9}$
    $5 = x$

11. $4(x - 1) = x + 2$
    $4x - 4 = x + 2$
    $-x -x$
    $3x - 4 = 2$
    $ +4 +4$
    $\dfrac{3x}{3} = \dfrac{6}{3}$
    $x = 2$

12. $2(x - 3) = x + 7$
    $2x - 6 = x + 7$
    $-x -x$
    $x - 6 = 7$
    $+6 +6$
    $x = 13$

13. $d = rt$
    $\dfrac{217}{62} = \dfrac{62t}{62}$
    $3\dfrac{1}{2}$ hr $= t$

14. $c = nr$ or $c = nr$
    $1.26 = 1\dfrac{1}{2} \cdot r$ $\$1.26 = 1.5r$
    $\dfrac{2}{3} \times \dfrac{\overset{0.42}{\cancel{1.26}}}{1} = \dfrac{3}{2}r \times \dfrac{2}{3}$ $\dfrac{\$1.26}{1.5} = \dfrac{1.5r}{1.5}$
    $\$0.84 = r$ or $\$0.84 = r$

15. $12\% = 0.12$ and 6 mo $= \dfrac{1}{2}$ yr
    $i = prt$
    $48 = p \cdot 0.12 \cdot \dfrac{1}{2}$
    $\dfrac{48}{0.06} = \dfrac{p \cdot 0.06}{0.06}$
    $\textbf{\$8 00} = p$
    $0.06 \overline{)48.00}$

16. $15\% = 0.15$
    $i = prt$
    $135 = 1200 \cdot 0.15 \cdot t$
    $\dfrac{135}{180} = \dfrac{180t}{180}$
    $\dfrac{3}{4} = t$
    $\dfrac{3}{\cancel{4}} \times \dfrac{\overset{3}{\cancel{12}}}{1} = \textbf{9 mo.}$

17. small no. $= x$
    large no. $= 5x$
    $5x - 9 = 2(x + 3)$
    $5x - 9 = 2x + 6$
    $-2x -2x$
    $3x - 9 = 6$
    $+9 +9$
    $\dfrac{3x}{3} = \dfrac{15}{3}$
    $x = 5$
    $5x = 5(5) = \textbf{25}$

18. boys = x
    girls = $x + 9$
    $x + x + 9 = 65$
    $2x + 9 = 65$
    $ -9 -9$
    $\dfrac{2x}{2} = \dfrac{56}{2}$
    $x = $ **28 boys**
    $x + 9 = 28 + 9 = $ **37 girls**

19. Tom's hours = x
    Bill's hours = $x + 10$
    Fred's hours = $2x$
    $x + x + 10 + 2x = 94$
    $4x + 10 = 94$
    $ -10 -10$
    $\dfrac{4x}{4} = \dfrac{84}{4}$
    $x = 21$
    Fred's hours = $2(21) = $ **42 hrs**

20.
	age now	age in 8 years
Rachel	x	$x + 8$
Sarah	$2x$	$2x + 8$

$3(2x + 8) = 5(x + 8) - 10$
$6x + 24 = 5x + 40 - 10$
$6x + 24 = 5x + 30$
$-5x -5x$
$x + 24 = 30$
$ -24 -24$
$x = 6$
Sarah = $2(6) = $ **12 yrs**

Chapter 7 GEOMETRY

Level 1 Preview, pg. 230

1. **acute**
2. **obtuse**
3. $90°$
   $-67°$
   23°
4. $180°$
   $-72°$
   108°
5. Because $\angle b$ and $\angle f$ are corresponding, $\angle f = $ **52°**.
6. $\angle f$
7. $180°$
   $-52°$
   128°
8. $56°180°$
   $+62°-118°$
   $\overline{118°}\overline{62°} = \angle F$
   Because two angles are equal, $\triangle DEF$ is **isosceles**.

9. $48°180°$
   $+42°-90°$
   $\overline{90°}\overline{90°} = \angle M$
   Because $\angle M = 90$, $\triangle KLM$ is a **right triangle**.
10. $46°180°$
    $+54°-100°$
    $\overline{100°}\overline{80°} = \angle Y$
    Because $\angle Y$ is the largest angle, **side WX is the longest**.

Lesson 1 Exercise, pg. 232

1. a. **acute**   b. **right**   c. **obtuse**   d. **reflex**
2. a. **straight**   b. **reflex**   c. **right**   d. **acute**
3. a. **acute**   b. **acute**   c. **obtuse**   d. **right**
4. a. **reflex**   b. **obtuse**   c. **straight**   d. **acute**

Lesson 2 Exercise, pg. 233

1. $90°$
   $-22°$
   68°
2. $180°$
   $-22°$
   158°
3. $180°$
   $-43°$
   137°
4. $180°$
   $-106°$
   74°
5. $\angle b$
6. $\angle c$
7. **106°**
8. **74°**
9. $\angle COD$
10. $\angle AOD$
11. $180°$
    $-58°$
    122°
12. **72°**
13. $180°\dfrac{3m}{3} = \dfrac{117°}{3}$
    $-63°$
    $\overline{117°}m = $ **39°**
14. small angle = x
    large angle = $x + 20$
    $x + x + 20 = 90$
    $2x + 20 = 90$
    $ -20 -20$
    $\dfrac{2x}{2} = \dfrac{70}{2}$
    $x = $ **35°**
    $x + 20 = 35 + 20 = $ **55°**

15. small angle = x
large angle = $4x$
$x + 4x = 180°$
$$\frac{5x}{5} = \frac{180}{5}$$
$x = 36°$
$4x = 4(36) = $ **144°**

Lesson 3 Exercise, pg. 235

1. ∠**t**
2. ∠**s**
3. ∠**q**
4. ∠**p**
5. **118°**
6. $180°$
   $- 49°$
   131°
7. **vertical**
8. **alternate interior**
9. **alternate exterior**
10. **corresponding**

Lesson 4 Exercise, pg. 237

1. $32°$ $180°$
   $+ 74°$ $- 106°$
   $106°$ $74° = ∠M$
   Because two angles are equal, **the triangle is isosceles.**

2. $35°$ $180°$
   $+ 55°$ $- 90°$
   $90°$ $90° = ∠F$
   Because there is one right angle, **the triangle is right.**

3. $42°$ $180°$
   $+ 42°$ $- 84°$
   $84°$ **96°**

4. $180°$ **51°**
   $- 78°$ $2\overline{)102}$
   $102°$

5. $90°$ $180°$
   $+ 33°$ $- 123°$
   $123°$ **57°**

6. $30°$ $180°$
   $+ 60°$ $- 90°$
   $90°$ $90° = ∠C$
   Because ∠C is the largest angle, **side AB is the longest side.**

7. 6 in.   24 in.
   $+ 8$ $- 14$
   14 in.   10 in. = side DF
   Because side DF is the longest, **∠E is the largest angle.**

8. $48°$ $180°$
   $+ 66°$ $- 114°$
   $114°$ $66° = ∠Z$
   Because ∠X is the smallest angle, **side YZ is the shortest side.**

9. Because side NO is opposite the right angle, **side NO is the hypotenuse.**

10. $60°$ $180°$
    $+ 60°$ $- 120°$
    $120°$ $60°$
    Because the three angles are equal, **the triangle is equiangular.**

Level 1 Review, pg. 238

1. **acute**
2. **reflex**
3. $180°$
   $- 84°$
   96°
4. $90°$
   $- 36°$
   54°
5. Because t and u are vertical, **∠u = 112°.**
6. ∠z
7. $180°$
   $- 112°$
   68°
8. $33°$ $180°$
   $+ 57°$ $- 90°$
   $90°$ $90° = ∠C$
   Because side AB is opposite the 90° angle, **side AB is called the hypotenuse.**
9. $49°$ $180°$
   $+ 82°$ $- 131°$
   $131°$ $49° = ∠z$
   Because two angles are equal, **the triangle is isosceles.**
10. $55°$ $180°$
    $+ 47°$ $- 102°$
    $102°$ $78° = ∠R$
    Because ∠R is the largest angle, **side PQ is the longest.**

Lesson 5 Exercise, pg. 242

1. **Yes.**   ∠A =   80°   180°
         ∠B = $+ 60°$ $- 140°$
                140°   40° = ∠C
         ∠D =   80°   180°
         ∠F = $+ 40°$ $- 120°$
                120°   60° = ∠E

   Because the angles in △ABC are the same as the angles in △DEF, the triangles are similar.

2. **No.** $\angle G = 65°$ $180°$
   $\angle H = +55°$ $-120°$
   $\quad\quad\ \ 120°$ $\ \ 60° = \angle I$
   $\angle J = 60°$ $180°$
   $\angle K = +50°$ $-110°$
   $\quad\quad\ \ 110°$ $\ \ 70° = \angle L$

   Because the angles in $\triangle GHI$ are not the same as the angles in $\triangle JKL$, the triangles are not similar.

3. **Yes.** The ratio of the short side to the long side in rectangle *MNOP* is $3:4$.

   The ratio of the short side to the long side in rectangle *QRST* is $9:12 = 3:4$.

   Because the ratios of the sides are the same, the rectangles are similar.

4. **No.** The ratio of the short side to the long side in the small rectangle is $5:6$.

   The ratio of the short side to the long side in the large rectangle is $10:11$.

   Because the ratios of the sides are not the same, the rectangles are not similar.

5. $\dfrac{\text{width}}{\text{length}} = \dfrac{4}{5} = \dfrac{20}{x}$

   $\dfrac{4x}{4} = \dfrac{100}{4}$

   $x = \mathbf{25}$ **in. long**

6. $\dfrac{\text{short leg}}{\text{long leg}} = \dfrac{18}{30} = \dfrac{x}{25}$

   $\dfrac{30x}{30} = \dfrac{450}{30}$

   $x = \mathbf{15}$

7. $\dfrac{\text{height}}{\text{shadow}} = \dfrac{5}{3} = \dfrac{x}{72}$

   $\dfrac{3x}{3} = \dfrac{360}{3}$

   $x = \mathbf{120}$ **ft**

8. $\dfrac{\text{long leg}}{\text{short leg}} = \dfrac{10}{3} = \dfrac{x}{24}$

   $\dfrac{3x}{3} = \dfrac{240}{3}$

   $x = \mathbf{80}$ **ft**

9. $\dfrac{\text{height}}{\text{width}} = \dfrac{6}{3} = \dfrac{7}{x}$

   $\dfrac{6x}{6} = \dfrac{21}{6}$

   $x = \mathbf{3\dfrac{1}{2}}$ **ft**

10. $\dfrac{\text{short leg}}{\text{long leg}} = \dfrac{5}{8} = \dfrac{x}{48}$

    $\dfrac{8x}{8} = \dfrac{240}{8}$

    $x = \mathbf{30}$ **cm**

Lesson 6 Exercise, pg. 245

1. **Yes.** The conditions satisfy the SAS requirement.

2. **Yes.** The conditions satisfy the ASA requirement.

3. **No.** Perimeter of triangle on the left is $6 + 8 + 10 = 24$ in.

   Missing side of triangle on the right is $23 - 6 - 10 = 7$ in.

   The conditions fail to satisfy the SSS requirement.

4. **No.** In the triangle on the left one leg is 4 inches.

   In the triangle on the right the hypotenuse is 4 inches. Because corresponding sides are not equal, the conditions fail to satisfy the SAS requirement.

5. **Yes.** The conditions satisfy the ASA requirement.

6. **Yes.** Perimeter of triangle at the left is $4 + 4 + 6 = 14$.

   Missing side of triangle at right is $14 - 4 - 6 = 4$. Because the sides are the same, the conditions satisfy the SSS requirement.

7. **(4)** $\angle B = \angle E$ This condition satisfies the ASA requirement.

8. **(1)** $GI = JL$ This condition satisfies the SSS requirement.

9. **(2) B only** This condition satisfies the SSS requirement.

10. **(5) A or C** Condition A satisfies the SAS requirement.

    Condition C satisfies the ASA requirement.

Lesson 7 Exercise, pg. 247

1. $c^2 = a^2 + b^2$
   $c^2 = 24^2 + 32^2$
   $c^2 = 576 + 1024$
   $c^2 = 1600$
   $c = \sqrt{1600}$
   $c = \mathbf{40}$ **ft**

2. $c^2 = a^2 + b^2$
   $c^2 = 36^2 + 48^2$
   $c^2 = 1296 + 2304$
   $c^2 = 3600$
   $c = \sqrt{3600}$
   $c = \mathbf{60}$ **in.**

3. $c^2 = a^2 + b^2$
   $34^2 = a^2 + 30^2$
   $1156 = a^2 + 900$
   $-\ 900 \quad\quad -\ 900$
   $\overline{256 = a^2}$
   $\sqrt{256} = a$
   $\mathbf{16} = a$

4. $c^2 = a^2 + b^2$
    $13^2 = 5^2 + b^2$
    $169 = 25 + b^2$
    $\underline{-25 \quad -25}$
    $144 = b^2$
    $\sqrt{144} = b$
    $12 \text{ ft} = b$

5. $c^2 = a^2 + b^2$
    $c^2 = 18^2 + 24^2$
    $c^2 = 324 + 576$
    $c^2 = 900$
    $c = \sqrt{900}$
    $c = 30 \text{ ft}$

6. $c^2 = a^2 + b^2$
    $c^2 = (1.2)^2 + (1.6)^2$
    $c^2 = 1.44 + 2.56$
    $c^2 = 4$
    $c = \sqrt{4}$
    $c = 2 \text{ cm}$

7. $c^2 = a^2 + b^2$
    $26^2 = a^2 + 24^2$
    $676 = a^2 + 576$
    $\underline{-576 \quad -576}$
    $100 = a^2$
    $\sqrt{100} = a$
    $10 = a$
    base $= 2 \cdot 10 = 20 \text{ in.}$

8. $c^2 = a^2 + b^2$
    $75^2 = 45^2 + b^2$
    $5625 = 2025 + b^2$
    $\underline{-2025 \quad -2025}$
    $3600 = b^2$
    $\sqrt{3600} = b$
    $60 = b$

9. $c^2 = a^2 + b^2$
    $17^2 = a^2 + 15^2$
    $289 = a^2 + 225$
    $\underline{-225 \quad -225}$
    $64 = a^2$
    $\sqrt{64} = a$
    $8 \text{ ft} = a$

10. $c^2 = a^2 + b^2$
    $c^2 = 15^2 + 20^2$
    $c^2 = 225 + 400$
    $c^2 = 625$
    $c = \sqrt{625}$
    $c = 25 \text{ mi}$

Lesson 8 Exercise, pg. 250

1. $I = (+6, +3)$ $N = (-7, 0)$
    $J = (+2, +7)$ $P = (-5, -8)$
    $K = (0, +5)$ $Q = (-3, -5)$
    $L = (-4, +6)$ $R = (0, -8)$
    $M = (-6, -1)$ $S = (+7, -4)$

2.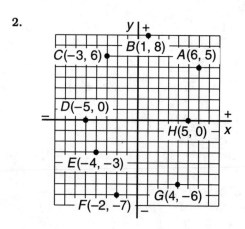

Lesson 9 Exercise, pg. 253

1. A is 5 spaces above x-axis.
    B is 3 spaces below x-axis.
    Distance AB is $5 + 3 = 8$.

2. B is 2 spaces left of y-axis.
    C is 4 spaces right of y-axis.
    Distance BC is $2 + 4 = 6$.

3. C is 3 spaces below x-axis.
    D is 1 space above x-axis.
    Distance CD is $3 + 1 = 4$.

4. $E = (x_1, y_1) = (+1, +2)$
    $F = (x_2, y_2) = (+13, +11)$
    $d = \sqrt{(x_2 - x_1)^2 + (y_2 - y_1)^2}$
    $d = \sqrt{(13 - 1)^2 + (11 - 2)^2}$
    $d = \sqrt{(12)^2 + (9)^2}$
    $d = \sqrt{144 + 81}$
    $d = \sqrt{225}$
    $d = 15$

5. $G = (x_1, y_1) = (+5, +15)$
    $H = (x_2, y_2) = (+21, +3)$
    $d = \sqrt{(x_2 - x_1)^2 + (y_2 - y_1)^2}$
    $d = \sqrt{(21 - 5)^2 + (3 - 15)^2}$
    $d = \sqrt{(16)^2 + (-12)^2}$
    $d = \sqrt{256 + 144}$
    $d = \sqrt{400}$
    $d = 20$

6. $I = (x_1, y_1) = (-5, -4)$
    $J = (x_2, y_2) = (+3, +2)$
    $d = \sqrt{(x_2 - x_1)^2 + (y_2 - y_1)^2}$
    $d = \sqrt{(3 - (-5))^2 + (2 - (-4))^2}$
    $d = \sqrt{(3 + 5)^2 + (2 + 4)^2}$
    $d = \sqrt{(8)^2 + (6)^2}$
    $d = \sqrt{64 + 36}$
    $d = \sqrt{100}$
    $d = 10$

Answers and Solutions

7. $K = (x_1, y_1) = (-2, +7)$
   $L = (x_2, y_2) = (+3, -5)$
   $d = \sqrt{(x_2 - x_1)^2 + (y_2 - y_1)^2}$
   $d = \sqrt{(3 - (-2))^2 + (-5 - (7))^2}$
   $d = \sqrt{(3 + 2)^2 + (-5 + (-7))^2}$
   $d = \sqrt{(5)^2 + (-12)^2}$
   $d = \sqrt{25 + 144}$
   $d = \sqrt{169}$
   d = 13

8. $N = (x_1, y_1) = (+3, +8)$
   $P = (x_2, y_2) = (+1, +6)$
   $M = \left(\dfrac{x_1 + x_2}{2}, \dfrac{y_1 + y_2}{2}\right)$
   $M = \left(\dfrac{3 + 1}{2}, \dfrac{8 + 6}{2}\right)$
   $M = \left(\dfrac{4}{2}, \dfrac{14}{2}\right)$
   M = (2, 7)

9. $Q = (x_1, y_1) = (-5, +2)$
   $R = (x_2, y_2) = (+3, +6)$
   $M = \left(\dfrac{x_1 + x_2}{2}, \dfrac{y_1 + y_2}{2}\right)$
   $M = \left(\dfrac{-5 + 3}{2}, \dfrac{+2 + 6}{2}\right)$
   $M = \left(\dfrac{-2}{2}, \dfrac{+8}{2}\right)$
   M = (-1, +4)

10. $S = (x_1, y_1) = (-3, -3)$
    $T = (x_2, y_2) = (+7, +5)$
    $M = \left(\dfrac{x_1 + x_2}{2}, \dfrac{y_1 + y_2}{2}\right)$
    $M = \left(\dfrac{-3 + 7}{2}, \dfrac{-3 + 5}{2}\right)$
    $M = \left(\dfrac{+4}{2}, \dfrac{+2}{2}\right)$
    M = (+2, +1)

11. $U = (x_1, y_1) = (-2, +4)$
    $V = (x_2, y_2) = (+6, -4)$
    $M = \left(\dfrac{x_1 + x_2}{2}, \dfrac{y_1 + y_2}{2}\right)$
    $M = \left(\dfrac{-2 + 6}{2}, \dfrac{+4 + (-4)}{2}\right)$
    $M = \left(\dfrac{+4}{2}, \dfrac{0}{2}\right)$
    M = (+2, 0)

Lesson 10 Exercise, pg. 257

1. For $y = x + 4$,
   when $x = 3$, $y = 3 + 4 = 7$
   when $x = -2$, $y = -2 + 4 = +2$
   when $x = -5$, $y = -5 + 4 = -1$

x	y
+3	+7
-2	+2
-5	-1

   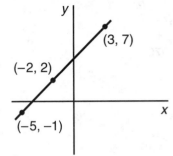

2. For $y = \dfrac{x}{2} + 1$,
   when $x = +8$, $y = \dfrac{+8}{2} + 1 = 4 + 1 = 5$
   when $x = +4$, $y = \dfrac{+4}{2} + 1 = 2 + 1 = 3$
   when $x = -6$, $y = \dfrac{-6}{2} + 1 = -3 + 1 = -2$

x	y
8	5
4	3
-6	-2

   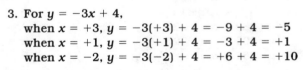

3. For $y = -3x + 4$,
   when $x = +3$, $y = -3(+3) + 4 = -9 + 4 = -5$
   when $x = +1$, $y = -3(+1) + 4 = -3 + 4 = +1$
   when $x = -2$, $y = -3(-2) + 4 = +6 + 4 = +10$

x	y
+3	-5
+1	+1
-2	+10

4. For $y = -2x - 3$,
   when $x = 2$, $y = -2(2) - 3 = -4 - 3 = -7$
   when $x = -3$, $y = -2(-3) - 3 = +6 - 3 = +3$
   when $x = -4$, $y = -2(-4) - 3 = +8 - 3 = +5$

x	y
2	−7
−3	+3
−4	+5

5. **Yes,** when $x = 1$, $y = 5x + 3 = 5(1) + 3 = 5 + 3 = 8$.
   Point $(1, 8)$ is a solution of $y = 5x + 3$.

6. **No,** when $x = 2$, $y = -3x + 1 = -3(2) + 1 = -6 + 1 = -5$.
   Point $(2, -4)$ is not a solution of $y = -3x + 1$.

7. **No,** when $x = 3$, $y = -x + 6 = -(3) + 6 = -3 + 6 = +3$.
   Point $(3, 5)$ is not on the graph.

8. **Yes,** when $x = -8$, $y = \frac{3}{4}x - 2 = \frac{3}{4}(-8) - 2 = -6 - 2 = -8$.
   Point $(-8, -8)$ is on the graph.

9. **(3) (2, −3)**
   For choice (1),
   when $x = 4$, $y = x - 5 = 4 - 5 = -1$.
     Point $(4, 0)$ is not on the graph.
   For choice (2),
   when $x = 3$, $y = x - 5 = 3 - 5 = -2$.
     Point $(3, -4)$ is not on the graph.
   For choice (3),
   when $x = 2$, $y = x - 5 = 2 - 5 = -3$.
     Point $(2, -3)$ is on the graph.
   For choice (4),
   when $x = -1$, $y = x - 5 = -1 - 5 = -6$.
     Point $(-1, -5)$ is not on the graph.

10. **(1) (3, −9)**
    For choice (1),
    when $x = 3$, $y = -2(3) - 3 = -6 - 3 = -9$.
      Point $(3, -9)$ is on the graph.
    For choice (2),
    when $x = 2$, $y = -2(2) - 3 = -4 - 3 = -7$.
      Point $(2, -4)$ is not on the graph.
    For choice (3),
    when $x = -1$, $y = -2(-1) - 3 = +2 - 3 = -1$.
      Point $(-1, +1)$ is not on the graph.
    For choice (4),
    when $x = -3$, $y = -2(-3) - 3 = +6 - 3 = +3$.
      Point $(-3, +4)$ is not on the graph.

Lesson 11 Exercise, pg. 260

1. $y = x^2 + 2$
   $y = 2^2 + 2 = 4 + 2 = 6$
   $y = 1^2 + 2 = 1 + 2 = 3$
   $y = 0^2 + 2 = 0 + 2 = 2$
   $y = (-1)^2 + 2 = 1 + 2 = 3$
   $y = (-2)^2 + 2 = 4 + 2 = 6$

x	y
2	6
1	3
0	2
−1	3
−2	6

   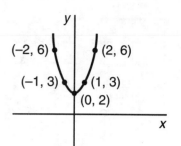

2. $y = x^2 - 2x$
   $y = 4^2 - 2(4) = 16 - 8 = 8$
   $y = 3^2 - 2(3) = 9 - 6 = 3$
   $y = 2^2 - 2(2) = 4 - 4 = 0$
   $y = 1^2 - 2(1) = 1 - 2 = -1$
   $y = 0^2 - 2(0) = 0 - 0 = 0$
   $y = (-1)^2 - 2(-1) = +1 + 2 = 3$
   $y = (-2)^2 - 2(-2) = +4 + 4 = 8$

x	y
4	8
3	3
2	0
1	−1
0	0
−1	3
−2	8

   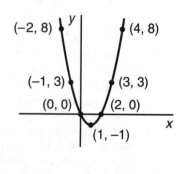

3. $y = x^2 + x - 2$
   $y = 2^2 + 2 - 2 = 4 + 2 - 2 = 4$
   $y = 1^2 + 1 - 2 = 1 + 1 - 2 = 0$
   $y = 0^2 + 0 - 2 = 0 + 0 - 2 = -2$
   $y = (-1)^2 + (-1) - 2 = 1 - 1 - 2 = -2$
   $y = (-2)^2 + (-2) - 2 = 4 + (-2) - 2 = 0$
   $y = (-3)^2 + (-3) - 2 = 9 + (-3) - 2 = 4$

x	y
2	4
1	0
0	−2
−1	−2
−2	0
−3	4

   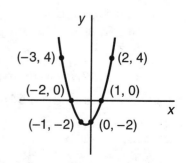

Answers and Solutions   353

4. $y = x^2 + x - 4$
   $y = 2^2 + 2 - 4 = 4 + 2 - 4 = 2$
   $y = 1^2 + 1 - 4 = 1 + 1 - 4 = -2$
   $y = 0^2 + 0 - 4 = 0 + 0 - 4 = -4$
   $y = (-1)^2 + (-1) - 4 = 1 - 1 - 4 = -4$
   $y = (-2)^2 + (-2) - 4 = 4 - 2 - 4 = -2$
   $y = (-3)^2 + (-3) - 4 = 9 - 3 - 4 = 2$

x	y
2	2
1	-2
0	-4
-1	-4
-2	-2
-3	2

Points: (-3, 2), (2, 2), (-2, -2), (1, -2), (-1, -4), (0, -4)

5. **No**, when $x = 5$, $y = 5^2 - 4(5) = 25 - 20 = 5$.
   Point (5, 4) is not on the graph.

6. **Yes**, when $x = 4$, $y = 4^2 + 4 + 3 = 16 + 4 + 3 = 23$.
   Point (4, 23) is on the graph.

7. **Yes**, when $x = -3$, $y = (-3)^2 - (-3) - 1 = 9 + 3 - 1 = 11$.
   Point (-3, 11) is on the graph.

8. **Yes**, when $x = -2$, $y = (-2)^2 + 3(-2) - 2 = 4 - 6 - 2 = -4$.
   Point (-2, -4) is on the graph.

9. **(2) (3, 15)**
   For choice (1), $y = 4^2 + 2(4) = 16 + 8 = 24$.
   Point (4, 20) is not on the graph.
   For choice (2), $y = 3^2 + 2(3) = 9 + 6 = 15$.
   Point (3, 15) is on the graph.
   For choice (3), $y = 1^2 + 2(1) = 1 + 2 = 3$.
   Point (1, 2) is not on the graph.
   For choice (4), $y = (-2)^2 + 2(-2) = 4 - 4 = 0$.
   Point (-2, 1) is not on the graph.

10. **(4) (-2, 17)**
    For choice (1),
    $y = 3^2 - 5(3) + 3 = 9 - 15 + 3 = -3$.
    Point (3, 3) is not on the graph.
    For choice (2),
    $y = 2^2 - 5(2) + 3 = 4 - 10 + 3 = -3$.
    Point (2, -4) is not on the graph.
    For choice (3),
    $y = (-1)^2 - 5(-1) + 3 = 1 + 5 + 3 = 9$.
    Point (-1, 8) is not on the graph.
    For choice (4),
    $y = (-2)^2 - 5(-2) + 3 = 4 + 10 + 3 = 17$.
    Point (-2, 17) is on the graph.

Lesson 12 Exercise, pg. 264

1. $C = (x_1, y_1) = (4, 5)$
   $D = (x_2, y_2) = (6, 9)$
   $m = \dfrac{y_2 - y_1}{x_2 - x_1} = \dfrac{9 - 5}{6 - 4} = \dfrac{4}{2} = +2$

2. $E = (x_1, y_1) = (2, 1)$
   $F = (x_2, y_2) = (10, 7)$
   $m = \dfrac{y_2 - y_1}{x_2 - x_1} = \dfrac{7 - 1}{10 - 2} = \dfrac{6}{8} = \dfrac{3}{4}$

3. $G = (x_1, y_1) = (4, 8)$
   $H = (x_2, y_2) = (9, 3)$
   $m = \dfrac{y_2 - y_1}{x_2 - x_1} = \dfrac{3 - 8}{9 - 4} = \dfrac{-5}{5} = -1$

4. $I = (x_1, y_1) = (-6, +10)$
   $J = (x_2, y_2) = (+8, +3)$
   $m = \dfrac{y_2 - y_1}{x_2 - x_1} = \dfrac{3 - (10)}{8 - (-6)} = \dfrac{3 - 10}{8 + 6} = \dfrac{-7}{14} = -\dfrac{1}{2}$

5. **Line b has positive slope.**

6. $y = 2x - 3$
   $y = 2(0) - 3$
   $y = 0 - 3 = -3$
   y-intercept = (0, -3)

7. $y = -2x + 1$
   $y = -2(0) + 1$
   $y = 0 + 1 = 1$
   y-intercept = (0, 1)

8. $y = 5x - 7$
   $y = 5(0) - 7$
   $y = 0 - 7 = -7$
   y-intercept = (0, -7)

9. $y = 3x - 9$
   $0 = 3x - 9$
   $+9 = + 9$
   $\dfrac{9}{3} = \dfrac{3x}{3}$
   $3 = x$
   x-intercept = (3, 0)

10. $y = 8x + 4$
    $0 = 8x + 4$
    $-4 = -4$
    $\dfrac{-4}{8} = \dfrac{8x}{8}$
    $-\dfrac{1}{2} = x$
    x-intercept $= \left(-\dfrac{1}{2}, 0\right)$

Level 2 Review, pg. 265

1. $\dfrac{\text{width}}{\text{length}}$ $\dfrac{3}{5} = \dfrac{15}{x}$
   $\dfrac{3x}{3} = \dfrac{75}{3}$
   $x = 25$ in. long

2. $\dfrac{\text{height}}{\text{base}}$ $\dfrac{9}{5} = \dfrac{h}{8}$
   $\dfrac{5h}{5} = \dfrac{72}{5}$
   $h = 14\dfrac{2}{5}$ ft

3. $\dfrac{12}{20} = \dfrac{15}{x}$

$\dfrac{12x}{12} = \dfrac{300}{12}$

$x = 25$

4. **(2)** $\angle D = \angle G$
   This satisfies the ASA requirement.

5. **(4)** $RT = UW$
   This satisfies the SAS requirement.

6. $c^2 = a^2 + b^2$
   $c^2 = (3.2)^2 + (2.4)^2$
   $c^2 = 10.24 + 5.76$
   $c^2 = 16$
   $c = \sqrt{16}$
   $c = \mathbf{4\ cm}$

7. $c^2 = a^2 + b^2$
   $35^2 = 28^2 + b^2$
   $1225 = 784 + b^2$
   $-784 \quad -784$
   $441 = b^2$
   $\sqrt{441} = b$
   $21 = b$

8. **(3)** $(5, 2)$

9. **(3)** $(4, -3)$

10. E is 7 spaces above x-axis.
    F is 3 spaces below x-axis.
    Distance EF is $7 + 3 = 10$.

11. $D = (x_1, y_1) = (-4, -3)$
    $E = (x_2, y_2) = (+6, +7)$
    $M = \left(\dfrac{x_1 + x_2}{2}, \dfrac{y_1 + y_2}{2}\right)$
    $M = \left(\dfrac{-4 + 6}{2}, \dfrac{-3 + 7}{2}\right)$
    $M = \left(\dfrac{+2}{2}, \dfrac{+4}{2}\right)$
    $M = (+1, +2)$

12. **(2)** $(1, 3)$
    For choice (1), $y = 0 + 2 = 2$.
      Point $(0, 5)$ is not on the graph.
    For choice (2), $y = 1 + 2 = 3$
      Point $(1, 3)$ is on the graph.
    For choice (3), $y = 2 + 2 = 4$.
      Point $(2, 7)$ is not on the graph.
    For choice (4), $y = 3 + 2 = 5$.
      Point $(3, 8)$ is not on the graph.

13. **(4)** $(4, 8)$
    For choice (1), $y = 3(-3) - 4 = -9 - 4 = -13$.
      Point $(-3, -10)$ is not on the graph.
    For choice (2), $y = 3(-1) - 4 = -3 - 4 = -7$.
      Point $(-1, -5)$ is not on the graph.
    For choice (3), $y = 3(2) - 4 = 6 - 4 = 2$.
      Point $(2, 3)$ is not on the graph.
    For choice (4), $y = 3(4) - 4 = 12 - 4 = 8$.
      Point $(4, 8)$ is on the graph.

14. **(1)** $(-3, 18)$
    For choice (1),
    $y = (-3)^2 - 3(-3) = +9 + 9 = 18$.
      Point $(-3, 18)$ is on the graph.
    For choice (2),
    $y = (-2)^2 - 3(-2) = +4 + 6 = 10$.
      Point $(-2, 9)$ is not on the graph.
    For choice (3),
    $y = (1)^2 - 3(1) = 1 - 3 = -2$.
      Point $(1, -1)$ is not on the graph.
    For choice (4),
    $y = (4)^2 - 3(4) = 16 - 12 = 4$.
      Point $(4, 8)$ is not on the graph.

15. **(2)** $(3, 14)$
    For choice (1),
    $y = (5)^2 + 2(5) - 1 = 25 + 10 - 1 = 34$.
      Point $(5, 36)$ is not on the graph.
    For choice (2),
    $y = (3)^2 + 2(3) - 1 = 9 + 6 - 1 = 14$.
      Point $(3, 14)$ is on the graph.
    For choice (3),
    $y = (-2)^2 + 2(-2) - 1 = 4 - 4 - 1 = -1$.
      Point $(-2, -2)$ is not on the graph.
    For choice (4),
    $y = (-4)^2 + 2(-4) - 1 = 16 - 8 - 1 = 7$.
      Point $(-4, 6)$ is not on the graph.

16. **(4)** $+\dfrac{3}{2}$
    $m = \dfrac{y_2 - y_1}{x_2 - x_1} = \dfrac{11 - 5}{7 - 3} = \dfrac{6}{4} = \dfrac{3}{2}$

17. **(2)** $(0, 3)$
    When $x = 0$, $y = 5(0) + 3$
    $= 0 + 3 = 3$.
    Coordinates of the y-intercept are $(0, 3)$.

18. **(4)** $\left(\dfrac{7}{2}, 0\right)$
    When $y = 0$, $0 = -2x + 7$
    $\dfrac{-7}{-2} = \dfrac{-2x}{-2}$
    $\dfrac{7}{2} = x$
    Coordinates of the x-intercept are $\left(\dfrac{7}{2}, 0\right)$

Lesson 13 Exercise, pg. 270

1. $A = lw$
   $132 = 22w$
   $\dfrac{132}{22} = \dfrac{22w}{22}$
   $\mathbf{6\ ft} = w$

2. $P = 2l + 2w$
   $124 = 2l + 2(16)$
   $124 = 2l + 32$
   $-32 \qquad -32$
   $\dfrac{92}{2} = \dfrac{2l}{2}$
   $\mathbf{46\ ft} = l$

3. $P = a + b + c$
$55 = 17 + 17 + c$
$55 = 34 + c$
$\underline{-34 \quad -34}$
$21 = c$
21 in. = c

4. $C = \pi d$
$15.7 = 3.14d$
$\dfrac{15.7}{3.14} = \dfrac{3.14d}{3.14}$
$3.14 \overline{)15.70} \quad \mathbf{5\ m = d}$

5. $A = s^2$
$81 = s^2$
$\sqrt{81} = s$
9 yd = s

6. $P = 4s$
$150 = 4s$
$\dfrac{150}{4} = \dfrac{4s}{4}$
37.5 ft = s

7. $A = \dfrac{1}{2}bh$
$108 = \dfrac{1}{2} \cdot \dfrac{12}{1} \cdot h$
$\dfrac{108}{6} = \dfrac{6h}{6}$
18 ft = h

8. $V = lwh$
$180 = l \cdot \dfrac{10}{1} \cdot \dfrac{1}{2}$
$\dfrac{180}{5} = \dfrac{5l}{5}$
36 ft = l

9. $P = 2l + 2w$
$46 = 2(13.4) + 2w$
$46.0 = 26.8 + 2w$
$\underline{-26.8 \quad -26.8}$
$\dfrac{19.2}{2} = \dfrac{2w}{2}$
9.6 cm = w

10. $A = \dfrac{1}{2}bh$
$84 = \dfrac{1}{2}b \cdot 14$
$\dfrac{84}{7} = \dfrac{7b}{7}$
12 in. = b

11. $P = 4s$ $P = 2l + 2w$
$P = 4 \cdot 16$ $64 = 2(12) + 2w$
$P = 64$ $64 = 24 + 2w$
$\underline{\qquad -24 \quad -24}$
$\dfrac{40}{2} = \dfrac{2w}{2}$
20 ft = w

12. $A = lw$ $A = s^2$
$A = 25 \cdot 9$ $225 = s^2$
$A = 225$ $\sqrt{225} = s$
                      15 in. = s

Lesson 14 Exercise, pg. 272

1. $w = x$ $P = 2l + 2w$
$l = x + 6$ $48 = 2(x + 6) + 2x$
                     $48 = 2x + 12 + 2x$
                     $48 = 4x + 12$
                     $\underline{-12 \qquad -12}$
                     $\dfrac{36}{4} = \dfrac{4x}{4}$
                     $9 = x$
                     $w = \mathbf{9\ ft}$
                     $l = 9 + 6 = \mathbf{15\ ft}$

2. $w = x$ $P = 2l + 2w$
$l = 3x$ $64 = 2(3x) + 2x$
                     $64 = 6x + 2x$
                     $\dfrac{64}{8} = \dfrac{8x}{8}$
                     $8 = x$
                     $w = \mathbf{8\ ft}$
                     $l = 3 \cdot 8 = \mathbf{24\ ft}$

3. $a = x$ $P = a + b + c$
$b = x + 2$ $36 = x + x + 2 + x + 4$
$c = x + 4$ $36 = 3x + 6$
                     $\underline{-6 \qquad -6}$
                     $\dfrac{30}{3} = \dfrac{3x}{3}$
                     $10 = x$
                     $a = \mathbf{10\ in.}$
                     $b = 10 + 2 = \mathbf{12\ in.}$
                     $c = 10 + 4 = \mathbf{14\ in.}$

4. $a = 2x$ $P = a + b + c$
$b = 2x$ $45 = 2x + 2x + x$
$c = x$ $\dfrac{45}{5} = \dfrac{5x}{5}$
                     $9 = x$
long side $= 2(9) = \mathbf{18\ m}$

5. $w = x$ $A = lw$
$l = 2x$ $98 = 2x \cdot x$
                     $\dfrac{98}{2} = \dfrac{2x^2}{2}$
                     $49 = x^2$
                     $\sqrt{49} = x$
                     $7 = x$
                     $w = \mathbf{7\ yd}$
                     $l = 2 \cdot 7 = \mathbf{14\ yd}$

6. $l = x$ $A = lw$
    $w = \frac{1}{2}x$ $50 = x \cdot \frac{1}{2}x$
                $2 \cdot 50 = \frac{x^2}{2} \cdot 2$
                $100 = x^2$
                $\sqrt{100} = x$
                $10 = x$
                $l = \mathbf{10\ in.}$
                $w = \frac{1}{2} \cdot 10 = \mathbf{5\ in.}$

7. $l = 4x$ $A = lw$
    $w = 3x$ $300 = 4x \cdot 3x$
                $\frac{300}{12} = \frac{12x^2}{12}$
                $25 = x^2$
                $\sqrt{25} = x$
                $5 = x$
                $l = 4 \cdot 5 = \mathbf{20\ ft}$
                $w = 3 \cdot 5 = \mathbf{15\ ft}$

8. $b = 2x$ $A = \frac{1}{2}bh$
    $h = 3x$ $108 = \frac{1}{2} \cdot \frac{2x}{1} \cdot 3x$
                $\frac{108}{3} = \frac{3x^2}{3}$
                $36 = x^2$
                $\sqrt{36} = x$
                $6 = x$
                $b = 2 \cdot 6 = \mathbf{12\ in.}$
                $h = 3 \cdot 6 = \mathbf{18\ in.}$

9. $b = x$ $A = \frac{1}{2}bh$
    $h = \frac{1}{3}x$ $24 = \frac{1}{2}x \cdot \frac{1}{3}x$
                $6 \cdot 24 = \frac{1}{6}x^2 \cdot \frac{6}{1}$
                $144 = x^2$
                $\sqrt{144} = x$
                $12 = x$
                $b = \mathbf{12\ m}$
                $h = \frac{1}{3} \cdot \frac{12}{1} = \mathbf{4\ m}$

10. $l = x$ $V = lwh$
     $h = 8x$ $1440 = x \cdot 5 \cdot 8x$
                 $\frac{1440}{40} = \frac{40x^2}{40}$
                 $36 = x^2$
                 $\sqrt{36} = x$
                 $6 = x$
                 $l = \mathbf{6\ in.}$
                 $h = 8 \cdot 6 = \mathbf{48\ in.}$

11. $w = 3x$ $V = lwh$
     $h = 4x$ $3000 = 10 \cdot 3x \cdot 4x$
                 $\frac{3000}{120} = \frac{120x^2}{120}$
                 $25 = x^2$
                 $\sqrt{25} = x$
                 $5 = x$
                 $w = 3 \cdot 5 = \mathbf{15\ ft}$
                 $h = 4 \cdot 5 = \mathbf{20\ ft}$

12. $l = x$ $V = lwh$
     $w = \frac{1}{2}x$ $64 = x \cdot \frac{1}{2}x \cdot 8$
                 $\frac{64}{4} = \frac{4x^2}{4}$
                 $16 = x^2$
                 $\sqrt{16} = x$
                 $4 = x$
                 $l = \mathbf{4\ in.}$
                 $w = \frac{1}{2} \cdot \frac{4}{1} = \mathbf{2\ in.}$

Lesson 15 Exercise, pg. 274

1. $c^2 = a^2 + b^2$
    $c^2 = 3^2 + 3^2$
    $c^2 = 9 + 9$
    $c^2 = 18$
    $c = \sqrt{18}$
    $c = \sqrt{9}\sqrt{2}$
    $c = \mathbf{3\sqrt{2}}$

2. $c^2 = a^2 + b^2$
    $c^2 = 2^2 + 3^2$
    $c^2 = 4 + 9$
    $c^2 = 13$
    $c = \mathbf{\sqrt{13}}$

3. $c^2 = a^2 + b^2$
    $c^2 = 4^2 + 4^2$
    $c^2 = 16 + 16$
    $c^2 = 32$
    $c = \sqrt{32}$
    $c = \sqrt{16}\sqrt{2}$
    $c = \mathbf{4\sqrt{2}}$

4. $c^2 = a^2 + b^2$
    $c^2 = 3^2 + 6^2$
    $c^2 = 9 + 36$
    $c^2 = 45$
    $c = \sqrt{45}$
    $c = \sqrt{9}\sqrt{5}$
    $c = \mathbf{3\sqrt{5}}$

5. $c^2 = a^2 + b^2$
    $c^2 = 8^2 + 8^2$
    $c^2 = 64 + 64$
    $c^2 = 128$
    $c = \sqrt{128}$
    $c = \sqrt{64}\sqrt{2}$
    $c = \mathbf{8\sqrt{2}\ mi}$

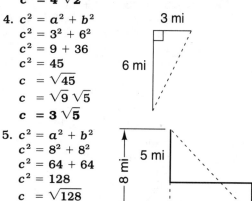

Level 3 Review, pg. 275

1. $P = a + b + c$
   $48 = a + 12 + 16$
   $48 = a + 28$
   $-28 -28$
   $20 = a$
   20 in. = a

2. $P = 2l + 2w$
   $40 = 2l + 2(9)$
   $40 = 2l + 18$
   $-18 -18$
   $\dfrac{22}{2} = \dfrac{2l}{2}$
   11 ft = l

3. $A = lw$
   $95 = 5l$
   $\dfrac{95}{5} = \dfrac{5l}{5}$
   19 ft = l

4. $V = lwh$
   $420 = 7 \cdot w \cdot 10$
   $\dfrac{420}{70} = \dfrac{70w}{70}$
   6 ft = w

5. $A = lw$ $A = s^2$
   $A = 12.5 \cdot 8$ $100 = s^2$
   $A = 100$ $\sqrt{100} = s$
                   10 m = s

6. $l = x$ $A = lw$
   $w = \dfrac{1}{3}x$ $75 = x \cdot \dfrac{1}{3}x$

   $3 \cdot 75 = \dfrac{1}{3}x^2 \cdot \dfrac{3}{1}$
   $225 = x^2$
   $\sqrt{225} = x$
   $15 = x$
   $\mathbf{l = 15\ ft}$
   $w = \dfrac{1}{3} \cdot \dfrac{15}{1} = \mathbf{5\ ft}$

7. $w = x$ $P = 2l + 2w$
   $l = x + 8$ $100 = 2(x + 8) + 2x$
                   $100 = 2x + 16 + 2x$
                   $100 = 4x + 16$
                   $-16 -16$
                   $\dfrac{84}{4} = \dfrac{4x}{4}$
                   $21 = x$
                   w = 21 in.
                   l = 21 + 8 = 29 in.

8. $l = 3x$ $A = lw$
   $w = 2x$ $294 = 3x \cdot 2x$
                   $\dfrac{294}{6} = \dfrac{6x^2}{6}$
                   $49 = x^2$
                   $\sqrt{49} = x$
                   $7 = x$
                   $l = 3 \cdot 7 = \mathbf{21\ ft}$
                   $w = 2 \cdot 7 = \mathbf{14\ ft}$

9. $b = x$ $A = \dfrac{1}{2}bh$
   $h = 2x$ $64 = \dfrac{1}{2}x \cdot \dfrac{2}{1}x$
                   $64 = x^2$
                   $\sqrt{64} = x$
                   $8 = x$
                   b = 8 m
                   $h = 2 \cdot 8 = \mathbf{16\ m}$

10. $w = x$ $V = lwh$
    $l = 2x$ $216 = 2x \cdot x \cdot 12$
                   $\dfrac{216}{24} = \dfrac{24x^2}{24}$
                   $9 = x^2$
                   $\sqrt{9} = x$
                   $3 = x$
                   w = 3 in.
                   $l = 2 \cdot 3 = \mathbf{6\ in.}$

11. $c^2 = a^2 + b^2$
    $c^2 = 6^2 + 6^2$
    $c^2 = 36 + 36$
    $c^2 = 72$
    $c = \sqrt{72}$
    $c = \sqrt{36}\sqrt{2}$
    $\mathbf{c = 6\sqrt{2}}$

12. $c^2 = a^2 + b^2$
    $c^2 = 4^2 + 8^2$
    $c^2 = 16 + 64$
    $c^2 = 80$
    $c = \sqrt{80}$
    $c = \sqrt{16}\sqrt{5}$
    $\mathbf{c = 4\sqrt{5}\ mi}$

13. $c^2 = a^2 + b^2$
    $c^2 = 7^2 + 7^2$
    $c^2 = 49 + 49$
    $c = \sqrt{98}$
    $c = \sqrt{49}\sqrt{2}$
    $\mathbf{c = 7\sqrt{2}}$

14. $c^2 = a^2 + b^2$
    $c^2 = 2^2 + 6^2$
    $c^2 = 4 + 36$
    $c^2 = 40$
    $c = \sqrt{40}$
    $c = \sqrt{4}\sqrt{10}$
    $\mathbf{c = 2\sqrt{10}}$

Posttest, pg. 277

1. **(5) 96**
   $80\% = 0.8$ 120
                    $\times 0.8$
                    $\overline{96.0} = 96$

2. **(4) 36**
   56 36
   22 $5\overline{)180}$
   41
   29
   $+\ 32$
   $\overline{180}$

358   Answers and Solutions

3. **(3) 12.5**
$$60 \overline{)750.0}$$
$$\underline{60}$$
$$150$$
$$\underline{120}$$
$$30\;0$$
$$\underline{30\;0}$$

4. **(1) 12($1800 − $400)**

5. **(2) 3 : 10**
   $\dfrac{\text{in favor}}{\text{total}}\quad \dfrac{360}{1200}=\dfrac{3}{10}$

6. **(5) 2(4.5) + 4(3.5)**

7. **(3) $1890**
   8% = 0.08
   $1750 × 0.08 = $140.00
   $1750 + 140 = $1890

8. **(5) $175**
   10% = 0.10
   $1750 × 0.10 = $175.00

9. **(5) Not enough information is given.**
   You do not know the amount that goes into the pension plan.

10. **(3) 67°**
    90° − 23° = 67°

11. **(1) $1\frac{3}{8}$ in.**
    $3\frac{1}{4} = 3\frac{2}{8} = 2\frac{2}{8} + \frac{8}{8} = 2\frac{10}{8}$
    $-1\frac{7}{8} \qquad\qquad\quad = 1\frac{7}{8}$
    $\phantom{-1\frac{7}{8}\qquad\qquad\quad =} 1\frac{3}{8}$ in.

12. **(4) 78.5**
    $V = \pi r^2 h$
    $V = 3.14 \times (2.5)^2 \times 4$
    $V = 3.14 \times 2.5 \times 2.5 \times 4$
    $V = 78.5\ \text{m}^3$

13. **(2) 80%**
    right 48
    wrong + 12
    total 60
    $\dfrac{\text{right}}{\text{total}}\ \dfrac{48}{60}=\dfrac{4}{5}=80\%$

14. **(5) 128 ft**
    $\dfrac{\text{height}}{\text{shadow}}\quad \dfrac{8}{3}=\dfrac{x}{48}$
    $\dfrac{3x}{3}=\dfrac{384}{3}$
    $x = 128$ ft

15. **(3) x^3**

16. **(2) 225**
    $A = s^2$
    $A = 15^2$
    $A = 15 \times 15$
    $A = 225$ sq ft

17. **(1) 36°**
    72° + 72° = 144°
    180° − 144° = 36°

18. **(2) 2n − 7**

19. **(5) 7 : 12**
    1 yd = 36 in.   21 in. : 36 in. = 7 : 12

20. **(1) 20**
    $w = x$ $A = lw$
    $l = 2x$ $800 = 2x \cdot x$
    $\dfrac{800}{2} = \dfrac{2x^2}{2}$
    $400 = x^2$
    $\sqrt{400} = x$
    $20 = x$

21. **(4) 75**
    $\dfrac{\text{height}}{\text{base}}\quad \dfrac{5}{2}=\dfrac{x}{30}$
    $\dfrac{2x}{2}=\dfrac{150}{2}$
    $x = 75$

22. **(2) 1**
    $8a − 2 = 6$
    $+ 2 + 2$
    $\dfrac{8a}{8} = \dfrac{8}{8}$
    $a = 1$

23. **(1) 6(3m + 4)**
    Divide 6 into both terms.

24. **(2) 12**
    $A = lw$ $A = s^2$
    $A = 20 \cdot 7.2$ $144 = s^2$
    $A = 144$ $\sqrt{144} = s$
    $\quad 12 = s$

25. **(4) D**

26. **(4) 415.9 million**
    487.1 million
    − 71.2
    415.9 million

27. **(1) more than 4 times the number sold in 1980**
    446.3 is close to 440
    110.2 is close to 110
    $110 \overline{)440}$ = 4 times

28. **(3) CD sales were nearly 3 times the sales of LPs.**
    71.2 to the nearest 10 = 70
    207.2 to the nearest 10 = 210
    $70 \overline{)210}$ = 3 times

29. **(2)** $t < 3$
$$5(t + 1) < 20$$
$$5t + 5 < 20$$
$$\quad\; -5 \quad\; -5$$
$$\frac{5t}{5} < \frac{15}{5}$$
$$t < 3$$

30. **(4)** $59°$
$$180°$$
$$-121°$$
$$\overline{59°}$$

31. **(4)** $2\frac{1}{16}$ in.^{2}
$$A = \tfrac{1}{2}bh$$
$$A = \tfrac{1}{2} \times 2\tfrac{3}{4} \times 1\tfrac{1}{2}$$
$$A = \tfrac{1}{2} \times \tfrac{11}{4} \times \tfrac{3}{2} = \tfrac{33}{16} = 2\tfrac{1}{16} \text{ in.}^2$$

32. **(1)** $\frac{5}{12}$

$$\begin{array}{r}60 \\ 48 \\ +36 \\ \hline 144\end{array}\quad \frac{\text{tomato}}{\text{total}}\quad \frac{60}{144} = \frac{5}{12}$$

33. **(3)** $\frac{5}{22}$

now: $36 - 6 = 30$ cans of chicken soup.
$144 - 12 = 132$ total
$$\frac{\text{chicken}}{\text{total}}\quad \frac{30}{132} = \frac{5}{22}$$

34. **(5)** $2(x - 5) = 12$

35. **(3)** 1970

36. **(4)** 1980

37. **(5)** The amount of freight carried by truck has increased steadily while the amount carried by railroad has declined or remained about the same.

38. **(3)** 15
$$c^2 = a^2 + b^2$$
$$17^2 = a^2 + 8^2$$
$$289 = a^2 + 64$$
$$-64 = \quad -64$$
$$\overline{225 = a^2}$$
$$\sqrt{225} = a$$
$$15 = a$$

39. **(4)** 9000

$$\begin{array}{rr}\text{voters} & 3 \\ \text{nonvoters} & +2 \\ \hline \text{registered} & 5\end{array}\quad \frac{\text{voters}}{\text{registered}}\;\frac{3}{5} = \frac{5400}{x}$$
$$\frac{3x}{3} = \frac{27{,}000}{3}$$
$$x = 9{,}000 \text{ registered}$$

40. **(1)** *A* only
This satisfies the SSS requirement.

41. **(5)** $+\frac{1}{2}$
$$A = (x_1, y_1) = (-2, -1)$$
$$B = (x_2, y_2) = (4, 2)$$
$$m = \frac{y_2 - y_1}{x_2 - x_1} = \frac{2 - (-1)}{4 - (-2)} = \frac{2 + 1}{4 + 2} = \frac{3}{6} = +\frac{1}{2}$$

42. **(3)** $\frac{3000}{15}$
$$\frac{\text{envelopes}}{\text{minutes}}\quad \frac{15}{1} = \frac{3000}{x}$$
$$15x = 3000$$
$$x = \frac{3000}{15}$$

43. **(5)** $(x + 10)(x - 10)$
$$x^2 - 100 = (x + 10)(x - 10)$$

44. **(1)** $\frac{9 \cdot 5}{2}$
$$x : 5 = 9 : 2$$
$$\frac{x}{5} = \frac{9}{2}$$
$$2x = 9 \cdot 5$$
$$x = \frac{9 \cdot 5}{2}$$

45. **(2)** $y = -4$ and $+3$
$$y^2 + y - 12 = 0$$
$$(y + 4)(y - 3) = 0$$
$$y + 4 = 0 \qquad y - 3 = 0$$
$$\;\;-4 \;\; -4 \qquad\;\; +3 \;\; +3$$
$$y = -4 \text{ and } y = +3$$

46. **(4)** $348
car payments $= x$
food $= 2x$
mortgage $= 3x$
$$x + 2x + 3x = \$696$$
$$\frac{6x}{6} = \frac{\$696}{6}$$
$$x = \$116$$
$$\text{mortgage} = 3(\$116) = \$348$$

47. **(4)** 2000
$75\% = 0.75$
$$0.75\overline{)1500.00.}\;\; = 20\,00.$$

48. **(2)** 9
H is 6 spaces above *x*-axis.
I is 3 spaces below *x*-axis.
Distance *HI* is $6 + 3 = 9$.

49. **(5)** 15
$$G = (x_1, y_1) = (-5, -3)$$
$$H = (x_2, y_2) = (7, 6)$$
$$d = \sqrt{(x_2 - x_1)^2 + (y_2 - y_1)^2}$$
$$d = \sqrt{(7 - (-5))^2 + (6 - (-3))^2}$$
$$d = \sqrt{(7 + 5)^2 + (6 + 3)^2}$$
$$d = \sqrt{(12)^2 + (9)^2}$$
$$d = \sqrt{144 + 81}$$
$$d = \sqrt{225}$$
$$d = 15$$

50. **(3)** $\left(1, 1\frac{1}{2}\right)$

$G = (x_1, y_1) = (-5, -3)$
$H = (x_2, y_2) = (7, 6)$
$M = \left(\dfrac{x_1 + x_2}{2}, \dfrac{y_1 + y_2}{2}\right)$
$M = \left(\dfrac{-5 + 7}{2}, \dfrac{-3 + 6}{2}\right)$
$M = \left(\dfrac{2}{2}, \dfrac{3}{2}\right)$
$M = \left(1, 1\dfrac{1}{2}\right)$

Simulated Test, pg. 285

1. **(2) 25**

   25 wk
   $180\overline{)4500}$
   $\underline{360}$
   900
   $\underline{900}$

2. **(5) 24**

   green 3   green   3   x
   white +4   ─── = ─ = ──
   total   7   total   7   56
   $\dfrac{7x}{7} = \dfrac{168}{7}$
   $x = 24$ gallons of green paint

3. **(4) $2.70**

   $15   6% = 0.06 $45
   ×3   ×0.06
   $45 $2.70

4. **(3)** $\dfrac{1}{2}c - 20$

5. **(3) 3.63 in.**

   3.60   3.63 in.
   2.45   4)14.52
   4.63
   +3.84
   14.52

6. **(1)** $\dfrac{1}{3}$

   15   large   20   1
   25   ───── = ── = ─
   +20   total   60   3
   60

7. **(3)** $\dfrac{3}{11}$

   now: 15 small   small   15   3
         23 medium   ───── = ── = ──
         +17 large   total   55   11
         55

8. **(2)** 7($6) + 3($9)

9. **(3) $484**

   $24 $384
   ×16   +100
   144          $484
   24
   $384

10. **(1) $53**

    $36 $432 $492
    ×12   + 60   −439
    72        $492 $ 53
    36
    $432

11. **(5) Not enough information is given.**
    You do not know the cost of the guarantee.

12. **(1) $418.45**

    5% = 0.05 $389 $389.00
                 ×0.05   19.45
                 $19.45   +10.00
                             $418.45

13. **(4)** 2(35) + 3(55)

14. **(1) 84**

    $90 \text{ ft} = \dfrac{90}{3} = 30 \text{ yd}$ $P = 2l + 2w$
    $36 \text{ ft} = \dfrac{36}{3} = 12 \text{ yd}$ $P = 2(30) + 2(12)$
    $\phantom{36 \text{ ft} = \dfrac{36}{3} = 12 \text{ yd}}$ $P = 60 + 24 = 84$ yd

15. **(2) 10.8**

    $A = lw$ 10.8
                    300)3240.0
    $A = 90 \times 36$ 300
                             240
    $A = 3240$ ft   0
                             240 0
                             240 0

16. **(2) 3.4**

    $V = s^3$
    $V = 1.5^3$
    $V = 1.5 \times 1.5 \times 1.5 = 3.375$ to the nearest tenth of a cubic meter = 3.4

17. **(4) 2:3**

    $40,000
    −16,000   paid:owed = 16,000:24,000 = 2:3
    $24,000 owed

18. **(4) 30**

    $b = x$ $A = \dfrac{1}{2}bh$
    $h = \dfrac{1}{3}x$ $150 = \dfrac{1}{2} \cdot x \cdot \dfrac{1}{3}x$
    $\phantom{h = \dfrac{1}{3}x}$ $6 \cdot 150 = \dfrac{1}{6}x^2 \cdot 6$
    $\phantom{h = \dfrac{1}{3}x}$ $900 = x^2$
    $\phantom{h = \dfrac{1}{3}x}$ $30 = x$

19. **(1) Fred gets $12, and Gordon gets $7.**
    Fred's wage $= x$
    Gordon's wage $= x - 5$

    $$\begin{aligned} 40x + 40(x-5) &= 760 \\ 40x + 40x - 200 &= 760 \\ 80x - 200 &= 760 \\ +200 &= +200 \\ \overline{80x} &= \overline{960} \\ \frac{80x}{80} &= \frac{960}{80} \\ x &= 12 \\ x - 5 = 12 - 5 &= 7 \end{aligned}$$

20. **(4) 9,975**
    $10^4 = 10 \times 10 \times 10 \times 10 = 10{,}000$
    $5^2 = 5 \times 5 = 25$
    $10^4 - 5^2 = 10{,}000 - 25 = 9{,}975$

21. **(1)** $\dfrac{3x+8}{6}$

22. **(2) 62.5°**
    $180°$
    $-55°$
    $\overline{125°}$ $2\overline{)125.0°}$ = $62.5°$

23. **(5) 1990**

24. **(1)** The number of newspapers generally increased, and the number of periodicals remained about the same.

25. **(2)** $\dfrac{12{,}000}{8 \times 100}$

    $\dfrac{12{,}000}{100}$ = number of hours for 12,000 parts

    $\dfrac{12{,}000}{100} \div 8 = \dfrac{12{,}000}{100} \times \dfrac{1}{8}$

    $= \dfrac{12{,}000}{8 \times 100}$ number of days for 12,000 parts

26. **(4)** $n \geq 6$
    $$\begin{aligned} 2n - 7 &\geq 5 \\ +7 & +7 \\ \overline{2n} &\geq \overline{12} \\ \frac{2n}{2} & \frac{12}{2} \\ n &\geq 6 \end{aligned}$$

27. **(5) 106°**
    Alternate exterior angles are equal.

28. **(5) 98**
    $\dfrac{\text{short leg}}{\text{long leg}}\quad \dfrac{3}{7} = \dfrac{42}{x}$
    $\dfrac{3x}{3} = \dfrac{294}{3}$
    $x = 98$

29. **(2)** $+10m^4$
    $(-5m^3)(-2m) = +10m^4$

30. **(5)** $2\dfrac{1}{2}$
    $$\begin{aligned} 8z - 1 &= 6z + 4 \\ -6z & -6z \\ \overline{2z - 1} &= \overline{4} \\ +1 & +1 \\ \overline{2z} &= \overline{5} \\ \frac{2z}{2} &= \frac{5}{2} \\ z &= 2\dfrac{1}{2} \end{aligned}$$

31. **(4)** $\dfrac{-3c}{5}$
    $\dfrac{15c^2d}{-25cd} = \dfrac{-3c}{5}$

32. **(5) 60**
    $c^2 = a^2 + b^2$
    $c^2 = 48^2 + 36^2$
    $c^2 = 2304 + 1296$
    $c^2 = 3600$
    $c = \sqrt{3600}$
    $c = 60$

33. **(2) (3, 4)**

34. **(5) $13.50**
    $18\% = 0.18$ 0.18 \$ 13.50
    $\times900$ $12\overline{)\$162.00}$
    $\overline{\$162.00}$ $\underline{12}$
    42
    $\underline{36}$
    $6\,0$
    $\underline{6\,0}$
    00

35. **(4) 24**
    $80\% = 0.8$ $12\,0$ 120 miles
    $0.8\overline{)96.0}$ $\underline{-96}$
    24 miles

36. **(1) 7**
    $15\dfrac{3}{4} \div 2\dfrac{1}{4} = \dfrac{63}{4} \div \dfrac{9}{4} = \dfrac{\cancel{63}^7}{\cancel{4}_1} \times \dfrac{\cancel{4}^1}{\cancel{9}_1} = \dfrac{7}{1} = 7$

37. **(3) $540.00**
    $6\% = 0.06$ \$ 5 40
    $0.06\overline{)\$32.40}$

38. **(2) 130 miles**
    $\dfrac{\text{inches}}{\text{miles}}\quad \dfrac{1}{40} = \dfrac{3\frac{1}{4}}{x}$

    $x = 40 \times 3\dfrac{1}{4} = \dfrac{\cancel{40}^{10}}{1} \times \dfrac{13}{\cancel{4}_1} = \dfrac{130}{1} = 130$

39. **(1) 1970**
    Both were a little below $25,000.

40. **(5) $25,000**
    a little below $125,000
    a little below $-100,000$
    $\overline{\$\ 25,000}$

41. **(4)** The price of new private houses increased more than the price of existing houses.

42. **(3) 9 × 3.25**
    $A = lw$
    $A = 9 \times 3.25$

43. **(4) (x + 8)(x − 6)**
    $x^2 + 2x - 48 = (x + 8)(x - 6)$

44. **(4) 65**
    $A = bh$
    $A = 10 \times 6\frac{1}{2} = \frac{\overset{5}{\cancel{10}}}{1} \times \frac{13}{\underset{1}{\cancel{2}}} = \frac{65}{1} = 65$

45. **(1) a(a − 12)**
    $a^2 - 12a = a(a - 12)$
    Divide both terms by a.

46. **(4) 120**
    Replace b with 5 and c with 9.
    $a = 4 \cdot 5(9 - 3)$
    $a = 20(6)$
    $a = 120$

47. **(1) 8 × $1.19**
    $c = nr$
    $c = 8 \times \$1.19$

48. **(5) 9 times**
    $\frac{83\%}{9\%}$ = approximately 9 times

49. **(3) $438,000**
    7.3% = .073
    $.073$
    $\underline{\times\ 6{,}000{,}000}$
    $\$438{,}000.000 = \$438{,}000$

50. **(4) 260,000**
    $10^5 = 10 \times 10 \times 10 \times 10 \times 10 = 100{,}000$
    2.6
    $\underline{\times 100{,}000}$
    $260{,}000.0 = 260{,}000$

51. **(2) $80 − 0.15($80)**
    $80 = original price.
    0.15($80) = discount.

52. **(5) (0,−4)**
    When $x = 0$, $y = +3(0) - 4$
    $y = 0 - 4$
    $y = -4$
    The coordinates of the y-intercept are $(0,-4)$.

53. **(2) c = −6 and +3**
    $c^2 + 3c - 18 = 0$
    $(c + 6)(c - 3) = 0$
    $c + 6 = 0 \qquad c - 3 = 0$
    $\underline{-6 \quad -6} \qquad \underline{+3 \quad +3}$
    $c = -6$ and $c = +3$

54. **(5) −10**
    $\frac{2}{\underset{1}{\cancel{3}}} \times \frac{\overset{-5}{\cancel{-15}}}{1} = \frac{-10}{1} = -10$

55. **(4) 16**
    S is 6 spaces left of y-axis.
    T is 10 spaces right of y-axis.
    Distance ST is $6 + 10 = 16$.

56. **(5) 20**
    $R = (x_1, y_1) = (-6, -4)$
    $T = (x_2, y_2) = (10, 8)$

    $d = \sqrt{(x_2 - x_1)^2 + (y_2 - y_1)^2}$
    $d = \sqrt{(10 - (-6))^2 + (8 - (-4))^2}$
    $d = \sqrt{(10 + 6)^2 + (8 + 4)^2}$
    $d = \sqrt{(16)^2 + (12)^2}$
    $d = \sqrt{256 + 144}$
    $d = \sqrt{400}$
    $d = 20$